SUPERCONDUCTING LEVITATION

SUPERCONDUCTING LEVITATION

Applications to Bearings and Magnetic Transportation

FRANCIS C. MOON
Cornell University

With selected sections by
Pei-Zen Chang
National Taiwan University

A Wiley-Interscience Publication
JOHN WILEY & SONS, INC.
New York / Chichester / Brisbane / Toronto / Singapore

This text is printed on acid-free paper.

Library of Congress Cataloging in Publication Data:

Moon, F. C., 1939–
  Superconducting levitation: applications to bearings and magnetic
  transportation / Francis C. Moon; with selected sections by Pei-Zen
  Chang.
    p. cm.
  ISBN 0-471-55925-3 (alk. paper)
  1. High temperature superconductors. 2. Magnetic levitation
  vehicles. 3. Magnetic bearings. I. Chang, Pei-Zen. II. Title.
  QC611.98.H54M66 1993
  621.34—dc20 93-31759
  CIP

Printed in the United States of America

10 9 8 7 6 5 4 3 2 1

To My Teacher, Colleague, and Friend,
Professor Yih-Hsing Pao

CONTENTS

PREFACE

Why a book on magnetic levitation? Since the discovery of the new, higher-temperature superconductors in 1987, the levitated magnet has become a symbol of the new technology of superconductivity. Also, magnetic levitation (Mag-Lev) transportation systems have ignited the public's imagination about levitated motion. Yet, in both the lay public as well as among active scientists and engineers, there is a general lack of understanding about magnetic levitation. This book presents the basic principles that govern levitation of material bodies by magnetic fields without too much formal theory. This lack of formal theory is due, in part, to the incomplete nature of the theory of magnetic levitation, particularly the fundamental nature of passive, stable levitation based on flux pinning in superconductors. The other reason is my own bias toward experiment and phenomenology. My goals for this book include the desire to inspire both students and practitioners to explore the fascinating phenomena of levitation in the hopes that the wider interest will lead to a more complete understanding of the physics. My second goal is to present enough knowledge and experience about levitation to engineers and applied scientists to encourage them to create and invent new devices based on the use of magnetic forces and superconducting materials. My belief is that as new materials develop, we shall see (in future decades) applications of magnetic levitation not imagined at this time.

It is also my goal to begin to define the technology of magnetic bearings, particularly those based on superconductivity. Bearings enable the creation of machines with movable parts. After two decades of development, active magnetic bearings are beginning to assume a visible role in machine engineering. I believe that in the next decade, passive superconducting bearings will also begin to provide a usable tool for engineers. I also hope that this book will illustrate that materials development alone, while a necessary investment toward application, is not sufficient to design and optimize levitation devices. This book attempts to show the important roles that magnetics, mechanics, and dynamics play in the complete understanding of magnetic levitation and its applications.

My personal interest in magnetic levitation goes back to an article I read on superconducting trains in 1967 by Jim Powell and Gordon Danby of Brookhaven National Laboratory. At Princeton University I built and studied numerous magnetic levitation devices using eddy current forces. As a mechanical and aerospace engineer, I was aware of the role that concepts of stability and dynamics played in the success of early flying machines, and I believed that a proper understanding of dynamic stability would become important to the development of magnetic transportation systems. The decade 1976–1986 was a lost opportunity for U.S. scientists and engineers interested in superconducting levitation, myself included, due to lack of funding. However, the discoveries of 1986–1987 renewed not only my interest, but also that of many others, in the development of magnetic levitation of rotating machines and superconducting bearings.

Acknowledgments must be given to many students, research collaborators, and funding agencies. Professor Rishi Raj was an early collaborator in producing material for our first superconducting bearings in 1987. Robert Ware, K.-C. Weng, and Margaret Yanoviak were especially helpful in the early days of this program. Dr. Pei-Zen Chang performed a sizable number of the experiments described in this book as part of his dissertation. Other Cornell collaborators include Donald Chu, David Kupperman, Michael Chiu, Dorothea Yeh, Caswell Rowe Jr., William Homes, and Dr. Czeslaw Golkowski. Recent contributors to our laboratory are Professor Takashi Hikihara of Kansai University, Osaka, and Professor Peter Schönhuber of Technische Universität Wien.

I must also acknowledge the help of John Hull and Thomas Mulcahy of Argonne National Laboratory, Hamid Hojaji of The Catholic University of America, Dr. Z. Yang of Dalhousie University, Nova Scotia, and S. Jin of ATT Bell Laboratories. We are especially grateful for the interaction with the Japanese Superconductivity Center, ISTEC, particularly Dr. S. Tanaka and Dr. M. Murakami, as well as, K. Matsuyama and R. Takahata of Koyo Seiko Co. Ltd, Osaka.

Research support has come from the National Science Foundation, NASA Goddard Space Center, and Argonne National Laboratory. I am especially grateful to Dr. Yury Flom of NASA for his enthusiasm and support of this field. Special thanks are given to Ms. Teresa Howley, who drew many of the figures in this book, and to Ms. Cora Lee Jackson and Ms. Judith Stage who typed this manuscript.

Finally, I want to dedicate this book to my long-time teacher and friend, Professor Yih-Hsing Pao of Cornell University, whose early encouragement for magnetomechanics supported my interest in levitation science and mechanics over the years.

FRANCIS C. MOON

Ithaca, New York
March 1994

CHAPTER 1

INTRODUCTION TO MAGNETIC LEVITATION

Levitation: The raising of the human body in the air without mechanical means or contact.

—*Encyclopedia Brittanica*

1-1 INTRODUCTION

The suspension of objects and people with no visible means of support is fascinating to most people, even in an age grown jaded with high-tech products. To deprive objects of the effects of gravity is a dream common to generations of thinkers from Benjamin Franklin to Robert Goddard, and even to mystics of the East. This modern fascination with magnetic levitation stems from two singular technical and scientific achievements: (i) the creation of high-speed vehicles to carry people at 500 km/hr and (ii) the discovery of new superconducting materials.

The modern development of magnetic levitation transportation systems, known as Mag-Lev, started in the late 1960s as a natural consequence of the development of low-temperature superconducting wire and the transistor and chip-based electronic control technology. In the 1980s, Mag-Lev had matured to the point where Japanese and German technologists were ready to market these new high-speed levitated machines (see Figure 1-1).

At the same time, C. W. Chu of the University of Houston and co-workers in 1987 discovered a new, higher-temperature superconductor, yttrium–barium–copper oxide (YBCO), which unleashed a

Figure 1-1 Photograph of Japanese superconducting Mag-Lev vehicle; rated speed 500 km/hr.

wild year of speculation amongst both scientists and the media about new applications of magnetic fields and forces. Those premature promises of superconducting materials have been tempered by the practical difficulties of development. First, bulk YBCO was found to have a low current density, and early samples were found to be too brittle to fabricate into useful wire. However, from the very beginning, the hallmark of these new superconductors was their ability to levitate small magnets (Figure 1-2). This property, captured on the covers of both scientific and popular magazines, inspired a group of engineers and applied scientists to envision a new set of levitation applications based on superconducting magnetic bearings.

In the past few years, the original technical obstacles of YBCO have gradually been overcome, and new superconducting materials

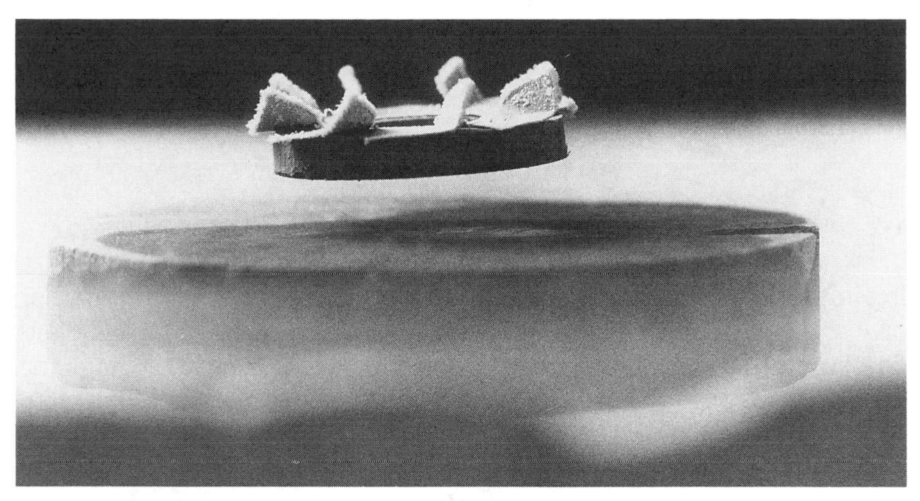

Figure 1-2 Photograph of a magnetically levitated rare earth magnet with a turbine disc above high-temperature superconducting material ($YBa_2Cu_3O_7$).

such as bismuth–strontium–calcium–copper oxide (BSCCO) have been discovered. Higher current densities for practical applications have been achieved, and longer and longer wire lengths have been produced with good superconducting properties. At this juncture of superconducting technology, we can now envisage, in the coming decade, the levitation of large machine components as well as the enhancement of existing Mag-Lev transportation systems with new high-temperature superconducting magnets.

Passive, simple, and defying intuition—these are the fascinating qualities of superconducting levitation. One can imagine the relative velocity of 100–200 m/sec between moving bodies with no contact, no wear, no need for fluid or gas intervention, and no need for active controls.

While the "explanation" of magnetic levitation is based on elementary principles of classical physics, I have found, in my travels, both technical and lay people alike amazed at the phenomenon and not completely understanding of the basic concepts. This book is written to encourage and inspire readers to play with and study magnetic levitation phenomena. However, this book will be most successful if it encourages engineers and applied scientists to create new magnetic levitation devices and applications based on the new discoveries of the modern superconductors.

The logic of describing both large-scale linear magnetic bearings for Mag-Lev transportation and rotary magnetic bearings in the same book is that they are based on the same physical principles, and often must meet the same technological constraints and challenges.

1-2 MAGNETIC LEVITATION SYSTEMS

Levitation is defined as the stable equilibrium of a body without contact with the solid earth. Excluding orbital motion, flying, and floating, levitation forces can be created by both electric or magnetic fields. The focus of this book will be the use of magnetic forces to equilibrate gravitational forces in a body in a manner such that all six degrees of freedom of the body are stable.

There are several types of total and partial magnetic levitation systems (see Figure 1-3):

- Permanent magnets (only partial stability)
- Diamagnetic materials (e.g., bismuth) in a magnetic field
- Electromagnets with feedback control
- Electromagnets with dynamic currents
- Eddy currents—alternating current (ac) devices
- Eddy currents—moving conductors
- Superconductors and permanent magnets
- Superconductors and superconducting magnets
- Hybrids—for example, permanent magnets with feedback-controlled electromagnet or controlled magnets with superconductors

Magnetic levitation requires two necessary subsystems:

(i) a primary system for generating the magnetic field and

(ii) a system for shaping or trapping the magnetic flux.

In the case of electromagnetic levitation, electric currents in a wire wound coil produce the primary field while the ferromagnetic coil holder and the ferromagnetic base create a means of shaping a magnetic circuit. In the case of eddy current levitation with a moving magnet over a conductor shown in Figure 1-3, the source of the field can be a permanent magnet or a normal or superconducting wire wound coil. The relative motion of the magnet and conductor provides the field-shaping system due to the induced eddy currents in the conductor. Finally, in the case of a passive superconducting levitator (Figure 1-3), a permanent magnet serves as the primary field source

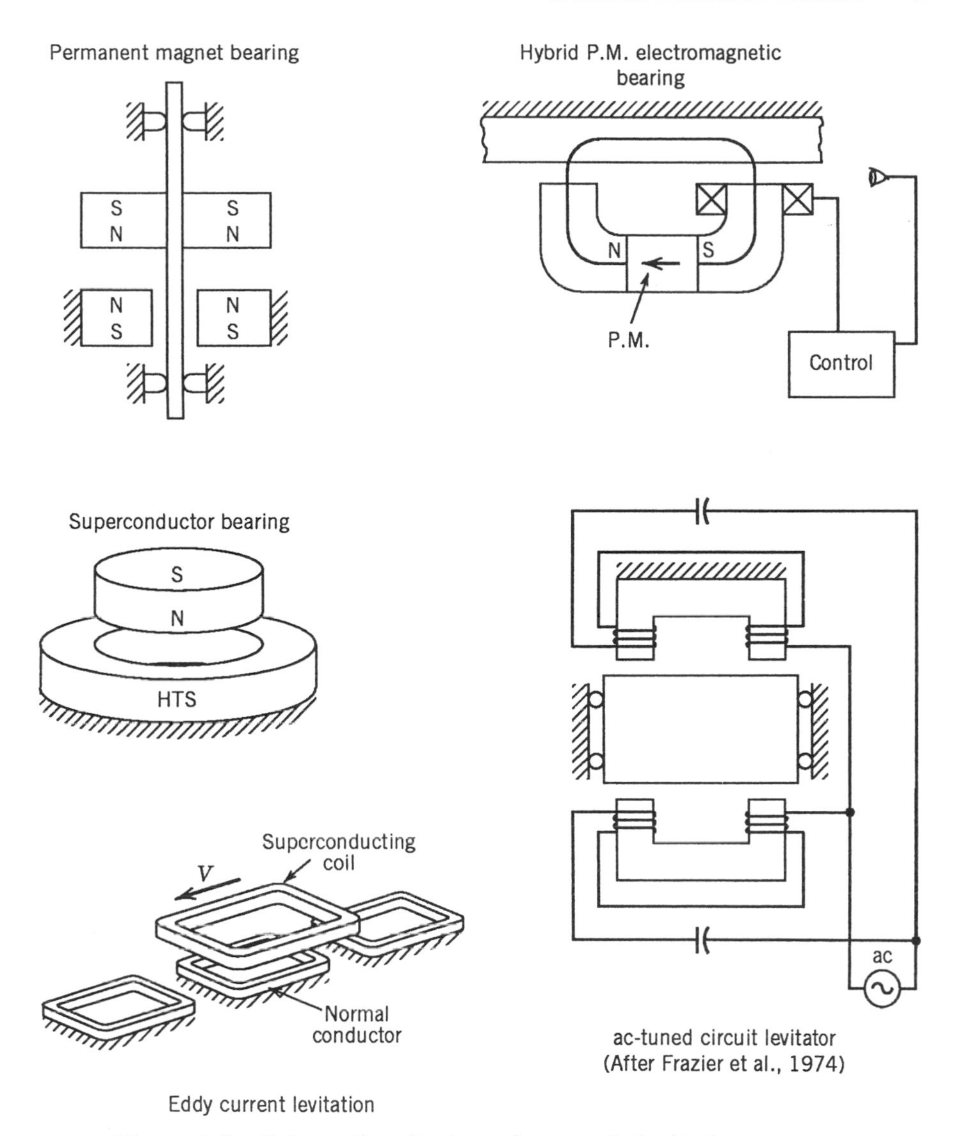

Figure 1-3 Schematics of selected magnetic levitation systems.

while the bulk or thin film superconductor provides the field shaping due to the induced supercurrents.

Active Magnetic Bearings Versus Passive Superconducting Bearings

Active-controlled electromagnetic levitation systems have become a maturing technology and offer the following advantages with a few significant disadvantages:

Advantages

- High stiffness
- Adaptive control to environment changes built in
- Low field leakage
- Industrial application proven

Disadvantages

- High cost
- Complexity-related reliability issues
- Small working gaps
- Total system weight penalty (e.g., power supply, controller, electronics)

The role of *superconducting levitation* systems in machines has yet to be defined, but some properties are evident:

Advantages

- Passive, no electronics or power supplies needed
- Potential high reliability (low complexity)
- Potential lower system weight
- Large or small working gaps
- Lower cost

Disadvantages

- Requires cryogen or cryogenic temperatures (cryocooler)
- Relatively low stiffness
- Magnetic field leakage
- Not a proven technology in service

1-3 STABILITY AND LEVITATION

Levitation is defined as the equilibrium of a body without solid or fluid contact with the earth. Levitation can be achieved using electric or magnetic forces or by using air pressure, though some purists would argue whether flying or hovering is levitation. However, the analogy of

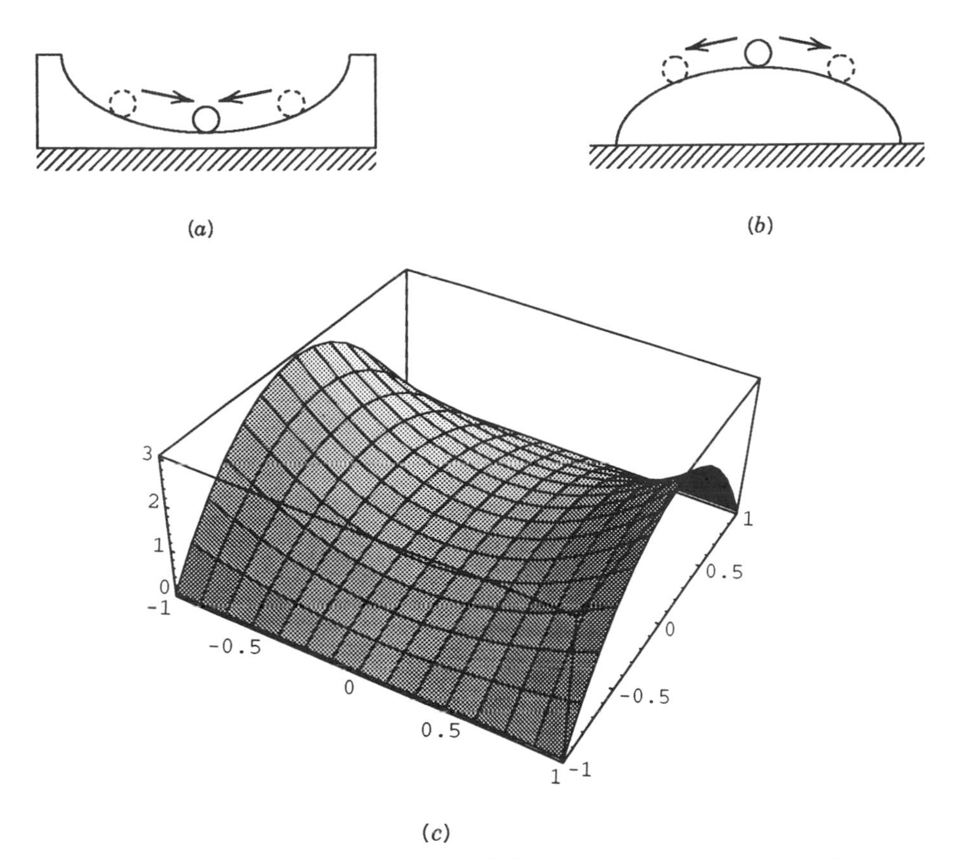

Figure 1-4 (*a*) Stable potential well. (*b*) Unstable potential hill. (*c*) Unstable potential saddle.

magnetic levitation with the suspension of aircraft provides insight into the essential requirements for levitation; that is, *lift alone is not levitation*. The success of the Wright machine in 1903 was based, in part, on the invention of a mechanism on the wings to achieve *stable* "levitated" flight. The same can be said of magnetic bearing design—namely, that an understanding of the nature of mechanical stability is crucial to the creation of a successful levitation device.

Simple notions of stability often use the paradigm of the ball in a potential well or on top of a potential hill as illustrated in Figure 1-4*a*, *b*. This idea uses the concept of potential energy, which states that physical systems are stable when they are at their lowest energy.

The minimum potential energy definition of stability is good to begin with, but is not enough in order to understand magnetic levitation. Not only must one consider the stability of the center of mass of the body, but one usually wants to achieve stability in the orientation or angular position of the body. If the levitated body is deformable, the stability of the deformed shape may also be important.

The second difficulty with the analogy with particles in gravitational potential wells is that we have to define what we mean by the magnetic or electric potential energy. This is straightforward if the sources of the levitating magnetic or electric forces are fixed. But when magnetization or electric currents are induced due to changes in the position or orientation of our levitated body, then the static concept of stability using potential energy can involve pitfalls that can yield the wrong conclusion regarding the stability of the system.

To really be rigorous in magneto-mechanics, one must discuss stability in the context of dynamics. For example, in some systems one can have static instability but dynamic stability. This is especially true in the case of time-varying electric or magnetic fields as in the case of actively controlled magnetic bearings. However, it is also important when the forces (mechanical or magnetic) depend on generalized velocities.

In general, the use of concepts of dynamic stability, rooted in modern nonlinear dynamics, must be employed to completely guarantee stability of magnetically levitated systems. This theory not only requires knowledge of how magnetic forces and torques change with position and orientation (i.e., magnetic stiffness), but also knowledge of how these forces change with both linear and angular velocities.

Earnshaw's Theorem

It is said that nothing is new under the sun, and this is particularly true in the case of stability and electromagnetic levitation. Early in the nineteenth century (1839) a British minister and natural philosopher, Samuel Earnshaw (1805–1888), examined this question and stated a fundamental proposition known as *Earnshaw's theorem*. The essence of this theorem is that a group of particles governed by inverse square law forces cannot be in stable equilibrium. The theorem naturally applies to charged particles and magnetic poles and dipoles. A modern statement of this theorem can be found in Jeans (1925) (see also Earnshaw, 1842): "*A charged particle in the field of a fixed set of charges cannot rest in stable equilibrium.*" This theorem can be extended to a set of magnets and fixed circuits with constant current sources. To the chagrin of many a would-be inventor, and contrary to the judgment of many a patent officer or lawyer, the theorem rules out many clever magnetic levitation schemes. This is especially the case of levitation with a set of permanent magnets as any reader can verify. Equilibrium is possible, but stability is not. [A short article on Earnshaw himself can be found in Scott (1959)].

Later we will address the question of how and why one can achieve stable levitation of ferromagnetic electromagnets using active feedback. However, here we will try to motivate why superconducting systems appear to violate or escape the consequences of Earnshaw's theorem. One of the first to show how diamagnetic or superconducting materials could support stable levitation was Braunbeck (1939a, b).

Earnshaw's theorem is based on the mathematics of inverse square force laws. Particles which experience such forces must obey a partial differential equation known as *Laplace's equation*. The solutions of this equation do not admit local minima or maxima as in Figure 1-4*a*, *b*, but only saddle-type equilibria Figure 1-4*c*. However, there are circumstances under which electric and magnetic systems can avoid the consequences of Earnshaw's theorem:

- Time-varying fields (e.g., eddy currents, alternating gradient)
- Active feedback
- Diamagnetic systems
- Ferrofluids
- Superconductors

The theorem is easily proved if the electric and magnetic sources are fixed in space and time, and one seeks to establish the stability of a single free-moving magnet or charged particle. However, in the presence of polarizable, magnetizable, or superconducting materials, the motion of the test body will induce changes in the electric and magnetic sources in the nearby bodies. In general magnetic flux attractors such as ferromagnetic materials still obey Earnshaw's theorem, whereas for flux repellers such as diamagnetic or Type I superconductors, stability can sometimes be obtained (Figure 1-5). Superconductors, however, have several modes of stable levitation;

- Type I or Meissner repulsive levitation based on complete flux exclusion
- Type II repulsive levitation based on both partial flux exclusion and flux pinning
- Type II suspension levitation based on flux pinning forces (see Figure 1-10).

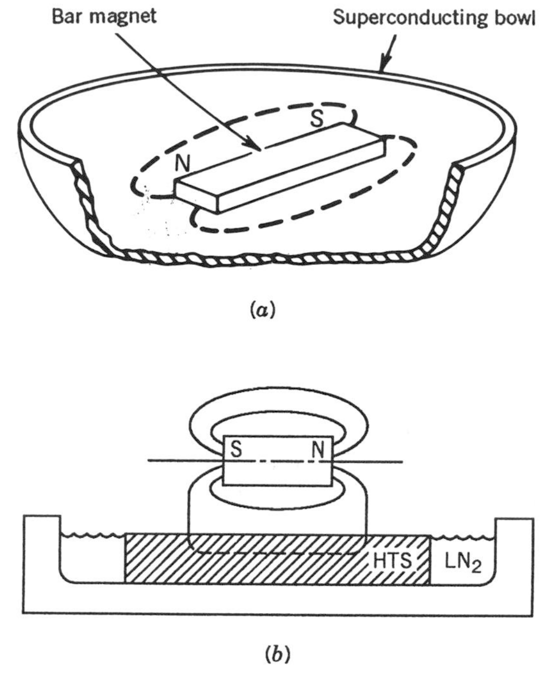

Figure 1-5 (*a*) Type I or Meissner superconducting levitation. (*b*) Type II superconducting levitation.

In the case of Meissner repulsive levitation shown in Figure 1-5*a*, superconducting currents in the bowl-shaped object move in response to changes in the levitated magnet. The concave shape is required to achieve an energy potential well.

In the case of Type II levitation, both repulsive (Figure 1-5*b*) and suspension (or attractive) (Figure 1-10) stable levitation forces are possible without shaping the superconductor. As explained in the next section and in Chapter 2, magnetic flux exclusion produces equivalent magnetic pressures which result in repulsive levitation whereas flux attraction creates magnetic tensions (similar to ferromagnetic materials) which can support suspension levitation. Flux penetration into superconductors is different from ferromagnetic materials, however.

In Type II superconductors, vortex-like supercurrent structures in the material create paths for the flux lines. When the external sources of these flux lines move, however, these supercurrent vortices resist motion or are pinned in the superconducting material. This so-called *flux-pinning* is believed to be the source of stable levitation in these materials [see, e.g., Brandt (1989a, b, 1990a, b, c)].

Finally, from a fundamental point of view, it is not completely understood why supercurrent-based magnetic forces can produce stable attractive levitation while spin-based magnetic forces in ferromagnetic materials produce unstable attractive or suspension levitation. Given the restricted assumptions upon which Earnshaw's theorem is based, the possibility that some new magnetic material will be discovered, which supports stable levitation, cannot be entirely ruled out.

1-4 MAGNETIC FIELDS AND FORCES

An intuitive feel for magnetic fields and magnetic forces on bodies is necessary to be able to understand levitation as well as to create new applications to machines. The paradox for many mechanical engineers is that their career path was motivated, in part, by an avoidance of electromagnetics. However, magnetic fields have common mathematical roots with fluid mechanics (magnetostatics) and heat transfer (magnetic field diffusion). Also, the dynamics and control of levitated bodies have analogies with mechanical vibrations and control.

A more detailed review of electromagnetics is presented in Chapter 2. For a more tutorial presentation, the reader is referred to classical texts on electromagnetics such as Smythe (1968), Stratton (1941), and Jackson (1962).

One of the confusing aspects of magnetics is the fact that the magnetic field really plays an intermediary role in determining magnetic forces. That is, electromagnetic forces occur between two different bodies which carry charge, current, or electrically polarized or magnetized materials. In this text we will only study forces between bodies carrying electric currents and magnetized material. When one wishes to focus on the force on just one body (#2), we replace the other body (#1) with a vector quantity called the *magnetic field*, namely, $\mathbf{B}_1$.

For example, the attractive force per unit length between two parallel current carrying wires is proportional to the product of the values of the currents in each wire and inversely proportional to the distance r between the wires (Figure 1-6a)

$$F = -\frac{\mu_0}{2\pi}\frac{I_1 I_2}{r} \qquad (\text{N/m}) \qquad (1\text{-}4.1)$$

where: $\mu_0 = 4\pi \times 10^{-7}$ N/A^2, I_1 and I_2 are measured in amperes; and r is the separation of the wires, measured in meters. If one

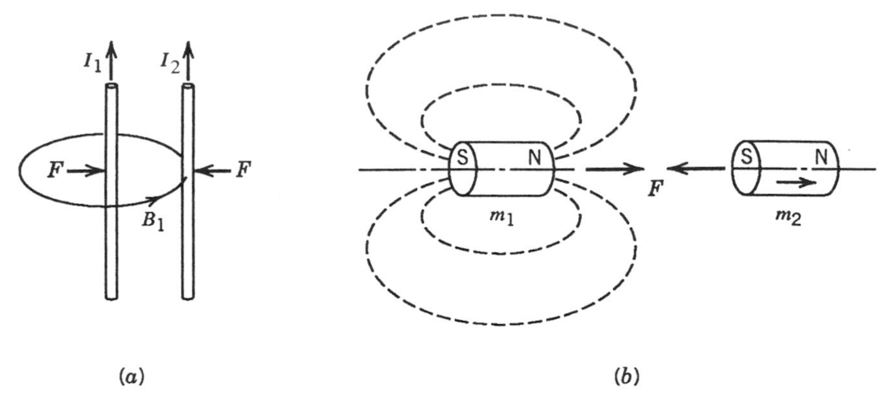

Figure 1-6 (*a*) Magnetic force between two parallel current filaments. (*b*) Magnetic force between two permanent magnet dipoles.

defines a quantity B_1 called the magnetic field produced by I_1, the force on I_2 takes the simple form

$$F = I_2 B_1 \tag{1-4.2}$$

and

$$B_1 = \frac{\mu_0 I_1}{2\pi r} \tag{1-4.3}$$

In MKS units, B_1 is in units called tesla (T). (The earth's field at the surface is around 0.5×10^{-4} T.)

Similarly, the force between two aligned point dipole magnets of strengths m_1 and m_2 is given by (see Figure 1-6*b*)

$$F = -\frac{\mu_0}{4\pi} \frac{6 m_1 m_2}{z^4} \tag{1-4.4}$$

where z is the axial separation between dipoles, and the minus sign indicates an attractive force when m_1 and m_2 both have the same sign. For finite size dipoles, Eq. (1-4.4) is valid when z is large.

Again, one can define a quantity called the magnetic field B_1 representing the effect of m_1, only this time the magnetic force is proportional to the *gradient* of B_1; that is,

$$F = m_2 \frac{dB_1}{dz} \qquad \text{(N)} \tag{1-4.5}$$

where m_1 and m_2 are in MKS units of A-m^2. Of course, all three variables F, m_2, and B_1 are vector quantities, and Eq. (1-4.5) can be generalized to

$$\mathbf{F} = \mathbf{m}_2 \cdot \nabla \mathbf{B}_1 \tag{1-4.6}$$

where ∇ is the vector gradient operation.

In addition to a force between dipoles, there is a couple or torque on each of them when they are not aligned. For example, the torque on dipoles in a plane oriented 90° to each other is proportional to

$$C \sim \mu_0 \frac{m_1 m_2}{z^3} \qquad (\text{N}-\text{m}) \tag{1-4.7}$$

and the torque axis is normal to the plane of the dipoles. Replacing the effect of dipole #1 by a magnetic field vector, the general expression for the torque is given by a vector cross-product operation

$$\mathbf{C} = \mathbf{m}_2 \times \mathbf{B}_1 \tag{1-4.8}$$

Induced Magnetic Forces

We have seen that electric currents and permanent magnets are sources of magnetic forces on other currents and magnets. However, the second body need not be an independent source of a magnetic field. For example, in ferromagnetic materials such as iron and its alloys, bringing a field source such as an electric current near the material will induce a distribution of magnetic dipoles which will, in turn, produce a force between the body carrying current and the ferromagnetic material (Figure 1-7). It can be shown, for example, that the force between a long wire with current I placed parallel to a ferromagnetic half-space produces an attractive force:

$$F = -\frac{\mu_r - 1}{\mu_r + 1} \frac{\mu_0 I^2}{4\pi h^2} \qquad (\text{N/m}) \tag{1-4.9}$$

where μ_r is called the *relative permeability* of the material. For ferromagnetic materials, μ_r is approximately 10^2 or greater, and the quotient in Eq. (1-4.9) can be set to unity. Note that for independent currents or dipoles, the forces are *linear* in I or m. However, for induced dipoles, the force is quadratic or *nonlinear* in the source current I. One way to envision the induced force is to replace the

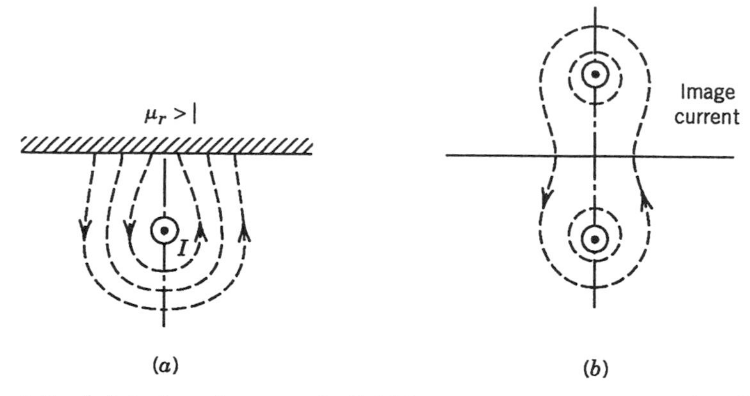

Figure 1-7 (*a*) Induced magnetic field between a current-carrying filament and a soft ferromagnetic half-space. (*b*) Image current replacing the ferromagnetic half-space.

ferromagnetic half-space with an "image current filament" as shown in Figure 1-7*b*. Then one can apply the force equation between two current filaments [Eq. (1-4.1)], where $I_1 = I_2 = I$.

Another example of an induced magnetic force is that between a long current-carrying filament and a thin-layer superconducting plane as shown in Figure 1-8*a*. As the current filament is brought up close to the superconducting layer, supercurrents are induced which act to produce forces on the current filament. In the so-called Type I regime, the total field below the superconducting layer is zero; that is, the induced currents produce just enough flux to cancel the flux of the source current below the layer. This flux screening is called the

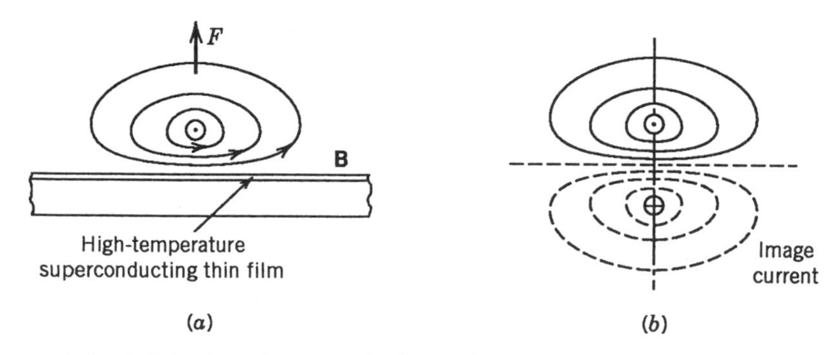

Figure 1-8 (*a*) Induced magnetic force between a current filament and a superconducting thin film. (*b*) Image current equivalent to the superconducting thin film.

Meissner effect. The magnetic force between the layer and the filament can be shown to be

$$F = \frac{\mu_0}{4\pi} \frac{I^2}{h} \qquad \text{(N/m)} \tag{1-4.10}$$

where in this case the force is repulsive. Again, an image current filament can be used to replace the effect of the layer, but in this case the image current is of opposite sense to the source current; that is, in Eq. (1-4.1), $I_1 = -I_2 = I$, and $r = 2h$ (Figure 1-8*b*).

The same effect of flux exclusion and a repulsive force can be obtained with a normal-conducting layer of thickness Δ and electric conductivity σ by moving the current filament parallel to the flat sheet with velocity v. The repulsive force normal to the sheet is given by

$$F = \frac{\mu_0}{4\pi} \frac{I^2}{h} \frac{v^2}{v^2 + w^2} \tag{1-4.11}$$

where $w = 2/\sigma \Delta \mu_0$ is a characteristic velocity. [See e.g., Moon (1984, Chapter 8).] This phenomenon, known as *eddy current levitation*, is the basis for electrodynamic levitation of high-speed vehicles (see Figure 1-17). The physics can be best understood by placing one's frame of reference with the moving current filament. Then the source field is stationary, but the sheet conductor moves with velocity v to the left. One then invokes Lenz's law, which states that a moving conductor in a magnetic field produces an electric field in the conductor. This electric field produces the electric currents in the sheet, which, in turn, produces forces on the source current filament.

Magnetic Stiffness

Stability of a mechanical system depends not only on the forces, but also on how those forces change due to small changes in the geometry. In the example of the levitated wire above a superconducting layer or a moving conducting layer, we look at the change in the height h due to a small perturbation; that is, $h = h_0 + z$, where $z/h_0 \ll 1$. Then one can approximate the magnetic force [Eq. (1-4.10)]:

$$F = \frac{\mu_0 I^2}{4\pi h_0}\left(1 - \frac{z}{h_0}\right) \tag{1-4.12}$$

or

$$F = F_0 - \kappa z \tag{1-4.13}$$

where $F_0 = \mu_0 I^2/4\pi h_0$ and κ is called the *magnetic stiffness*; that is,

$$\kappa = \frac{F_0}{h_0} \tag{1-4.14}$$

has units of a mechanical spring. The negative sign indicates that a small change in height decreases the force, which tends to restore the wire to its original position. In general, when the force is written as a function of a geometric coordinate h, we obtain

$$F(h) = F(h = h_0) + \frac{dF}{dh}(h - h_0) \tag{1-4.15}$$

where the derivative is evaluated at $h = h_0$. The magnetic stiffness is defined as

$$\kappa = -\left.\frac{dF}{dh}\right|_{h=h_0} \tag{1-4.16}$$

For stability with respect to change in h, κ *must be positive*.

A calculation of κ for the wire near a ferromagnetic half-space shows that $\kappa < 0$ and the force is unstable.

Magnetic Stresses

We have seen how the magnetic force between two bodies can be calculated from knowledge of the magnetic field produced by one body, and the current or magnetization in the other. However, the observed magnetic field between two bodies is the sum of fields produced by each. Thus, one can ask how does one determine the force of one body on the other using the total magnetic field vector? This problem was solved by both Faraday and Maxwell using the concept of *magnetic pressure* or *magnetic tension* or, more generally, *magnetic stresses*. More will be said about this concept in Chapter 2. For now assume that a body screens out a magnetic field. Then we can imagine a pressure force on a surface element dA of the body with normal vector $\mathbf{n}$ directed outward (Figure 1-9a). This magnetic pressure P is proportional to the square of the tangential field component,

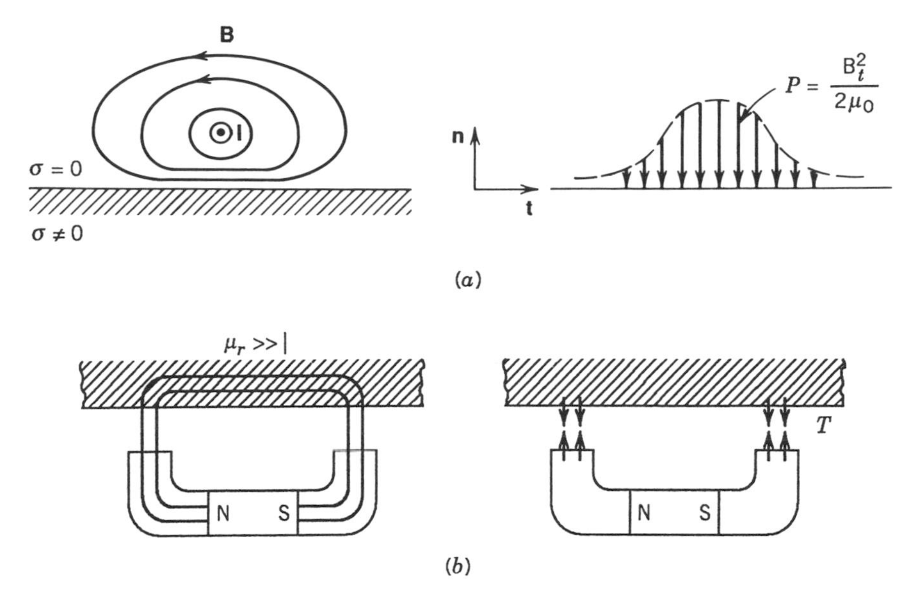

Figure 1-9 (*a*) Magnetic pressure distribution (*right*) on a superconducting half-space due to magnetic flux exclusion of a current-carrying filament and a Type I superconducting half-space. (*b*) Magnetic tension on a ferromagnetic half-space due to flux penetration.

or B_t; that is,

$$P = \frac{B_t^2}{2\mu_0} \tag{1-4.17}$$

Likewise, assume that a body attracts magnetic flux as in Figure 1-9*b*. Then we imagine a tensile stress T acting normal to the surface of the body proportional to the square of the magnetic field component, or B_n; that is,

$$T = \frac{B_n^2}{2\mu_0} \tag{1-4.18}$$

To get an order of magnitude of these stresses, a 1-T field produces either a pressure or tension of 40 N/cm^2 (57 psi). The use of this concept is illustrated in Figure 1-9*b* for a U-shaped electromagnet. The high-permeability ferromagnetic material creates a magnetic flux circuit through which the flux lines in the gap are almost completely normal to the horizontal rail. If the area of the U section is A_0, then the levitation force is given by

$$F = 2TA_0 = \frac{B_n^2 A_0}{\mu_0} \qquad (1\text{-}4.19)$$

Thus, a 1-cm^2 cross section with a 1-T field is capable of lifting 80 N. Note that *doubling the field quadruples the force.*

The use of a wire wound coil and ferromagnetic material to produce lift in an electromagnet goes back to W. Sturgeon in 1825 and J. Henry in 1832. Henry was able to create lift magnets (with zero gap distance) with a capacity of 15 kN (3500 lbf), which was remarkable in its day.

Thus, concepts of magnetic pressure and tension, when applied to the levitation of magnets by bulk superconducting materials, lead to the fact that repulsive levitation requires more flux exclusion from the superconductor (B_t is large) than flux trapping, whereas suspension levitation of a magnet by a superconductor requires more flux trapping B_n is large) as illustrated in Figure 1-10.

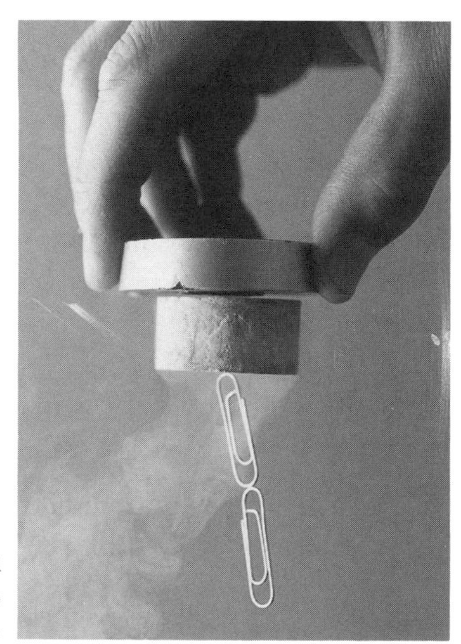

Figure 1-10 Suspension of a rare earth magnet below a high-temperature superconducting disk. (Courtesy M. Murakami, ISTEC, Tokyo)

1-5 BEARINGS AND LEVITATION

Bearings make relative motion possible. From the wheel to the air-cushion vehicle, man has sought to avoid the consequences of producing relative motion between two solid bodies, namely, friction, wear, heat, and energy loss. Conventional nonmagnetic bearing systems including the following (see Figure 1-11*a*, *b*):

- Rolling elements; ball and cylinder elements
- Fluid-based systems; hydrostatic and hydrodynamic
- Gas-based systems; aerostatic and aerodynamic

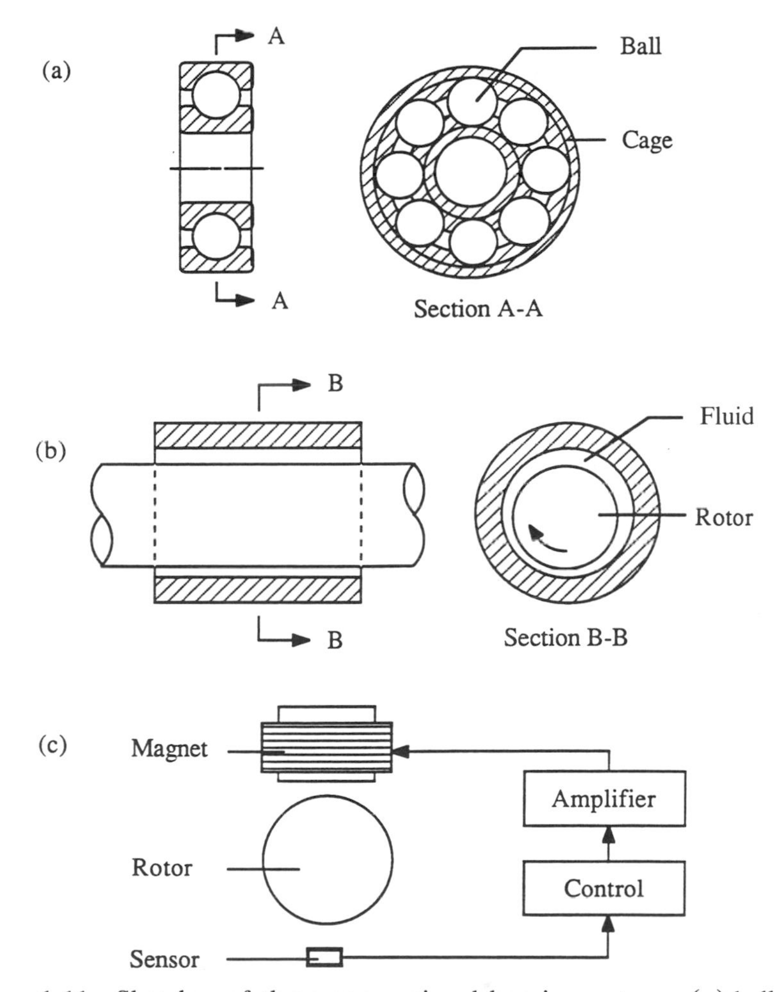

Figure 1-11 Sketches of three conventional bearing systems: (*a*) ball bearing, (*b*) fluid bearing, and (*c*) active control magnetic bearing.

Conventional electromagnetic bearing systems that are not based on superconductivity include: (Figure 1-11*c*):

- Permanent magnets
- Active controlled electromagnets
- Ferrofluid systems
- Electric field devices
- Hybrids based on some of the above systems

It is the thesis of this book that superconducting bearings based on new high-temperature superconductors can replace some of these systems.

The potential applications of linear and rotary magnetic bearings span a wide range of technologies, as illustrated by the following list:

- Gyroscopes
- High-speed machine tools
- Energy storage flywheels
- Angular momentum wheels for spacecraft
- Rotary scanners for optical and infrared devices
- High-speed optical shutters
- Centrifuges
- Micromachine bearing
- Cryocooler turbines
- Cryopumps: rockets, space vehicles, ground-based systems
- Aircraft engine bearings
- Underground gas pipeline pumps
- High-speed spindles for textile manufacturing
- Particle beam choppers
- Computer disk storage devices
- Linear conveyor pallets for clean room or vacuum processing system

The most natural application areas for superconducting bearings are those which have a cryogenic environment such as liquid oxygen or hydrogen cryopumps and cryocoolers. Applications that create a high-value-added product per hour such as high-speed machine tools, textile spindles, or computer chip conveyor systems could also justify the added expense of providing a cryogenic environment for superconducting bearings.

Active Magnetic Bearings Using Normal Conductors

To understand the advantages of superconductor-based levitation systems, it is necessary to place the existing magnetic bearing technology in perspective—in particular, active magnetic bearings for rotating machinery. There are now more than a dozen producers of magnetic bearings using normal conductor electromagnets and based on feedback control. One of the world leaders in this technology is the French company called S2M, or Société de Méchanique Magnétique. Over the past decade it has produced magnetic bearings in over 1400 rotating machines. These devices can produce lift forces of over 1500 kg and can produce axial forces of over 5000 kg, running at speeds of up to 5000 rpm. A short review of the current technology of active bearings has recently appeared (O'Connor, 1992) which lists a number of current applications;

- High-speed machine tools
- Turbomolecular pumps
- Centrifugal compressors for natural gas transmission
- Utility boiler feed pumps
- High-speed printing press
- Satellite control moment gyros
- Circulating fan for gas-laser system

Applications under development include energy storage flywheels for electric auto systems and bearings for advanced gas turbine engines for jet aircraft.

A typical magnetic bearing (see Figures 1-12 and 1-13) consists of the following subsystems:

- Ferromagnetic rotor
- Multipole slotted stator with normal conductor windings in the slots
- Position-sensitive proximity gages (e.g., inductance impedance sensors) to measure the position of the shaft in two planes

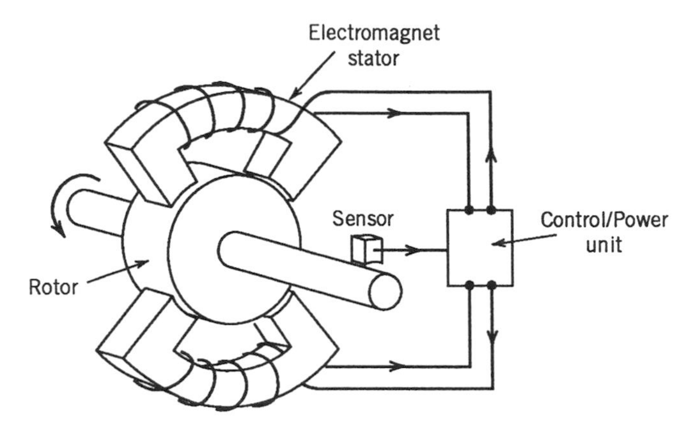

Figure 1-12 Sketch of an active control magnetic bearing.

Figure 1-13 Photograph of commercial high-speed magnetic bearing for machine tools (30,000–180,000 rpm). (Courtesy of Koyo Seiko Corp., Osaka)

- Electronic analog or digital control board
- Power supply to feed current to the stator windings
- Silicon–iron or cobalt–vanadium steel to provide magnetic circuit paths

The typical working gap is on the order of 0.5 mm. The operating temperature ranges have been as low as 60 K to above 100°C. However, development is underway to push operating temperatures to near 300°C for operation in gas turbine engines.

One of the more spectacular applications of S2M's magnetic bearings has been in natural gas pipeline systems. These bearings, retrofitted to centrifugal gas compressors, support a rotor weight of 1500 kgm, provide better reliability, and prevent leakage of bearing oil contamination in the pipeline, which was a problem with conventional mechanical bearings.

Some designs of active magnetic bearings use a hybrid permanent magnet levitation with active electromagnet control [see e.g., Weh (1989)]. One such system designed by Avcon Advanced Controls Technology, Inc. of Northridge, California uses rare earth magnets to provide the primary lift force between the stator and the rotor and uses separate feedback-controlled electromagnets to provide stability and positive magnetic stiffness. In an application for NASA, Avcon has designed a 450-kg lift system for a cyropump that operates at 15,000 rpm.

Passive Superconducting Bearings

Earlier proposals and experiments in superconducting levitated bearings began in the 1950s and 1960s. For example, Harding and Tuffias (1960) of the Jet Propulsion Laboratory in California built a levitated niobium sphere at liquid helium temperatures (4.2 K) for a superconducting gyro (see Section 6-1 for other references). However, the necessity to work with liquid helium discouraged further work in superconducting bearings. After the discovery of YBCO in 1987, many laboratories around the world began to build prototype high-temperature superconducting bearings for temperatures near that of liquid nitrogen (78 K). One of the early prototypes was built at Cornell University in 1987, and it was spun up to 10,000 rpm. A later version in 1988 (Figures 1-14 and 1-15) levitated a 5-g rotor with two YBCO journal bearings up to speeds of 120,000 rpm. Subsequently, the levitation of small rotors (10 g) with speeds of 500,000 rpm (have been

Figure 1-14 Photograph of a prototype passive magnetic bearing using high-temperature superconducting material, $YBa_2Cu_3O_7$ (50,000–120,000 rpm) at Cornell University.

reported by Allied Signal Corporation in 1992, and larger rotors of 1–10 kg on thrust bearings have been rotated to speeds of 5000–30,000 rpm at other laboratories.

Initial application programs by several companies in the United States (e.g., Allied Signal Corp. and Creare, Inc.) have focused on small levitated turbines for small cryocoolers for long-time space applications. These applications have natural cryogenic environments, and the bearing lifetimes of 5–15 years without maintenance gives superconducting bearings a potential advantage over conventional rolling elements or gas bearings.

As material improvements have progressed through dramatic increases in critical currents and magnetization in superconductors, the sights of the superconducting bearing community have been raised to consider higher-load machines such as heavier cryopumps and energy storage flywheels. Initially, magnetic bearing pressures were low, in the range below 1 N/cm^2. However, new material processing techniques, especially the use of the melt-quench method of preparing YBCO, has dramatically improved lift capability. Lift pressures in superconducting bearings depend not only on the material, but also

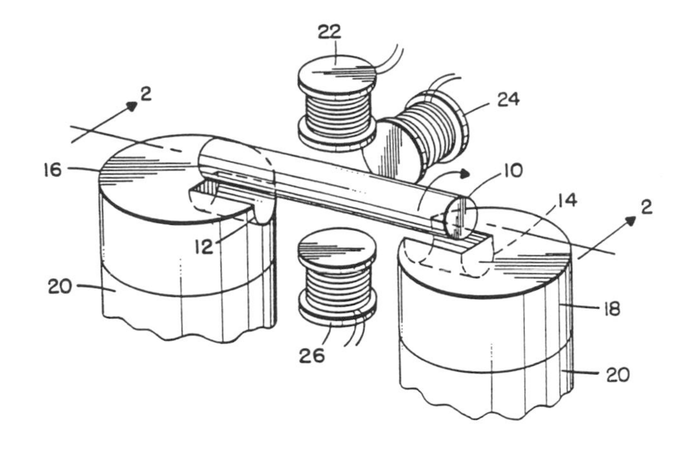

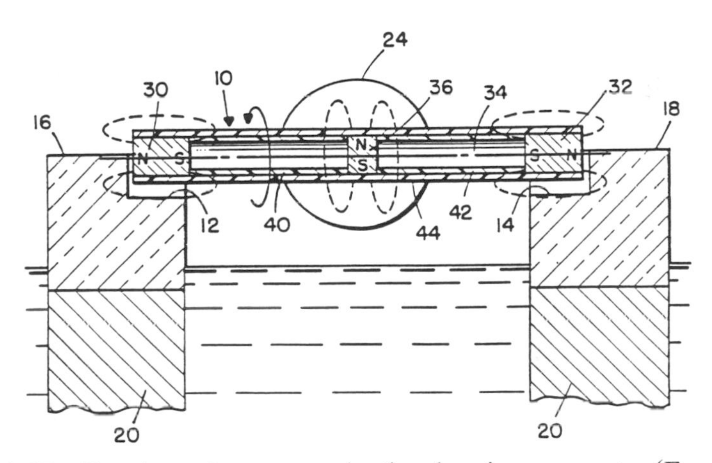

Figure 1-15 Sketches of superconducting bearing concepts. (From Moon and Raj, 1989)

on the field source. In almost all studies to date, this source has been rare earth magnets such as samarium–cobalt or neodymium–iron–boron which can only produce fields in an open magnetic circuit of around 0.5 T. Using the concepts of magnetic stresses in Section 1-4, a 0.5-T field will produce a stress on the order of 10 N/cm^2 (14 psi). (Flux compression in some bearing devices can increase this value.) However, if the permanent magnet is replaced with a higher field source such as a superconducting permanent magnet [see e.g., Weinstein and Chen (1992)] or a superconducting wire coil [Moon et al. (1993)], then higher magnetic pressures of over 100 N/cm^2 are possible.

Another potential limitation of passive superconducting bearings is the magnetic stiffness. Active magnetic bearings now have stiffnesses of over 10^6 N/m. Passive bearings have had measured stiffnesses of 10^3–10^4 N/m. However, recent measurements at Cornell using a 2-T superconducting coil as a field source have produced stiffnesses of over 10^5 N/m. It is known, however, that because magnetic forces depend on field gradients, it may be possible to significantly increase magnetic stiffness through clever magnetic field design. Promising work along this line has begun at the University of Houston [see, e.g., Chu et al. (1992)].

1-6 MAGNETICALLY LEVITATED VEHICLES

To skim over the ground with no visible means of suspension has long been a dream of science fiction writers. Now this dream is reality, at least at a few research centers. People-carrying vehicles can be suspended, guided, and propelled by electromagnetic fields. Designs for revenue systems range from low-speed airport to city-center service, to intercity vehicles that travel at 500 km/hr [see e.g., Rhodes and Mulhall (1981)]. While the steel-wheeled vehicle on steel rails has long been the most efficient way to travel on land, at high speeds dynamic problems can cause increased noise, vibration, and high maintenance costs.

There are two principal methods for magnetically levitating vehicles. The first, shown in Figure 1-16*a*, called electromagnetic levitation (EML) or attracting levitation, uses nonsuperconducting electromagnets which suspend the vehicle below ferromagnetic rails. The natural tendency of magnets to slam up against the steel rail is overcome by feedback control of the currents. The suspension gap is of the order of 1 cm (see Chapter 7). A recent design by the Grumman Corp. has proposed using a superconducting electromagnet and a much larger gap.

The second method, shown in Figure 1-16*b*, called electrodynamic levitation (EDL) or repulsive levitation, employs large superconducting magnets on the vehicle which generate eddy currents in a conducting track below the vehicle. Lift is developed when the vehicle moves, reaching an asymptotic limit at high speeds (see Figure 1-17). The associated magnetic drag increases at first with speed and then decreases with increasing velocities. The magnetic lift-to-drag force ratio for passenger-vehicle designs has ranged from 20 to 100. The dependence of lift and drag forces on speed can be put in nondimensional

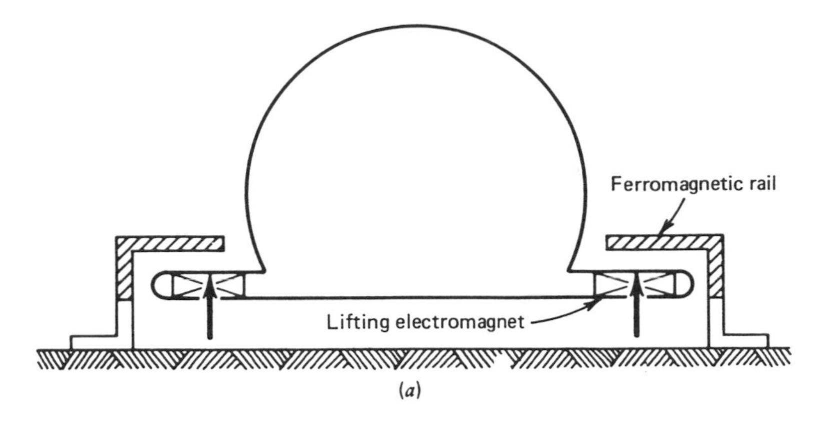

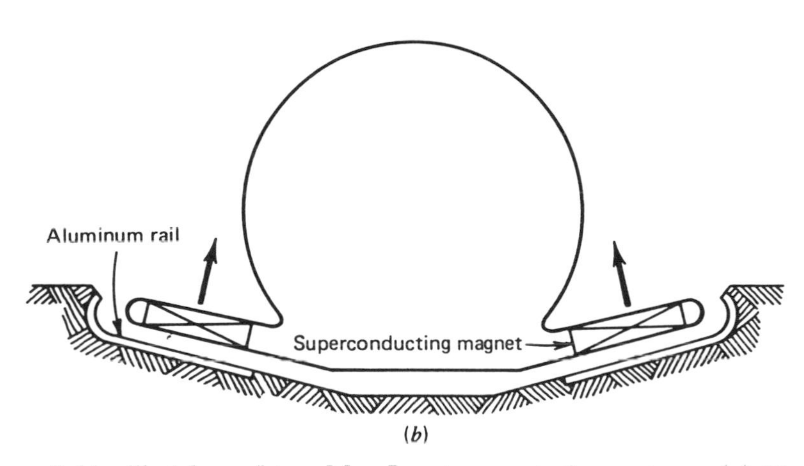

Figure 1-16 Sketches of two Mag-Lev transportation systems. (*a*) Electromagnetic levitation (EML) or attractive method. (*b*) Electrodynamic levitation (EDL) or repulsive method.

form by plotting

$$\frac{F}{\mu_0 I^2} \quad \text{versus} \quad R_m = \frac{\mu_0 \sigma \Delta}{2} v \qquad (1\text{-}6.1)$$

where I is the total current in the magnet, R_m is the magnetic Reynolds number, and Δ is the track thickness.

The magnets proposed for the EDL method are made from superconducting wire. The large fields generated by superconducting magnets ($\sim$ 2–3 T) can create a levitation gap of up to 30 cm. Permanent

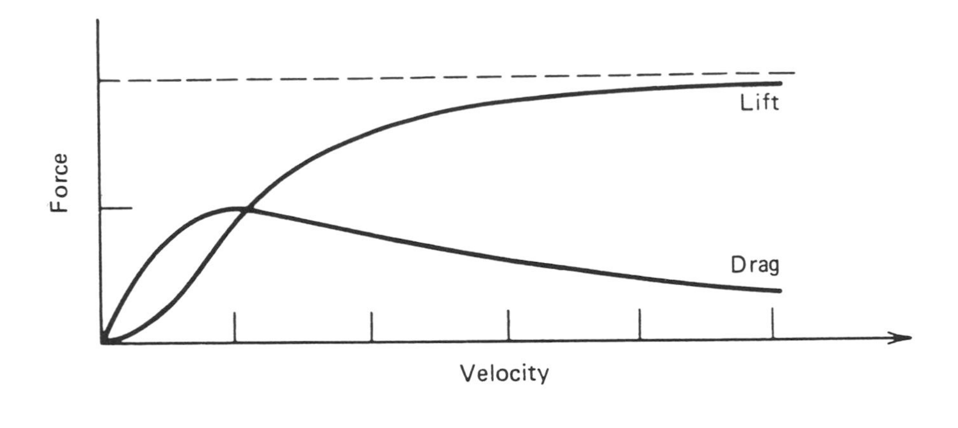

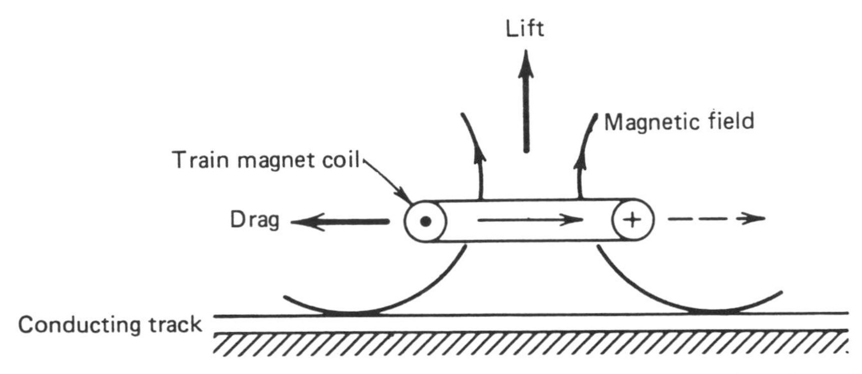

Figure 1-17 Magnetic lift and drag forces for the electrodynamic levitation method.

magnets can also be used on the vehicle for EDL, but they have a large weight penalty for full-sized vehicles.

Low-speed, people-moving Mag-Lev transportation using electromagnets have been employed in Birmingham (England) and in Berlin (called the M-Bahn). Another low-speed prototype, HSST, built by a group sponsored by Japan Air Lines, has carried over three million passengers at three different expositions in Japan and Canada.

At the high-speed level, two different demonstration Mag-Lev systems have been built. The Transrapid-07, an EML system built in Germany, is being marketed to run at speeds of up to 400–500 km/hr (Figure 1-18*a*). A test track in Emsland, Germany has been used for a decade to test several prototypes of this vehicle. The vehicle carries nonsuperconducting electromagnets which suspend the vehicle under a ferromagnetic rail. The currents in the electromagnet are constantly

(*a*)

(*b*)

Figure 1-18 (*a*) German EML vehicle, Transrapid-07. (*b*) Japanese EML vehicle, HSST.

adjusted using gap sensors and feedback control. Active coils in the guideway are used to propel the vehicle forward. A revenue system is planned for Orlando, Florida, the site of Disney World. A longer line was envisioned between Hamburg and Berlin in Germany in 1992, but funds have yet to be allocated. Another EML system designed for lower speeds called HSST, was initially sponsored by Japan Air Lines (Figure 1-18*b*).

A superconducting EDL Mag-Lev prototype system called the Linear Motor Car has been built on a 7.0-km test track in Miyazaki, Japan. Since 1978, this center has developed several prototypes

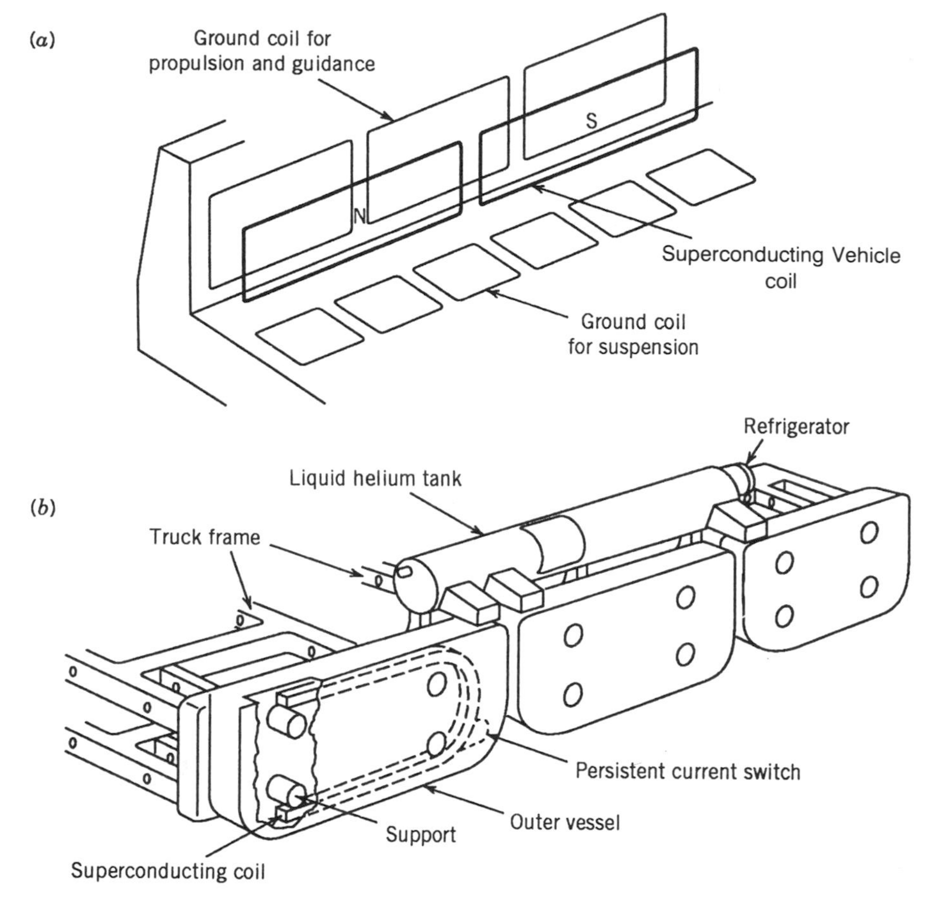

Figure 1-19 Japanese superconducting EDL vehicle magnets and guideway. (After Kyotani, 1988)

including one that achieved 500 km/hr. The latest version (Figure 1-1) uses low-temperature superconducting magnets on the vehicle and discrete copper coils in the guideway. Additional coils in the guideway carry active currents which are used to propel the vehicle. The superconducting magnets are wound with niobium–titanium wire in racetrack-shaped coils that generate magnetic fields of 2–3 T (Figure 1-19). The movement of this field source past the guideway coils generates eddy currents in the coils. The interaction between the currents in the guideways coils and the superconducting coils on the vehicle creates lift, drag, and guidance forces. The Japanese government is now building a demonstration line 43 km long on a section between Tokyo and Osaka in Yamanishi Prefecture.

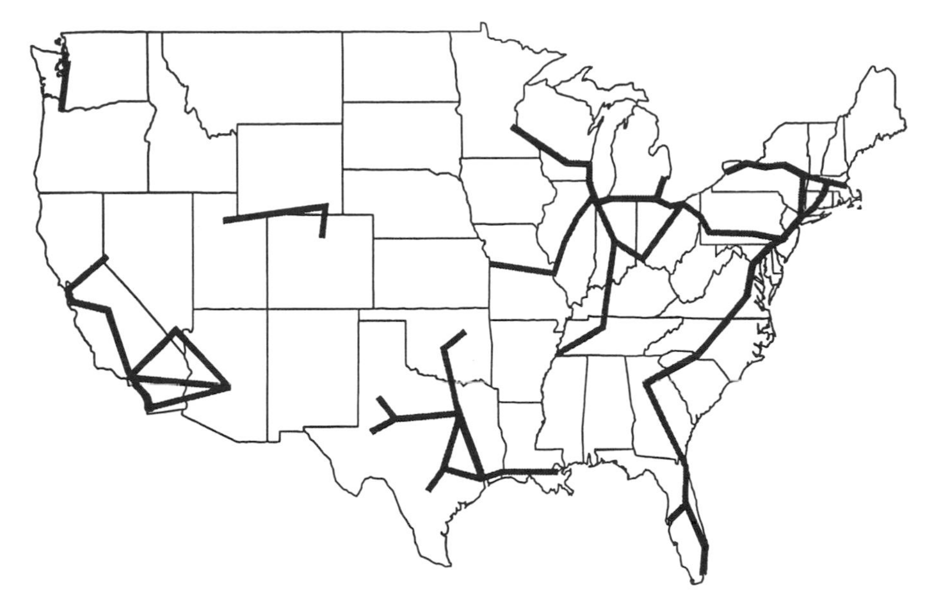

Figure 1-20 Map of potential Mag-Lev network in the United States.

In the early 1980s, Canada developed a superconducting Mag-Lev EDL design for a system between Montreal and Toronto, but at present there are no plans to build such a system [see e.g., Rudback et al. (1985)]. (See Figure 7-9)

Recently (1991–1992) the U.S. Department of Transportation initiated a National Mag-Lev Initiative (NMI) and sponsored four design studies for a superconducting Mag-Lev system. These designs are summarized in Chapter 7. Several corridors in the United States for potential application have been identified as shown in Figure 1-20. In 1992 the U.S. Congress passed a transportation bill called the Intermodal Surface Transportation Act which called for building a Mag-Lev prototype vehicle to compete with European and Japanese technologies. At the time of this writing (early 1994), continued research and development funding of this program had received approval by the U.S. Congress.

CHAPTER 2

PRINCIPLES OF MAGNETICS

The nation that controls magnetism will control the universe.

—Dick Tracy

2-1 BASIC LAWS OF ELECTROMAGNETISM

Socrates is reputed by Plato to have been aware of the attractive properties of the lodestone for pieces of iron. The Romans were also believed to have observed magnetic repelling forces between lodestones. However, the first extensive treatise on magnetism did not appear until 1600 when Gilbert, a court physician to Queen Elizabeth I, wrote *De Magnete*.

For 200 years, knowledge of magnetism did not advance much beyond Gilbert's work. In 1820, Oersted discovered that electric currents can produce magnetic fields, and Ampere found that coils of current-carrying wire would act as a magnet. Ampere also discovered the laws of force between currents; that is, parallel currents attract and antiparallel currents repel.

Finally, in the 1830s, Faraday in England and Henry in the United States discovered electromagnetic induction, namely, that a changing magnetic field would induce a voltage or current in a nearby conductor not in physical contact with the source of magnetic field.

Maxwell credits Faraday with the concept of magnetic tension and pressure, thus endowing the magnetic field with a more physical reality. Finally, Maxwell added the missing displacement current to the laws discovered by Faraday, Ampere, and Henry which resulted in his celebrated "Maxwell's equations," from which the wave nature of the electric and magnetic fields results.

There are many excellent classical treatments of electromagnetics, and we shall not attempt to do more than summarize the basic equations for low-frequency, low-velocity electromagnetics [see, e.g., Stratton (1941) or Jackson (1962)]. In this chapter we shall review the basic relations between the electric charges, currents, and magnetization and the electromagnetic fields, forces, and moments.

In electromagnetic problems, one uses physical quantities of charge Q (in coulombs), current I (in amperes), voltage V (in volts) and magnetic flux Φ (in webers).

In a continuum description of physical material, we use charge density q (C/m^3), current density $\mathbf{J}$ (A/m^2), and magnetization density $\mathbf{M}$ (A/m), where the MKSC system of units is used. The long-range interactions between material objects, separated in space, are described with the aid of auxiliary variables $\mathbf{E}$, $\mathbf{B}$, $\mathbf{D}$, and $\mathbf{H}$, which one assumes permeates all space inside and outside of the material objects of interest.

The difference between the $\mathbf{B}$ and $\mathbf{H}$ is related to the physical magnetization density in material bodies; that is,

$$\mathbf{B} = \mu_0(\mathbf{H} + \mathbf{M}) \tag{2-1.1}$$

where $\mu_0 = 4\pi \times 10^{-7}$ in MKS units and is called the *permeability of a vacuum*. Thus, outside a magnetizable body, $\mathbf{M} = 0$ and $\mathbf{B} = \mu_0\mathbf{H}$. When $\mathbf{M} = 0$, there is no real difference in the two fields $\mathbf{B}$ and $\mathbf{H}$ except the constant μ_0.

The magnetic dipole model is often represented as the limit of a small circuit of radius r and current I as $r \to 0$ and $I \to \infty$. The magnetic dipole $\mathbf{m}$ is defined by the equation (see Figure 2-1)

$$\mathbf{m} = \frac{1}{2}\oint \mathbf{r} \times \mathbf{I}\, ds \tag{2-1.2}$$

or for a circular circuit

$$\mathbf{m} = AI\mathbf{n}$$

where $A = \pi r^2$ and $\mathbf{n}$ is the normal in the direction of $\mathbf{r} \times \mathbf{I}$. Another magnetic dipole model is that of a small bar magnet. In analogy to the electric dipole, we have two magnetic poles of equal magnitude and opposite sense separated by a distance $\mathbf{d}$. However, in ferromagnetic materials, neither circulating microcurrents nor small bar magnets are responsible for the magnetization, but instead the magnetization is caused by a quantity called *spin* which resides in the electrons.

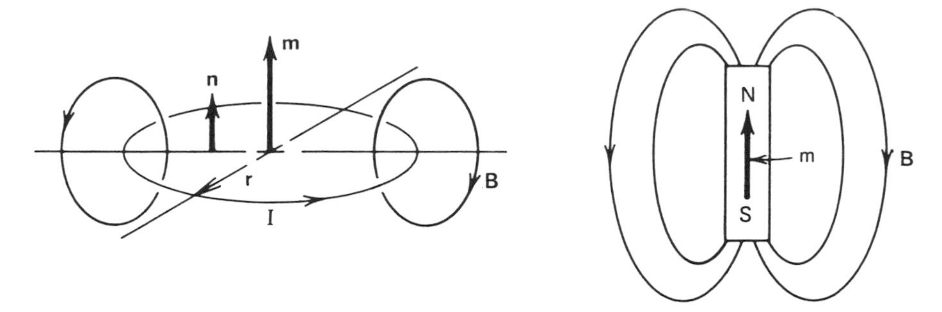

Figure 2-1 *Left*: Electric current dipole. *Right*: Magnetic dipole.

Although the current and pole models for elementary magnetism are physically artificial, they are useful concepts for mathematical models of magnetic materials.

Magnetic Stresses

An idea of great importance in magnetomechanics is that of magnetic stresses. The mathematical definition of these stresses is presented in Section 2-2. The electromagnetic stress concept is attributed to Faraday, who imagined that lines of tension act along field lines and that pressure forces act normal to field lines. For example, consider the attraction of north and south poles of two magnets as shown in Figure 2-2. Faraday imagined that the effect of each magnet on the other could be represented by lines of magnetic force (i.e., magnetic field lines) and that the attractive force was equivalent to tension forces acting along these lines. Thus one replaces the forces acting between the bodies by "stresses" acting in the field between them. The total force on the one pole face is given by the integration of these "stresses" over the pole face; that is,

$$F = \int_A \frac{B_n^2}{2\mu_0} \, da \tag{2-1.3}$$

where B_n is the normal component of the field to the surface.

Another illustration is shown in Figure 2-3, where current in a long cylinder with circumferential currents, called a *solenoid*, produces very high axial magnetic fields inside the cylinder and produces low-density fields outside the cylinder. The body forces on the currents near the

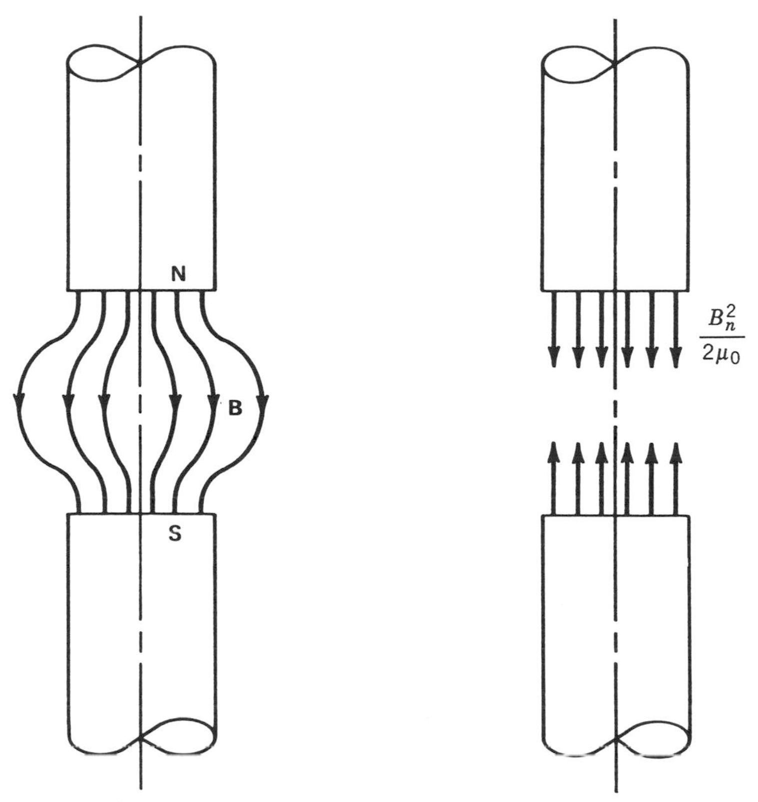

Figure 2-2 Magnetic field lines and magnetic stresses near magnetic poles.

center of the solenoid produce radial magnetic forces. However, these forces are replaced by a magnetic "pressure" acting on the inside surface of the cylinder of strength

$$P_m = \frac{B_t^2}{2\mu_0} \tag{2-1.4}$$

where B_t is the tangential component of the magnetic field to the cylinder surface. Thus a solenoid which produces a 1-T magnetic field will produce $10^3/8\pi$ N/cm^2 (57.7 psi) pressure on the inside wall of the cylinder. Of course, the actual forces are distributed over the current filaments in the cylinder wall. However, where the details of the stress distribution in the wall are not important, as in a thin-walled solenoid, the magnetic pressure concept is useful.

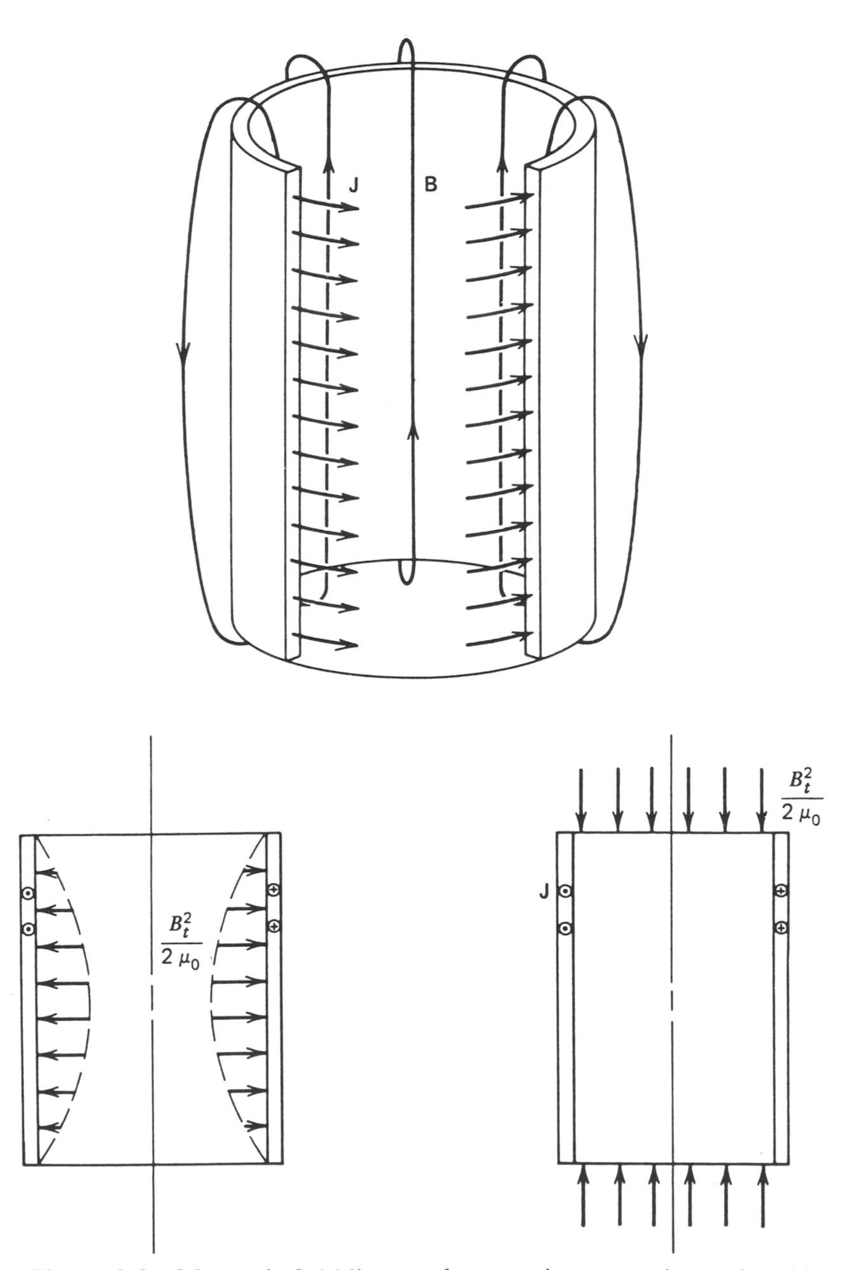

Figure 2-3 Magnetic field lines and magnetic stresses in a solenoid.

Nondimensional Groups

In any parameter study of a magnetomechanical problem, it is useful to reduce the number of independent variables by introducing nondimensional groups such as is done in thermal-fluid mechanics. One important nondimensional group is the ratio of magnetic to mechanical stresses. As noted in the previous section, B^2/μ_0 has units of stress or energy density. This can be nondimensionalized by the elastic modulus Y or by one of the six stress components t_{ij} forming the groups

$$\frac{B^2}{\mu_0 Y} \quad \text{or} \quad \frac{B^2}{\mu_0 t_{ij}}$$

These groups will appear where applied magnetic fields induce either currents or magnetization in the solid. For these problems the stresses and strains are proportional to the *square* of the applied magnetic field.

A nondimensional group that appears in electric current problems is

$$\frac{\mu_0 I^2}{Y\Delta^2}$$

where $\mu_0 I^2$ has units of force and Δ is a thickness or length variable. When more than one current appears, such as a coil with current I_1 in the field of another coil with current I_2, the following group will occur:

$$\frac{\mu_0 I_1 I_2}{Y\Delta^2}$$

When time enters the problem, such as frequency of an oscillating field ω or the length of a current pulse τ, the concept of *skin depth* becomes important. The skin depth is the depth to which an oscillating magnetic field can penetrate a conductor [see e.g., Reitz and Milford (1960)]. This length is given by

$$\delta_m = \left(\frac{2}{\mu_0 \sigma \omega}\right)^{1/2} \tag{2-1.5}$$

A natural nondimensional group is the ratio of δ_m to some geometric length such as the thickness of a plate—that is, δ_m/Δ. When $\delta_m/\Delta \ll 1$, we can sometimes assume that the magnetic field is zero inside the conductor.

When electric currents generate heat in a solid, one can define a thermal skin depth:

$$\delta_t = \left(\frac{2k}{c\omega}\right)^{1/2} \tag{2-1.6}$$

where k is the thermal conductivity and c is the heat capacity. Another natural nondimensional group is the ratio

$$\frac{\delta_m}{\delta_t} = \left(\frac{c}{\mu_0 \sigma k}\right)^{1/2} \tag{2-1.7}$$

When both magnetic field and temperature oscillate outside a good conductor, the magnetic field penetrates much deeper into the solid than does the temperature.

When a conductor moves in a magnetic field with velocity v, as in a superconducting Mag-Lev vehicle as it moves past a conducting guideway, there exists a nondimensional group similar to the Reynolds number in viscous-fluid mechanics. Hence we call this group a *magnetic Reynolds number*,

$$R_m = \frac{v\mu_0\sigma\Delta}{2} = \frac{v}{\omega} \tag{2-1.8}$$

where Δ is a length parameter. The quantity $\omega = 2/\mu_0\sigma\Delta$ represents a characteristic velocity parameter. For oscillating field problems, we can replace v by $\omega\Delta$ or

$$R_m = \frac{\omega\mu_0\sigma\Delta^2}{2} = \frac{\Delta^2}{\delta_m^2}$$

Maxwell's Equations

The differential forms of the laws of electromagnetics are known as *Maxwell's equations* [see e.g., Reitz and Milford (1960)]:

Conservation of charge

$$\nabla \cdot \mathbf{J} + \frac{\partial q}{\partial t} = 0 \tag{2-1.9}$$

Conservation of flux

$$\nabla \cdot \mathbf{B} = 0 \tag{2-1.10}$$

Gauss' law

$$\nabla \cdot \mathbf{D} = q \tag{2-1.11}$$

Maxwell's generalization of Ampere's law

$$\nabla \times \mathbf{H} = \mathbf{J} + \frac{\partial \mathbf{D}}{\partial t} \tag{2-1.12}$$

Faraday's law of induction

$$\nabla \times \mathbf{E} + \frac{\partial \mathbf{B}}{\partial t} = 0 \tag{2-1.13}$$

The preceding relations, known generally as Maxwell's equations, can be written in the form of balance laws:

Conservation of charge

$$\int_S \mathbf{J} \cdot d\mathbf{a} = -\frac{\partial}{\partial t} \int_{\mathscr{V}} q\, dv \tag{2-1.14}$$

Conservation of flux

$$\int_S \mathbf{B} \cdot d\mathbf{a} = 0 \tag{2-1.15}$$

Gauss' law

$$\int_S \mathbf{D} \cdot d\mathbf{a} = \int_{\mathscr{V}} q\, dv \tag{2-1.16}$$

Ampere's law

$$\oint_c \mathbf{H} \cdot d\mathbf{l} = \int_S \mathbf{J} \cdot d\mathbf{a} + \frac{\partial}{\partial t} \int_S \mathbf{D} \cdot d\mathbf{a} \tag{2-1.17}$$

Faraday's law

$$\oint_c \mathbf{E} \cdot d\mathbf{l} = -\frac{\partial}{\partial t} \int_S \mathbf{B} \cdot d\mathbf{a} \tag{2-1.18}$$

In the last two expressions, the area S is defined as that enclosed by the closed curve C. (See Figure 2-4.)

Low-Frequency Electromagnetics

The full set of Maxwell's equations leads to propagating wave solutions. Such wave-type solutions are important in the study of

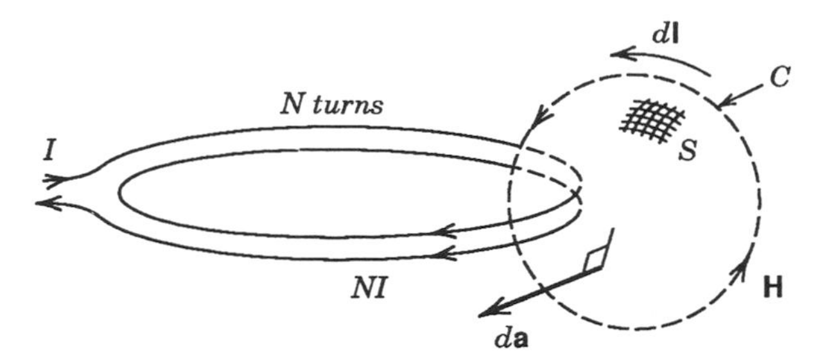

Figure 2-4 Sketch illustrating the relationship between the magnetic field and the electric current [Eq. (2-1.20)].

waveguides, antennas, and electromagnetic wave propagation and scattering problems. However, for most magnetomechanics problems, wave-type solutions in electromagnetic field variables are not required. This is because for frequencies of less than 10^7 Hz the wavelengths associated with such wave solutions are much longer than the circuits we are interested in. For example, the wavelength of a 10^6-Hz wave in air is 300 m ($c = 3 \times 10^8$ m/sec). The essential term in Maxwell's equations which leads to wave propagation is the displacement current $\partial \mathbf{D}/\partial t$. When this term is dropped, the equations take on the characteristics of either a diffusion equation or an elliptic equation. Neglect of $\partial \mathbf{D}/\partial t$ in Ampere's law is sometimes called the *quasi-static approximation*.

In low-frequency electromagnetics, Ampere's law in local form becomes

$$\nabla \times \mathbf{H} = \mathbf{J} \tag{2-1.19}$$

and in the form of an integral expression is

$$\oint_c \mathbf{H} \cdot d\mathbf{l} = \int_s \mathbf{J} \cdot d\mathbf{a} = NI \tag{2-1.20}$$

where NI is the total current penetrating the area enclosed by the circuit C in Figure 2-4.

Electromagnetic Constitutive Relations

Equations (2-1.10)–(2-1.18) are not sufficient to determine all the fields, because there are more unknowns than equations. Additional equations are needed which relate to material properties. These extra equations are analogous to stress–strain relations in mechanics. (A descriptive book on magnetic phenomena may be found in Burke, 1986.)

In a vacuum, $q = \mathbf{J} = 0$ and **B** and **H** are essentially the same fields; that is,

$$\mathbf{B} = \mu_0 \mathbf{H}$$

In the simplest theory of electromagnetics, material bodies are assumed to possess electric polarization **P** and magnetization **M**. These are defined by the equations

$$\mathbf{P} = \mathbf{D} - \varepsilon_0 \mathbf{E}$$

$$\mathbf{M} = \frac{\mathbf{B}}{\mu_0} - \mathbf{H}$$

Stationary Media

For a stationary, rigid body where **E** and **B** are considered to be independent, constitutive equations of the following form must be prescribed:

$$\begin{aligned} \mathbf{P} &= \mathbf{P}(\mathbf{E}, \mathbf{B}) \\ \mathbf{M} &= \mathbf{M}(\mathbf{E}, \mathbf{B}) \\ \mathbf{J} &= \mathbf{J}(\mathbf{E}, \mathbf{B}) \end{aligned} \qquad (2\text{-}1.21)$$

In the classical *linear* theory of isotropic rigid, stationary, electromagnetic materials these equations take the form

$$\mathbf{P} = \varepsilon_0 \eta \mathbf{E} \quad \text{or} \quad \mathbf{D} = \varepsilon_0 (1 + \eta) \mathbf{E} \qquad (2\text{-}1.22)$$

$$\mathbf{M} = \chi \mathbf{H} \quad \text{or} \quad \mathbf{B} = \mu_0 (1 + \chi) \mathbf{H} \qquad (2\text{-}1.23)$$

and

$$\mathbf{J} = \sigma \mathbf{E} \qquad (2\text{-}1.24)$$

The constant η is called the *electric susceptibility*, and χ is called the *magnetic susceptibility*. These constants as well as the electric conductivity σ can have a strong dependence on the temperature.

At low temperatures (between 0 and 200 K) many materials become superconducting; that is, $\sigma \to \infty$. In this state, steady closed currents can persist indefinitely. Thus voltage drops across superconducting circuits become zero for steady currents. However, for time-varying currents, an electric force is required to change the momentum of the electrons, which is proportional to **J**. Thus Ohm's law [Eq. (2-1.24)] is replaced by a relation of the form

$$\mathbf{E} = \mu_0 \lambda^2 \dot{\mathbf{J}}$$

However, this relation does not completely describe the macroscopic behavior of superconductors. Of equal or greater importance are the values of critical temperature T_c, magnetic field H_c, and current density J_c, at which the normal material becomes superconducting. The set of values of T, H, and J for which the material is superconducting is bounded by a surface $J_c = f(H_c, T_c)$. Further discussion of the properties of superconducting materials is given in Chapter 3.

Moving Media

It is known from elementary physics that the motion of a conductor in a steady magnetic field can create an electric field or voltage that can induce the flow of current in the conductor. Thus the electric field in a moving frame of reference ($\mathbf{E}'$) relative to that in a stationary frame ($\mathbf{E}$) differs by a term proportional to the velocity and the magnetic field (see Figure 2-5). This effect is one of the principal interactions between mechanics and electromagnetics, and is expressed mathematically by the relation

$$\mathbf{E}' = \mathbf{E} + \mathbf{v} \times \mathbf{B} \tag{2-1.25}$$

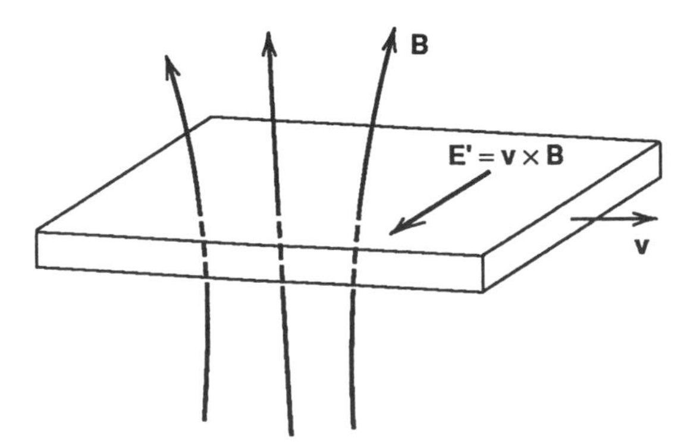

Figure 2-5 Induced electric field $\mathbf{E}'$ due to the motion of a conductor in a magnetic field [see Eq. (2-1.25)].

Similar relations hold for the other electromagnetic variables [see, e.g., Sommerfeld (1952)] but are not as important in magneto-mechanics problems as Eq. (2-1.25).

In the absence of net charge density in a conductor, the constitutive relation for current density in a moving medium becomes

$$\mathbf{J} = \sigma(\mathbf{E} + \mathbf{v} \times \mathbf{B}) \tag{2-1.26}$$

This expression is valid only for normal conductors. Thus, a moving conductor in a stationary magnetic field will have induced currents even if the initial electric field **E** and charge q are both zero.

2-2 MAGNETIC FORCES

Although isolated magnetic poles have not been found in nature, magnetic poles are a useful mathematical construct. The force between magnetic poles is analogous to the force between electric charges; that is, the force is proportional to the product of the strength of the two charges and is inversely proportional to the distance between them. If we represent the strength of the magnetic poles as P_1 and P_2, then in suitable units the magnitude of the magnetic force between these two poles is

$$F_{12} = \frac{P_1 P_2}{r^2} \tag{2-2.1}$$

where r is the distance between the two poles. The pole strength of a bar magnet with high aspect ratio is proportional to the product of the average normal magnetic field on the field on the end and the area of the pole force. Forces are, however, vector quantities; in the case of the force between two poles, it is repulsive if $P_1 P_2 > 0$ and lies along the vector between them. The vector nature of the magnetic pole–pole force can easily be expressed in terms of position vectors to each of the poles, $\mathbf{r}_1$ and $\mathbf{r}_2$. Then the vector force of pole P_1 on pole P_2 is given by

$$\mathbf{F}_{12} = \frac{P_1 P_2(\mathbf{r}_2 - \mathbf{r}_1)}{|\mathbf{r}_1 - \mathbf{r}_2|^3} \tag{2-2.2}$$

Note that $(\mathbf{r}_2 - \mathbf{r}_1)/|\mathbf{r}_1 - \mathbf{r}_2|$ is a unit vector from P_1 to P_2 and that $|\mathbf{r}_1 - \mathbf{r}_2|^2 = r^2$.

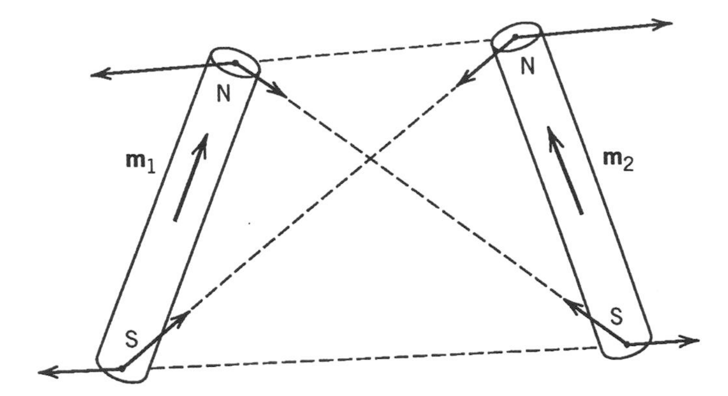

Figure 2-6 Magnetic forces on a dipole using the forces between magnetic poles.

Of course, all known magnetic bodies to date come in pairs of poles or dipoles. However, one can use Eq. (2-2.2) to calculate the forces between two dipoles, provided that we add all possible combinations of pairs of poles as illustrated in Figure 2-6.

Dipole–Dipole Forces

The dipole is a vector quantity because it has both magnitude and direction. A dipole is a mathematical construct where the magnetic body has equal and opposite pole strengths (i.e., $P_1 = -P_2$) and a direction along the vector from the negative to the positive pole, as illustrated in Figure 2-7. The magnetic dipole is represented by the vector symbol

$$\mathbf{m} = P\mathbf{l} \tag{2-2.3}$$

In the idealized case we have $\mathbf{l} \to 0$, and $\mathbf{P} \to \infty$ such that $P|\mathbf{l}|$ remains fixed.

The force between two dipoles can be calculated using the pole model in Figure 2-6. The results of this calculation is a complicated vector formula given by

$$\mathbf{F}_{12} = \frac{3\mu_0}{4\pi r^5}\left\{(\mathbf{m}_1 \cdot \mathbf{m}_2)\mathbf{r} + (\mathbf{m}_1 \cdot \mathbf{r})\mathbf{m}_2 + (\mathbf{m}_2 \cdot \mathbf{r})\mathbf{m}_1 - \frac{5(\mathbf{m}_2 \cdot \mathbf{r})(\mathbf{m}_1 \cdot \mathbf{r})\mathbf{r}}{r^2}\right\} \tag{2-2.4}$$

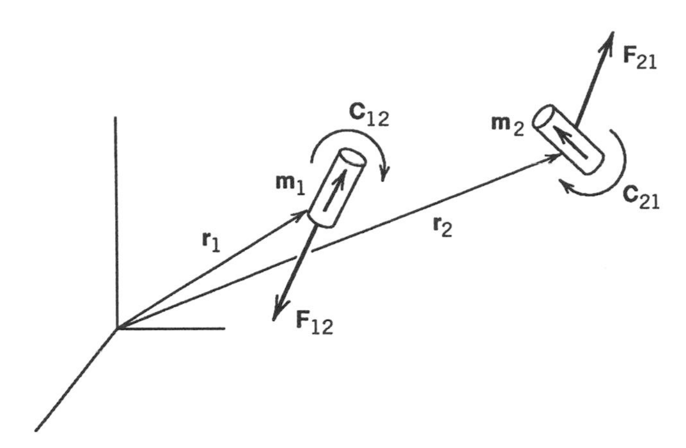

Figure 2-7 Magnetic forces and magnetic couples on two dipoles $\mathbf{m}_1$ and $\mathbf{m}_2$.

where $\mathbf{F}_{12}$ is the force on dipole #2 due to dipole #1, and $\mathbf{r}$ is the vector from $\mathbf{m}_1$ to $\mathbf{m}_2$.

In addition to a magnetic force, magnetic dipoles can experience torques or couples that try to change the vector orientation of the dipole. The torque on dipole #2 due to dipole #1 is given by

$$\mathbf{C}_{12} = \frac{\mu_0}{4\pi}\left(\frac{3(\mathbf{m}_1 \cdot \mathbf{r})\mathbf{m}_2 \times \mathbf{r}}{r^5} - \frac{\mathbf{m}_2 \times \mathbf{m}_1}{r^3}\right) \tag{2-2.5}$$

The torque $\mathbf{C}_{12}$ is a vector where the direction represents the axis about which the torque is applied (positive torque is given by the right-hand rule; see Figure 2-7).

Special Cases

1. Co-aligned Dipoles: $\mathbf{m}_1 \times \mathbf{m}_2 = 0$, $\mathbf{m}_2 \times \mathbf{r} = 0$ (Figure 2-8*a*). In this case the torque is zero. Also, the force is co-linear with the line between the two dipoles:

$$F = -\frac{3\mu_0}{2\pi} m_1 m_2 \frac{1}{r^4} \tag{2-2.6}$$

We see that the force varies as $1/r^4$, which is different than that for charges or poles which are inverse-square-dependent.

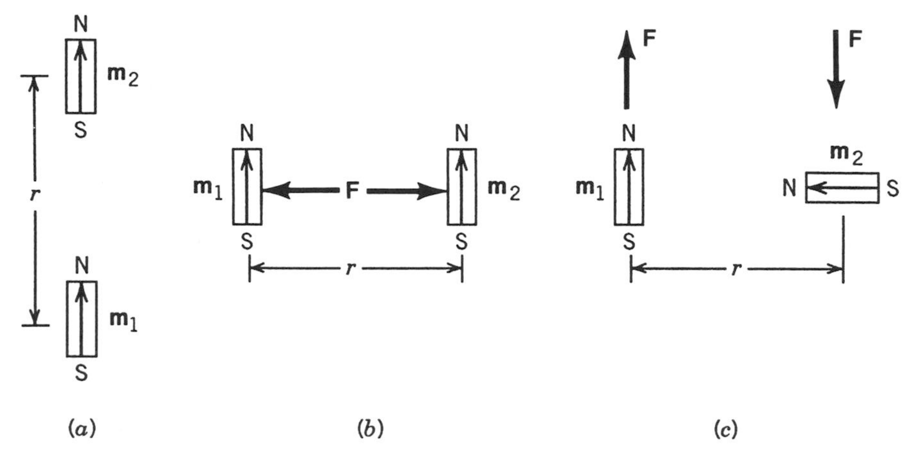

Figure 2-8 Three special configurations of pairs of magnetic dipoles [see Eqs. (2-2.6), (2-2.7), and (2-2.8)].

2. Parallel Dipoles: $\mathbf{m}_1 \cdot \mathbf{r} = 0$, $\mathbf{m}_2 \cdot \mathbf{r} = 0$ (Figure 2-8*b*). Here again, $\mathbf{C} = 0$. Also the force expression simplifies to an inverse fourth power law:

$$F = \frac{\mu_0}{4\pi} \frac{3m_1 m_2}{r^4} \tag{2-2.7}$$

3. Transverse Dipoles: $\mathbf{m}_1 \cdot \mathbf{m}_2 = 0$, $\mathbf{m}_1 \cdot \mathbf{r} = 0$ (Figure 2-8*c*). The force in this case is given by

$$\mathbf{F}_{12} = \frac{\mu_0}{4\pi} \frac{3(\mathbf{m}_2 \cdot \mathbf{r})\mathbf{m}_1}{r^5} \tag{2-2.8}$$

Thus, the force on dipole #2 due to dipole #1 is in the direction of $\mathbf{m}_2$. The torque on dipole #2, $\mathbf{C}_{12}$, is nonzero in this case and acts about an axis normal to the plane defined by $\mathbf{m}_1, \mathbf{m}_2$ with a magnitude

$$C_{12} = \frac{\mu_0}{2\pi} \frac{m_1 m_2}{r^3} \tag{2-2.9}$$

These formulas are sometimes useful for small permanent magnets whose physical dimensions are much smaller than the separation distance r.

Current – Current Forces

These formulas are based on the Biot–Savart law for steady currents. The simplest cases are treated in the classic physics textbooks. For example, consider the force between two parallel current filaments of infinite length with currents $\mathbf{I}_1$ and $\mathbf{I}_2$. The force per unit length of the wire is inversely proportional to the separation distance (Figure 1-6*a*):

$$F = \frac{\mu_0 I_1 I_2}{2\pi r} \tag{2-2.10}$$

This force is attractive if the sense of the two currents is the same.

The force on two filaments crossed at a right angle can be shown to be zero. In general the Biot–Savart law between two current-carrying circuits C_1 and C_2 (Figure 2-9) must be carried out numerically using the integral formula

$$\mathbf{F} = \frac{\mu_0 I_1 I_2}{4\pi} \oint_{C_1} \oint_{C_2} \frac{d\mathbf{s}_1 \times (d\mathbf{s}_2 \times \mathbf{R})}{R^3} \tag{2-2.11}$$

where $d\mathbf{s}_1$ and $d\mathbf{s}_2$ are differential vectors directed along the direction

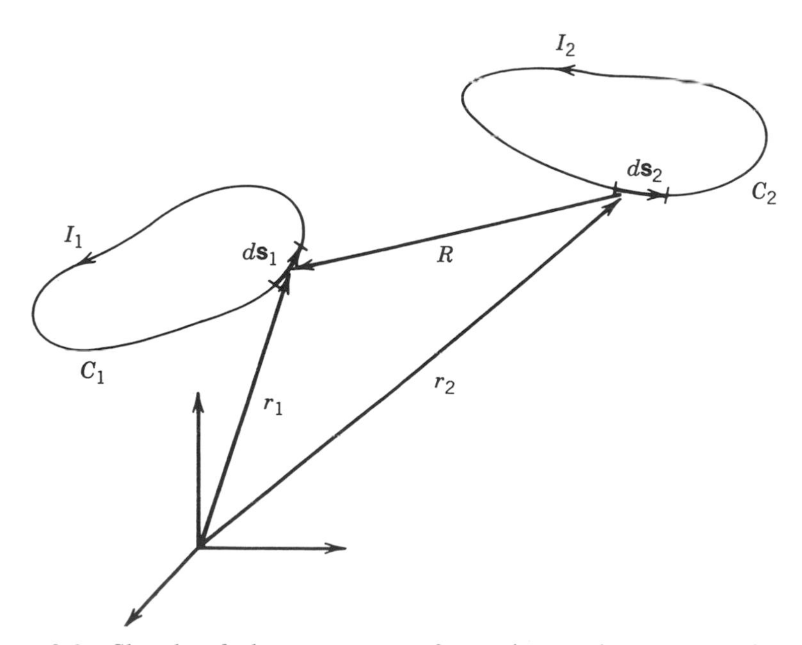

Figure 2-9 Sketch of the geometry of two interacting current filaments [Eq. (2-2.11)].

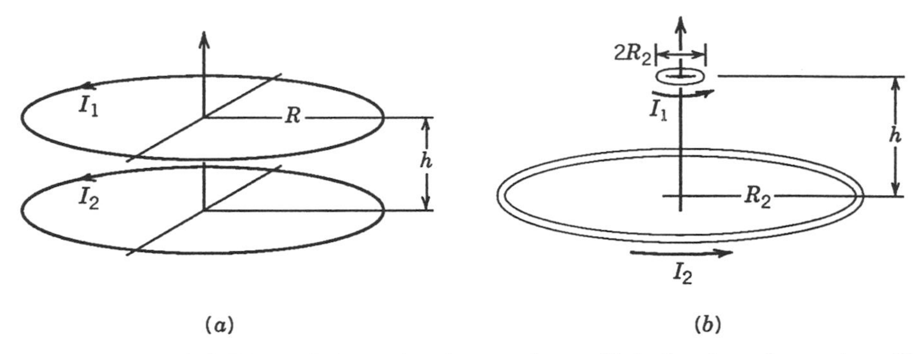

Figure 2-10 (*a*) Two planar, circular, and parallel circuits of equal radii [Eq. (2-2.12)]. (*b*) Two planar, circular, and parallel circuits with $R_2/R_1 \gg 1$ [Eq. (2-2.14)].

of the currents I_1 and I_2, respectively, and $\mathbf{R}$ is a position vector from $d\mathbf{s}_2$ to $d\mathbf{s}_1$.

One may also calculate the torque produced on one circuit due to the current in another circuit. The application of Eq. (2-2.11) to find the force between two circular rings whose axes are colinear can be found in terms of elliptic integrals of the first and second kind. When the radii of the two circuits are equal and the separation between the two planes of the circuits, h, is small compared with the radius, R, the total force is similar to the force between two parallel straight wires (see Figure 2-10a):

$$F = \mu_0 I_1 I_2 \frac{R}{h} \tag{2-2.12}$$

where the force acts along the axis normal to the circuits. Note that $\mu_0 I_1 I_2$ has units of force, so that in terms of nondimensional groups we obtain

$$\left(\frac{F}{\mu_0 I_1 I_2}\right) = \left(\frac{R}{h}\right) \tag{2-2.13}$$

Another special case is when the radius of one circuit R_1 is much smaller than the other R_2: $R_1 \ll R_2$ (Figure 2-10b). In this case the force is given by

$$F = -\frac{3}{2}\mu_0 I_1 I_2 \frac{\pi R_1^2 R_2^2 h}{\left(R_2^2 + h^2\right)^{5/2}} \tag{2-2.14}$$

We shall show in the next section that this can be thought of as the force between the circuit C_2 and a dipole of strength $I_2 \pi r^2$.

Current-Filament – Magnetized-Body Forces

Magnetic materials are often classified as soft and hard (see Section 2-4). Those that are soft do not retain magnetic flux or magnetization when the field sources are removed. However, the field generated by an electric circuit placed in the vicinity of soft magnetic material will induce magnetization or alignment of spins and thereby produce mutual forces between the circuit and the ferromagnetic body. An example is shown in Figure 2-11 where a current-carrying wire filament is placed parallel to a soft linear ferromagnetic material of relative permeability μ_r. The calculation of the magnetic attractive force can be performed by using an image method whereby the ferromagnetic half-space is replaced by a suitable current-carrying filament or by using the magnetic stress method [Eq. (2-1.3)].

Body-Field Forces

One can see from the preceding that the calculation of magnetic forces between material bodies by the direct method can be very complicated. When the magnetic sources are distributed, one must perform a double summation or double integral over both material bodies as in the case of two circuits [Eq. (2-2.11)]. By far the most popular method for calculating magnetic forces is the use of the

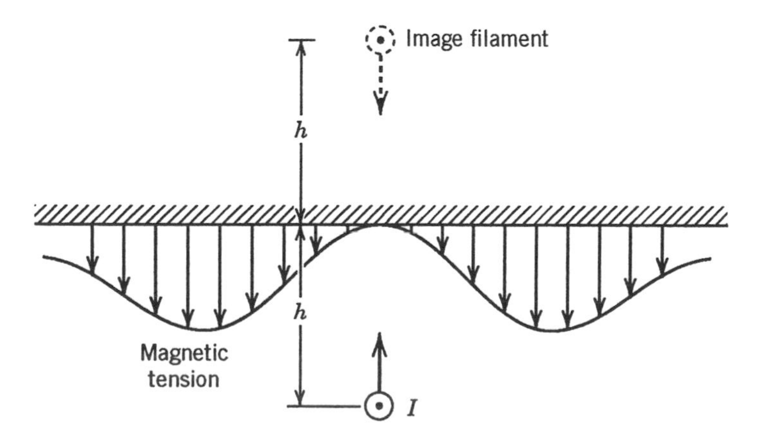

Figure 2-11 Magnetic force due to a current filament I near a ferromagnetic surface.

magnetic field distribution $\mathbf{B}(\mathbf{r})$. Earlier researchers in electromagnetism such as Faraday replaced the sources of magnetic effects by the force it would produce on a test unit magnetic pole. Thus, the conventional pictures of the magnetic field distribution actually give the direction and magnitude of the magnetic force on a unit pole as one moves the test monopole around in space; that is, for a pole of strength P we obtain

$$\mathbf{F}(\mathbf{r}) = P\mathbf{B}_0(\mathbf{r})$$

where $\mathbf{r}$ represents the position vector to some point in space, and the subscript in B_0 indicates that the field is due to all other magnetic sources excluding the test monopole.

Magnetic Dipole – Field Forces

With the above interpretation of the magnetic field as the force on a test monopole, it is easy to show that the force on a magnetic dipole of strength $\mathbf{m}$ due to other magnetic sources that are represented by the field $\mathbf{B}_0(\mathbf{r})$ is given by the gradient of $\mathbf{B}_0(\mathbf{r})$ (see Figure 2-12):

$$\mathbf{F} = \mathbf{m} \cdot \nabla \mathbf{B}_0(\mathbf{r}) \qquad (2\text{-}2.15)$$

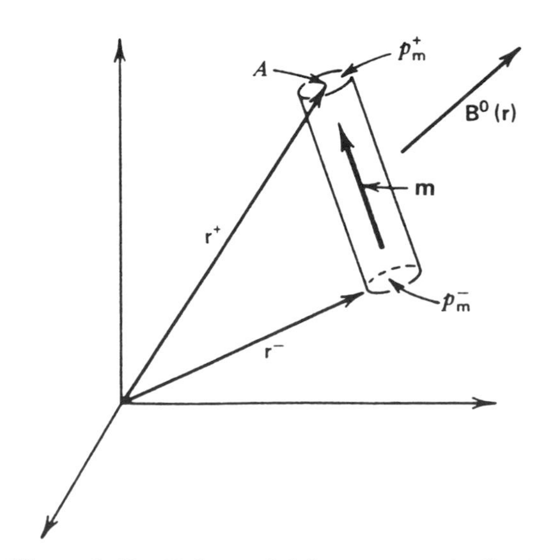

Figure 2-12 Pole model for a magnetic dipole.

Thus, if $\mathbf{r} = (x, y, z)$ in Cartesian coordinates, then

$$F_x = m_x \frac{\partial B_{0x}}{\partial x} + m_y \frac{\partial B_{0x}}{\partial y} + m_z \frac{\partial B_{0x}}{\partial z}$$

Current – Field Forces

Every student of basic physics learns that a current filament of length L directed along the x axis with strength I_x in a magnetic field along the y axis B_{0y} (due to other field sources) produces a force in the z direction (Figure 2-13a):

$$F_z = I_x B_{0y} L \tag{2-2.16}$$

This troika relation can be written in coordinate-independent form by using the vector cross-product operation

$$\mathbf{F} = (\mathbf{I} \times \mathbf{B}_0) L \tag{2-2.17}$$

When the current is not a straight-line element, one must integrate over the circuit elements (Figure 2-13b):

$$\mathbf{F} = I \oint d\mathbf{s} \times \mathbf{B}_0 \tag{2-2.18}$$

where $d\mathbf{s}$ lies along the filament direction.

For a volume distribution of current, where $\mathbf{J}$ represents the current density, the force is given by

$$\mathbf{F} = \int_{\mathscr{V}_1} \mathbf{J} \times \mathbf{B}_0 \, dv \tag{2-2.19}$$

Again note that the field $\mathbf{B}_0(\mathbf{r})$ is due to all other sources excluding that produced by $\mathbf{J}$ itself. The magnetic field $\mathbf{B}_1$ due to $\mathbf{J}$ can be calculated from the law [Eq. (2-1.19)]

$$\mathbf{B}_1 = \frac{\mu_0}{4\pi} \int \frac{\mathbf{J} \times (\mathbf{r} - \mathbf{r}') \, dv}{|\mathbf{r} - \mathbf{r}'|^3} \tag{2-2.20}$$

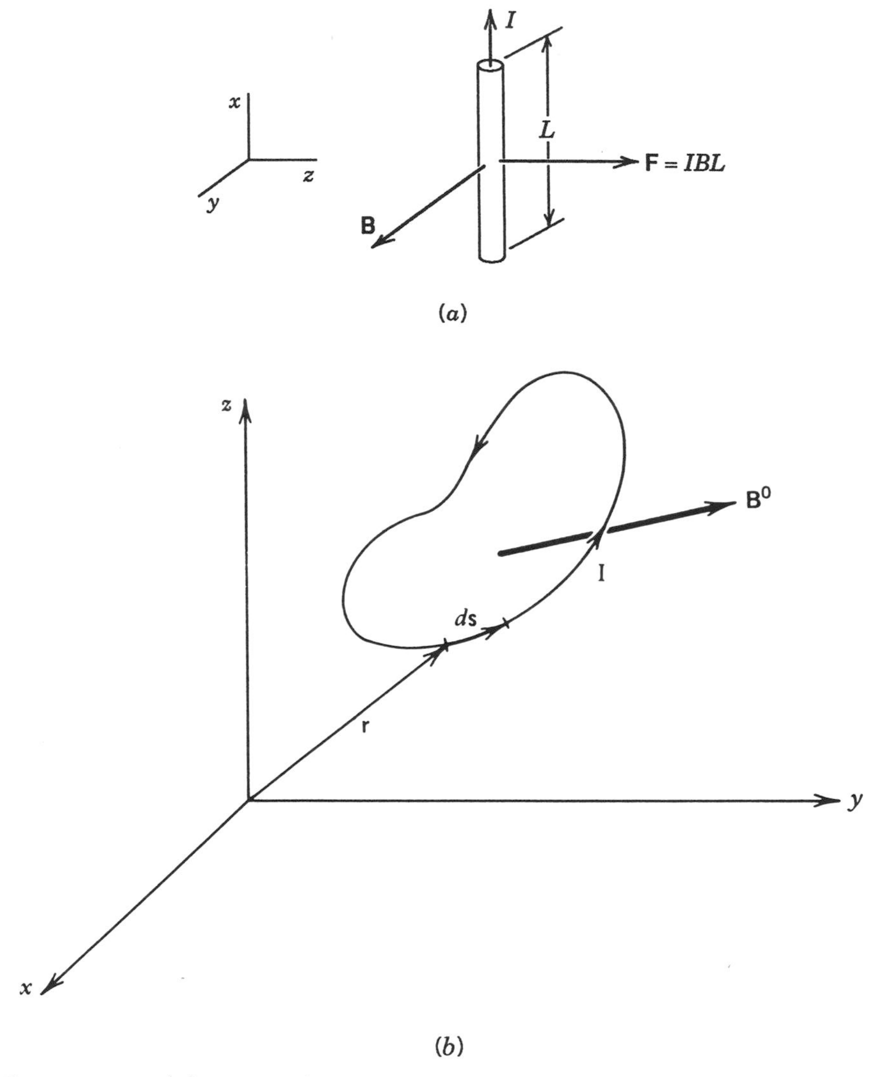

Figure 2-13 (*a*) Magnetic force on a current element in a transverse magnetic field [Eq. (2-2.16)]. (*b*) Sketch of a closed-current circuit in an external magnetic field **B**.

When the currents are contained inside a finite volume $\mathcal{V}$, one can show that the self-force is zero:

$$\int_{\mathcal{V}_1} \mathbf{J} \times \mathbf{B}_1(\mathbf{r})\, dv = 0$$

The Magnetic Stress Method

In the body-field method of calculating magnetic forces, one must separate out the field $B_0(\mathbf{r})$ due to all sources external to the test body. Sometimes, however, one only knows the total magnetic field $\mathbf{B} = \mathbf{B}_0 + \mathbf{B}_1$, where $\mathbf{B}_1$ is the self-field [Eq. (2-2.20)]. In this case, one can sometimes use the method of Faraday and Maxwell based on "magnetic stresses." This method uses the total field. A complete description of this method is beyond the scope of this monograph. The reader interested in more details should consult books such as Jackson (1962), Brown (1966), or Moon (1984).

In our discussion we will only talk about magnetic stresses in free space. The basic idea of the magnetic stress method is to calculate the force on the sources in $\mathscr{V}_1$ by integrating surface forces on an imaginary surface Σ surrounding $\mathscr{V}_1$. By analogy to forces on solid bodies, on each surface element of Σ, we can have magnetic tension or compression and magnetic shear. This is illustrated in Figure 2-14. The total magnetic field vector at the surface element dA with unit normal $\mathbf{n}$ is written in terms of a normal component B_n and the surface component B_s; that is,

$$\mathbf{B} = B_n\mathbf{n} + B_s\mathbf{s}$$

Then the magnetic force on this surface element consists of three terms, a magnetic tension σ_m proportional to the normal component, a magnetic pressure p_m proportional to the tangential field component, and a magnetic shear τ_m; that is,

$$\sigma_m = \frac{B_n^2}{2\mu_0}, \qquad p_m = \frac{B_s^2}{2\mu_0}, \qquad \tau_m = \frac{B_n B_s}{\mu_0}$$

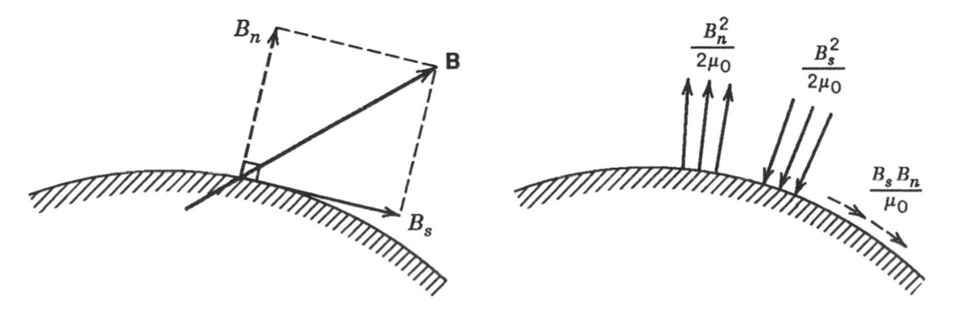

Figure 2-14 Normal and tangential components of the magnetic field vector at a surface and the corresponding magnetic stresses.

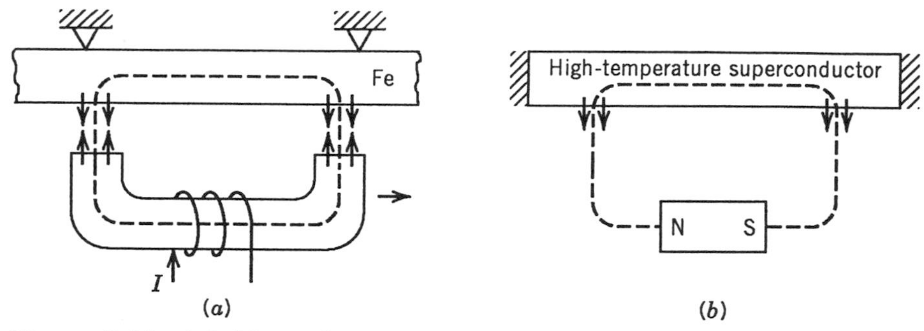

Figure 2-15 (*a*) Magnetic stresses between a ferromagnetic rail and an electromagnet in an electromagnetic levitation (EML) transportation system. (*b*) Magnetic stresses on a high-temperature superconductor due to a rare earth magnet suspended underneath.

or

$$\mathbf{dF} = \left[(\sigma - p_m)\mathbf{n} + \tau_m \mathbf{s}\right] dA$$

The total force on the magnetic sources in $\mathscr{V}_1$ due to all other sources external to $\mathscr{V}_1$ is then given by

$$\mathbf{F} = \frac{1}{2\mu_0}\int_{\Sigma_1}\left[(B_n^2 - B_s^2)\mathbf{n} + 2B_nB_s\mathbf{s}\right] dA \qquad (2\text{-}2.21)$$

where $\mathbf{n}$ is the outward normal from $\mathscr{V}_1$. This method is useful in calculating the force between an electromagnet and a ferromagnetic keeper or between a rare earth magnet and a superconductor, as shown in Figure 2-15.

2-3 MAGNETIC STIFFNESS

Magnetic levitation or suspension implies the existence of a magnetic force on the body which equilibrates the force of the earth's gravity—hence, the focus on calculation of magnetic forces. However, in the world of machine engineering, the ability of the magnetic forces to respond to changes in the position and velocity of the levitated body is sometimes of equal or greater importance than the gravity-canceling magnetic forces. In mechanical systems the change in force with position is usually proportional to the displacement. The proportional constant is known as the *elastic stiffness*. By analogy, the magnetic stiffness is the change in the magnetic force in a body due to a unit change in position of the body.

To be more precise, suppose that $\mathbf{r} = (x, y, z)$ is the position vector in space and that the magnetic force varies continuously with $\mathbf{r}$. Furthermore, suppose at $\mathbf{r}_0$ the magnetic force equals the gravity force on the body $m\mathbf{g}$. Then we can write

$$F(\mathbf{r}) = F(\mathbf{r}_0) + \left.\frac{\partial F}{\partial x}\right|_0 (x - x_0) + \left.\frac{\partial F}{\partial y}\right|_0 (y - y_0) + \left.\frac{\partial F}{\partial z}\right|_0 (z - z_0) \tag{2-3.1}$$

If we define displacements $\mathbf{u} = (u_x, u_y, u_z) = \mathbf{r} - \mathbf{r}_0$, then we define the magnetic stiffness associated with displacement u_x as

$$\kappa_x = -\left.\frac{\partial F}{\partial x}\right|_0 \tag{2-3.2}$$

A positive magnetic stiffness means that the perturbation magnetic force is restoring as in an elastic spring and in most cases means that the levitation in that direction is stable. A negative stiffness, $\kappa < 0$ or $\partial F / \partial u > 0$, implies that the body will be unstable due to disturbances associated with that particular displacement. An example is shown in Figure 2-16. The magnetic force on the current element I_1 due to the long wire I_2 produces a vertical force. Small displacement of the coil in the vertical direction results in a restoring force or positive stiffness, $\kappa = -F_0/h$.

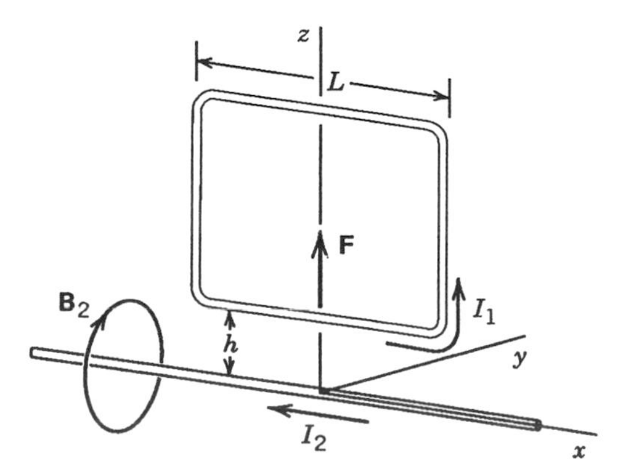

Figure 2-16 Magnetic force between a long magnetic filament with current I_2 and a rectangular circuit with current I_1.

It should be noted that, in general, each component of the magnetic force may depend on all three displacements (u_x, u_y, u_z), so that in principle there would be a magnetic stiffness matrix $\kappa_{\alpha\beta}$ $(\alpha, \beta = 1, 2, 3)$ with $\kappa_{\alpha\beta} = \kappa_{\beta\alpha}$.

Likewise the magnetic force on the body may depend on the rigid body orientation which may be described by three angles (say θ, ϕ, and ψ) and will likely depend on both the relative position and the angular orientation. Thus, one may extend the idea of magnetic stiffness to angular magnetic stiffness (e.g., $-\partial C/\partial\theta$).

The idea of magnetic stiffness has been generalized further to deal with deformable magnetic systems and has been used to understand the possible elastic collapse or buckling of arrays of superconducting magnets [see, e.g, Moon (1984)].

Magnetic Stiffness and Magnetic Energy

In mechanical systems the forces can sometimes be derived from a scalar energy function $\mathscr{V}(\mathbf{r})$; that is,

$$\mathbf{F} = -\nabla\mathscr{V} \quad \text{or} \quad F_x = -\frac{\partial\mathscr{V}}{\partial x} \tag{2-3.3}$$

Such systems are often called *conservative systems*. In magneto-mechanics the magnetic energy function serves a similar role. As an example, consider the interaction of two circuits shown in Figure 2-17. The magnetic energy can be shown to be a quadratic function of the

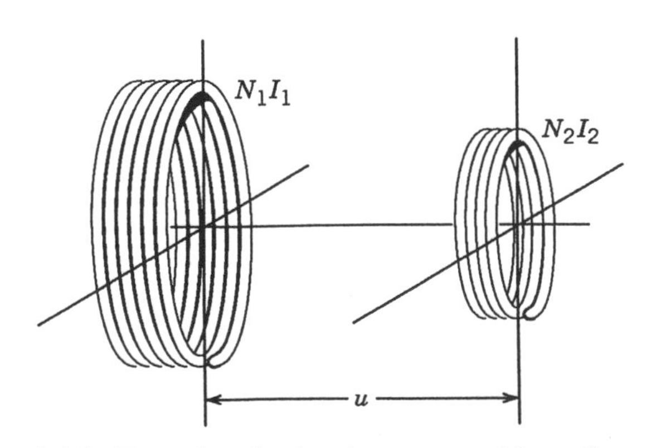

Figure 2-17 Two electric circuits separated by a distance u.

currents I_1 and I_2:

$$\mathscr{W} = \tfrac{1}{2}L_{11}I_1^2 + L_{12}I_1I_2 + \tfrac{1}{2}L_{22}I_2^2 \tag{2-3.4}$$

where L_{11} and L_{22} are called the *self-inductances* and L_{12} is called the *mutual inductance*. L_{12} gives the amount of magnetic flux threading circuit "two" due to a unit current in circuit "one". Hence L_{12} depends on the mutual displacement u in Figure 2-17. When the currents are independent of the relative position of the two circuits, it can be shown that the magnetic force between the circuits is given by [see, e.g., Moon (1984)]

$$F_x = \frac{\partial \mathscr{W}}{\partial u_x} \tag{2-3.5}$$

where the currents are held fixed when taking the derivative. In general, the self-inductances do not depend on the relative displacement u; that is,

$$\frac{\partial L_{11}}{\partial u} = 0, \qquad \frac{\partial L_{22}}{\partial u} = 0$$

Therefore, the magnetic force depends only on L_{12}; that is,

$$F_x = I_1 I_2 \frac{\partial L_{12}}{\partial u_x}, \ldots, \text{etc.} \tag{2-3.6}$$

If we can write L_{12} in a Taylor series in u_x, we obtain

$$L_{12} = L_0 + L_1 u_x + L_2 u_x^2$$

or

$$F_x = I_1 I_2 (L_1 + 2L_2 u_x) \tag{2-3.7}$$

Then the magnetic stiffness associated with the displacement u_x is given by

$$\kappa_x = -I_1 I_2 \left.\frac{\partial^2 L_{12}}{\partial u_x^2}\right|_0$$

or

$$\kappa_x = -2I_1 I_2 L_2 \tag{2-3.8}$$

If we consider all three force components F_i and displacements u_j, $i, j = \{x, y, z\}$, then one can show that the magnetic stiffness matrix is symmetric when there exists a magnetic force potential; that is,

$$\kappa_{ij} = -I_1 I_2 \frac{\partial L_{12}}{\partial u_i \, \partial u_j} = \kappa_{ji} \tag{2-3.9}$$

This means that there are, in general, six independent magnetic stiffnesses associated with a change in relative displacements of the two circuits. Similar symmetry properties hold if we consider the stiffness associated with relative angular positions of the two circuits.

Nonlinear Effects

The above discussion about magnetic stiffness implies that one can use the constructs of linear stability and vibration of mechanical systems by replacing the elastic springs by magnetic springs. However, caution must be exercised. Magnetomechanical systems are highly nonlinear and sometimes hysteretic (see Chapters 3, 5). This means that for larger departures from equilibrium, the magnetic stiffnesses can change drastically and can even change sign. In some systems, the complete determination of stability depends on dynamic effects, so that the dependence of magnetic forces on velocity may also be important (see Chapter 5).

2-4 MAGNETIC MATERIALS AND MAGNETIC CIRCUITS

The most familiar source of magnetic flux are permanent magnets. In the last two decades, dramatic improvements in the flux density of these materials have been made. The magnetics designer can now choose off-the-shelf magnets in many shapes and sizes. The most powerful magnets are usually based on the elements of iron, nickel, and cobalt—hence the term *ferromagnetic materials*. The highest field magnets to date are the so-called rare earth magnets (REMs) such as samarium–cobalt or neodymium–boron–iron. [For a discussion of the physics of magnetic materials see Cullity (1972).]

The most familiar flux-shaping magnetic materials are, of course, the iron-based alloys such as silicon-iron used in transformers. Such materials become magnetized only in the presence of an external field source such as a current-carrying coil. These so-called "soft" magnetic

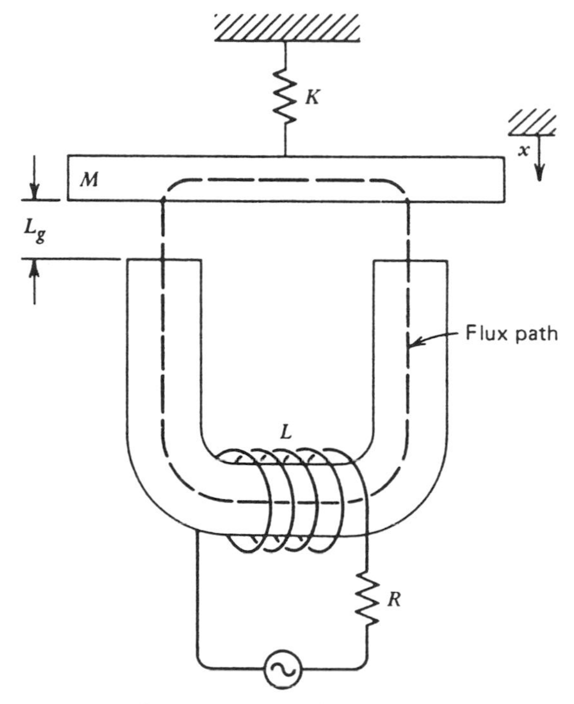

Figure 2-18 Electromagnet with ferromagnetic keeper.

materials create induced flux when magnetized such that the total magnetic flux is confined to some limited volume magnetic flux circuit as shown in Figure 2-18. The goal of the magnetics designer is to optimize the combination of flux source and flux shapers to meet some specification such as maximum flux density or magnetic force or flux gradient. This process involves a choice of materials, knowledge of material properties, and the specific geometric shapes and dimensions.

In this section we will briefly review some of the basic properties of magnetic materials and their use in magnetic circuits, especially as these properties relate to magnetic levitation. This review can only be given from a phenomenological perspective. The physics, chemistry, and materials processing of ferromagnetic material comprises a fascinating subject, which is too lengthy for this short treatise on levitation. For further reading on the physics and processing of permanent magnet materials, the reader is referred to Parker (1990). For further study on magnetic circuits and applications of permanent magnets, the reader should consult McCaig and Clegg (1987).

Magnetization Properties

The force on a magnetized body in the field of another source depends on the magnetization **M** or magnetic dipole strength as shown in Section 2-3. We also know that there are two measures of the magnetic field, namely, the flux density **B** and the magnetic field strength **H**. In MKS units **B** and $\mu_0\mathbf{H}$ have the same units. From Maxwell's equations the change in **B** is related to the electric field **E**, while the integral of **H** around a circuit is proportional to the current encircling the path of integration. In a vacuum, or nonmagnetic medium, **B** and $\mu_0\mathbf{H}$ are identical. Inside a magnetic body, however, these two fields are different, and the difference between them is the dipole magnetization per unit volume:

$$\mu_0\mathbf{M} = \mathbf{B} - \mu_0\mathbf{H}$$

or

$$\mathbf{B} = \mu_0(\mathbf{H} + \mathbf{M}) \tag{2-1.1}$$

For linear magnetic material,

$$\mathbf{M} = \chi\mathbf{H} \quad \text{and} \quad \mathbf{B} = \mu_0(1 + \chi)\mathbf{H} \tag{2-1.23}$$

In general, however, the relationship between **M** and **H** is nonlinear and hysteretic; that is, **M** depends on the time history of the magnetizing field strength **H**. This relationship can be measured by an experiment on a toroidal-shaped specimen shown in Figure 2-19 in which **H** is proportional to the ampere-turns circling the torus and **B** is proportional to the voltage in the second coil. In this experiment, one usually plots **B** versus **H** as shown in Figure 2-20, where the path or history of

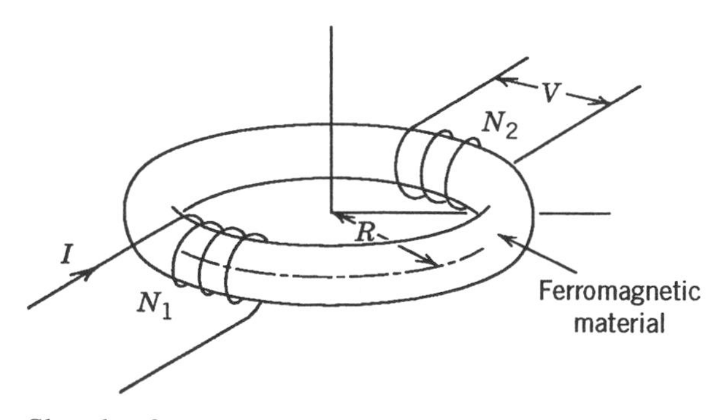

Figure 2-19 Sketch of experiment to measure magnetic properties of ferromagnetic materials.

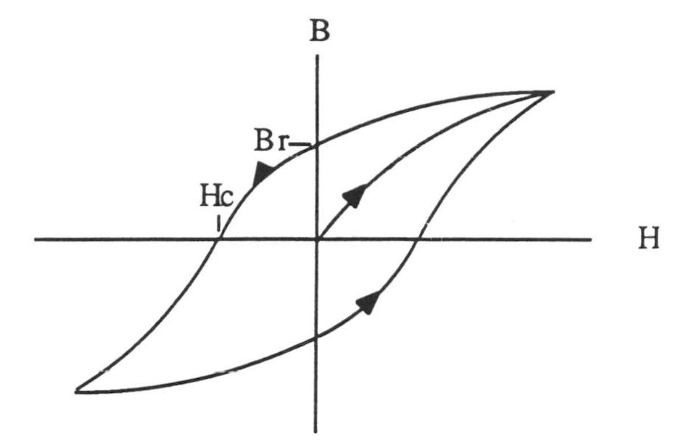

Figure 2-20 Magnetic flux density **B** versus magnetic field strength **H** for a typical ferromagnetic material.

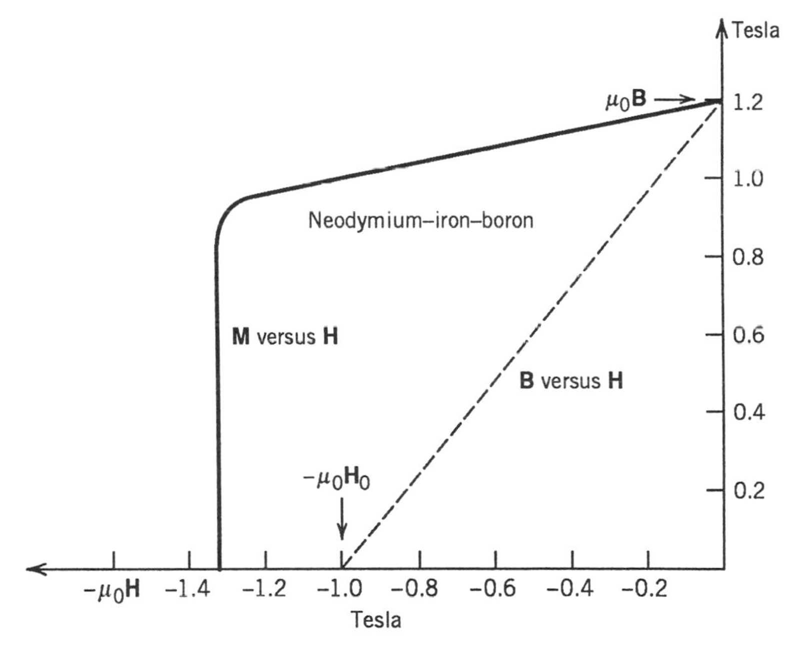

Figure 2-21 **B–H** curve and magnetization curve for commercial rare earth magnet neodymium–iron–boron.

the applied field **H** is shown by the arrows. Another plot is obtained by plotting **M** versus **H** as shown in Figure 2-21. In some books,

$$\mathbf{J} \equiv \mu_0 \mathbf{M} \equiv \mathbf{B}_i \tag{2-4.1}$$

is sometimes referred to as the *intrinsic magnetization* or the *magnetic polarization*. In typical ferromagnetic materials, **M** or **J** or $\mathbf{B}_i$ reaches a saturation value as the magnetic field strength **H** becomes very large, whereas **B** versus **H** continues to increase as **H** increases. The saturation of **M** in these materials is attributed to the nearly complete alignment of all the magnetic domains or spins of the ferromagnetic material when placed in a large magnetizing force **H**.

In the traditional **B–H** curve, B_r is known as the *remanence* or *residual magnetization*, and H_c is known as the *coercivity*. In typical rare earth ferromagnetic materials, $\mathbf{B}_r$ and $\mu_0\mathbf{M}_{\text{sat}}$ are between 1 and 2 T, with values closer to 1 T for most materials. [See e.g., Permag (1986).]

To obtain a B–H curve, one takes a toroidal specimen (e.g., a short length of a hollow cylinder) and wraps two pairs of coils around the torus as shown in Figure 2-19. Through the N_1 turns of coil #1, one puts a current I; and across the N_2 turns of coil #2, one measures the induced voltage $V(t)$, produced by the changing flux in the torus. The integral forms of Ampere's law and Faraday's law give the following relationships:

$$H = \frac{N_1 I}{2\pi R}$$
$$B = \frac{1}{N_2 A}\int_0^t V(\tau)\,d\tau \tag{2-4.2}$$

where R is the major radius of the torus, A is the cross-sectional area of the torus, and V is the voltage.

These properties, however, only hold for the closed toroidal specimen. If the torus is split or opened up, as in Figure 2-22, then the

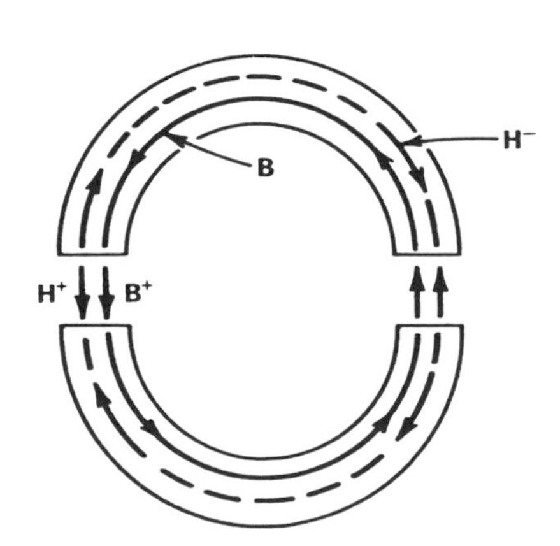

Figure 2-22 Split permanent magnetic toroid with air gaps.

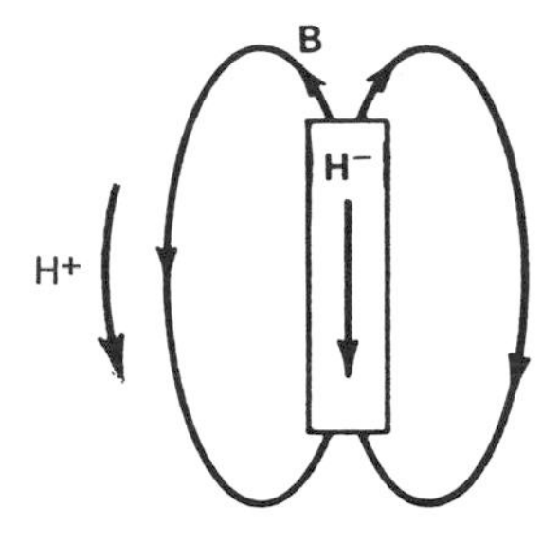

Figure 2-23 Permanent magnetic material showing demagnetization field H.

magnetic field just outside the material in the gap drops dramatically. To understand this nonintuitive behavior, one must understand the nature of a magnetic circuit.

It is important to note that if the flux lines leave the material, as in the bar magnet in Figure 2-23, then H inside the material cannot be zero and is negative. This follows from the fact that when $I = 0$ we obtain

$$\oint \mathbf{H} \cdot d\mathbf{s} = 0$$

Because $\mathbf{H}_{\text{out}} = \mathbf{B}_{\text{out}}/\mu_0 \neq 0$, H_{in} must be opposite the sense of B_{in}. Hence the importance of the demagnetization section of the B–H curve.

Another important feature of the B–H curve for permanent magnets is the *recoil curve* shown in Figure 2-24. Suppose our torus is split in two (see Figure 2-22), with $I = 0$ and no gap between the halves; then $\mathbf{H} = 0$ and we start at point a on the B–H curve in Figure 2-24. As the two halves are pulled apart by an amount u, the field $\mathbf{H}$ inside

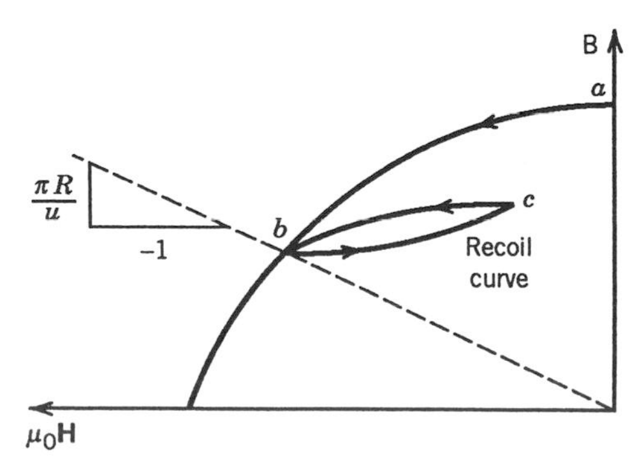

Figure 2-24 Demagnetization curve and recoil curve for a typical permanent magnetic material.

the material, $\mathbf{H}^-$, must satisfy the relation

$$\mu_0 H^- = -\frac{u}{\pi R} B(H) \tag{2-4.3}$$

H is found from the intersection of the curves $B = -\mu_0(\pi R/u)H$ and from the demagnetization curve (point b in Figure 2-24). However, if the two halves are moved back together, B does not retrace the demagnetization curve but follows another curve called the *recoil curve*, shown in Figure 2-24 as point c. In many ferromagnetic materials the slope of the recoil curve is approximately equal to the slope of the B–H curve at $H = 0$. This again shows the importance of the magnetization history of the material in determining the field B. In this respect, ferromagnetic constitutive behavior is analogous to plasticity in solid mechanics.

Domains

The source of spontaneous magnetization lies in a property of atomic particles called *spin*. Spin is a concept which endows atom particles with intrinsic angular momentum. Between the atomic scale of electron spin and macroscopic magnetization lies another substructure called *domains*. Below the Curie temperature (Table 2-1), electron

TABLE 2-1 Properties of Ferromagnetic Materials

Material	Curie Temperature (°C)	μ_r Initial	μ_r Maximum	Density (kg/m^3)	Saturation Field (tesla)
Fe (purified)	770	10^4	2×10^5	7800	2.15
Co	1131				
Ni	358				
Grain-oriented Fe-Si (3% Si)	740	7500	55,000	7670	2.0
78 Permalloy (78.5% Ni)	600	8000	10^5	8600	1.08
Supermalloy (5% Mo, 79% Ni)	400	10^5	10^6	8770	0.79
		Residual Flux Density, B_r (tesla)		Coercivity $\mu_0 H_c$ (tesla)	
Sintered Alnico	860	1.09		0.062	
SmCo 26	825	1.05		0.92	
Nd Fe B 35	310	1.23		1.16	

Source: American Institute of Physics Handbook, McGraw-Hill, New York.

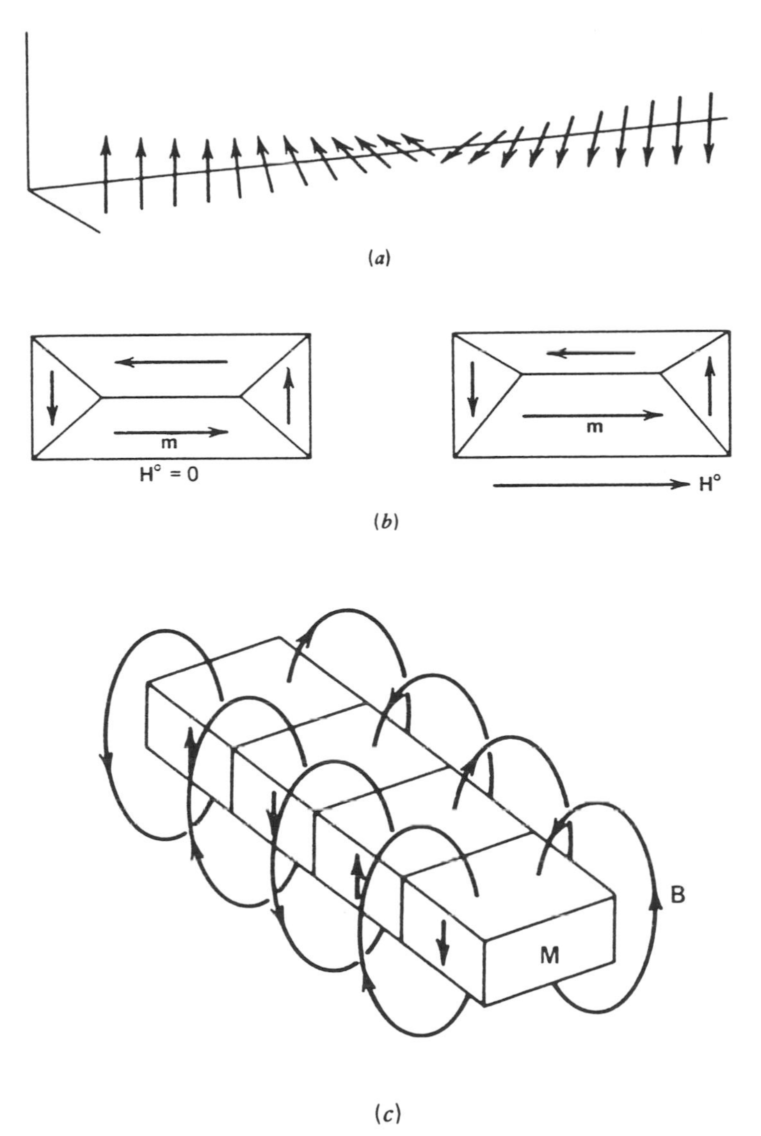

Figure 2-25 (*a*) Alignment of spins or magnetic moments in a linear array of atoms. (*b*, *c*) Alignment of magnetic moments in domains.

spins of neighboring atoms will spontaneously align themselves only over a certain number of atoms beyond which the alignment of magnetic moments will change direction as shown in Figure 2-25. Each region of aligned spins is called a *domain*. In an applied magnetic field, domains aligned with the field can grow at the expense of those that are not, thereby creating a net magnetization. Thus, while domains owe their microscopic magnetization to spin alignment, macroscopic magnetization **M** is an average of **m** over many domains.

Magnetic Circuits

In certain problems the magnetic flux is contained within well-defined flux paths, especially when ferromagnetic materials are present (see Figure 2-26). In such cases the concept of a magnetic circuit is sometimes useful. The flux in each tube of a magnetic circuit is defined by

$$\Phi(s) = \int_A = \mathbf{B} \cdot d\mathbf{a} \tag{2-4.4}$$

where s is a coordinate along the tube and $d\mathbf{a}$ is normal to the cross-sectional area of the tube. The conservation-of-magnetic-flux law [Eq. (2-1.15)] applied to a tube of varying cross section requires that

$$\Phi(s) = \text{constant}$$

If several flux tubes intersect, each carrying flux Φ_k, then the conservation of flux becomes

$$\sum \Phi_k = 0 \tag{2-4.5}$$

where positive flux is directed out of the junction.

In many applications the magnetic circuit is encircled by an electric circuit, as in a transformer (see Figure 2-26). To find the relation between the current in the electric circuit and the flux in the magnetic circuit, one uses the low-frequency form of Ampere's law (2-1.20). If

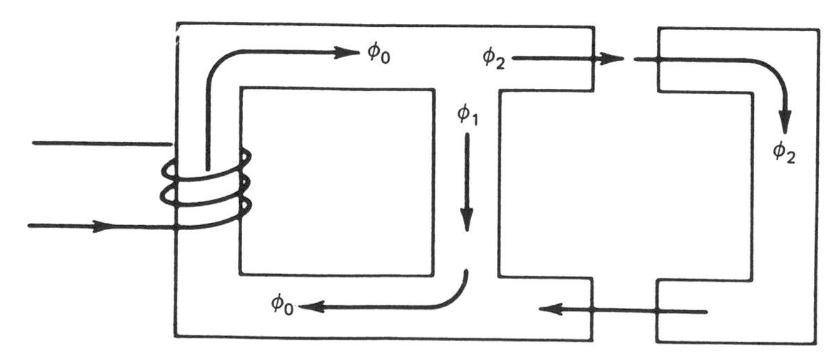

Figure 2-26 Ferromagnetic magnetic flux circuit.

we write **H** in the form $\mathbf{H} = \mathbf{B}/\mu$ and assume that the electric current I encircles the magnetic circuit N times, then we obtain

$$NI = \Phi\oint\frac{ds}{\mu A}, \quad \text{or} \quad \Phi = \frac{NI}{\Re} \tag{2-4.6}$$

The integral in Eq. (2-4.6) is called the *reluctance* $\Re$, where

$$\Re \equiv \oint\frac{ds}{\mu A} \tag{2-4.7}$$

and is analogous to the resistance in electric circuits.

In linear circuits the relation between flux and current is written in the form

$$\Phi = LI \tag{2-4.8}$$

where L is called the *inductance of the magnetic circuit*. Thus the voltage across an electric circuit which creates a flux Φ is given by

$$V = \frac{dLI}{dt} \tag{2-4.9}$$

Demagnetization

In many applications, permanent magnetic materials are used in a magnetic circuit with a gap as shown in Figure 2-27. For a given gap volume and gap magnetic field, there is an optimum circuit reluctance $\Re$ which will minimize the volume of permanent magnetic material required to generate a given flux density B_g in the gap. The magnetic intensity in the permanent magnet H_m has an opposite sense to the flux in the air gap, B_g. If λ_m and λ_g are the path lengths in the magnetic material and air gap, respectively, then

$$H_m\lambda_m + H_g\lambda_g = 0$$

where H in the soft magnetic material is assumed to be approximately zero (i.e., $\mu_r \gg 1$). If we multiply this expression by the flux in the circuit, $\Phi = B_g A_g = B_m A_m$, we obtain the following expression for

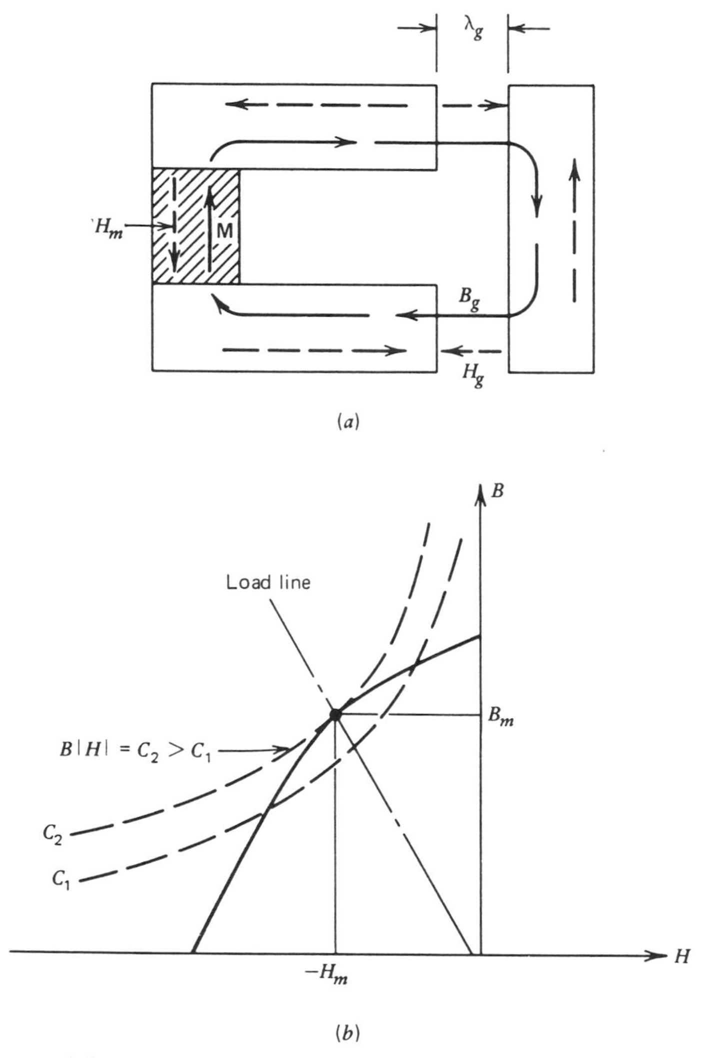

Figure 2-27 (*a*) Ferromagnetic circuit with permanent magnet field source (shaded element). (*b*) Demagnetization curve for permanent magnetic material, and the load line curve for the flux circuit.

the volume of the permanent magnet, $V_m = A_m \lambda_m$; that is,

$$V_m = -\frac{V_g B_g^2}{\mu_0 B_m H_m} \tag{2-4.10}$$

If B_m is assumed to be positive, then H_m is negative and B_m and H_m must lie on the demagnetization curve of the B–H diagram of the permanent magnetic material (see Figure 2-27). Thus, for a minimum volume of material, the product $|B_m H_m|$ should be a maximum. This defines a unique point on the B–H curve. Because the reluctance of the circuit determines the product $B_m H_m$, $\mathfrak{R}$ can be chosen to maximize $|B_m H_m|$ and minimize V_m.

Demagnetization of an Ellipsoid Self-demagnetization is also a concept applicable to soft magnetic materials where **M** is linearly proportional to **H**: $\mathbf{M} = \chi \mathbf{H}$. In the special case of an ellipsoid placed in a uniform magnetic field $\mathbf{B}^0$, the induced field inside the magnetized ellipsoid is uniform. If a, b, and c are the lengths of the principal semiaxis in the x, y, and z directions, respectively, then the induced magnetization may be related to the external field averaged over the particle volume:

$$\mathbf{M} = \frac{\chi}{\mu_0}\left(\frac{B_x^0}{1+\chi n_1}, \frac{B_y^0}{1+\chi n_2}, \frac{B_z^0}{1+\chi n_3}\right) \tag{2-4.11}$$

The numbers n_1, n_2, and n_3 are called *demagnetization factors* and are given by elliptic integrals.

In the special case of $b = c$, $a > b$, we have

$$n_1 = \frac{1-e^2}{2e^3}\left(\ln \frac{1+e}{1-e} \quad 2e\right) \tag{2-4.12}$$

where

$$e_2 = 1 - \frac{b^2}{a^2} \tag{2-4.13}$$

This function is shown plotted in Figure 2-28.

For a sphere ($a = b = c$), $n_1 = n_2 = n_3 = \frac{1}{3}$, and for a cylinder, $n_1 = 0$ and $n_2 = n_3 = \frac{1}{2}$. For a long needlelike particle, the couple acting on the particle will orient the particle such that $\mathbf{M} \times \mathbf{B}^0 = 0$, or the long axis will lie along the field lines.

Forces on Magnetized Material ($\mathbf{J} = 0$)

The law of conservation of flux $\nabla \cdot \mathbf{B} = 0$ implies that there are no isolated magnetic poles or monopoles; that is, a source of a magnetic field line must be accompanied by a sink. It is natural then that the

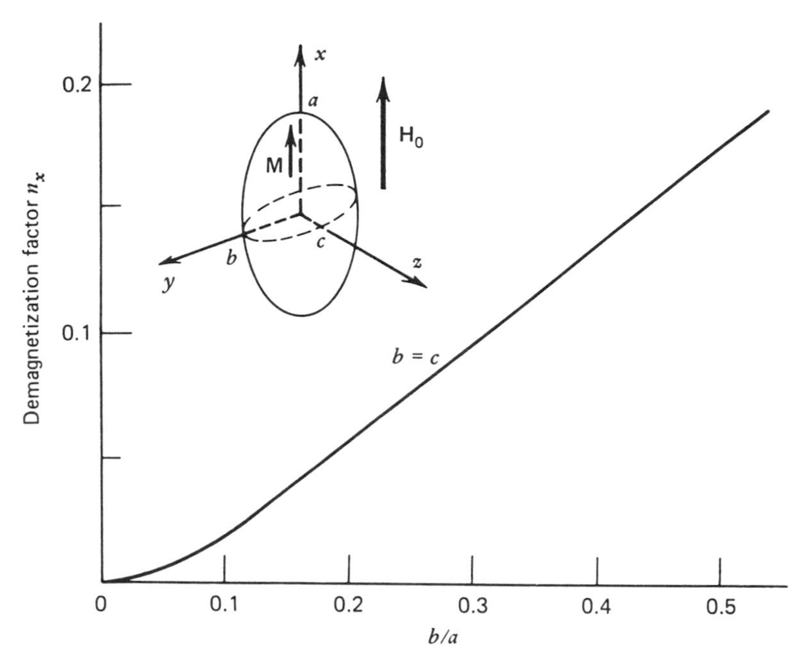

Figure 2-28 Demagnetization factor for an ellipsoid.

concept of a magnetic dipole is often used as the basic element of the continuum theory of magnetic materials because it contains its own source and sink. If a magnetic body can be thought of as a distribution of magnetic dipoles, then the magnetic field outside the body is a linear superposition of the fields produced by this dipole distribution. Thus, it can be shown that outside the body the magnetic field is given by

$$\mathbf{B} = \frac{\mu_0}{4\pi} \int_V \left(-\frac{\mathbf{M}}{r^3} + \frac{3\mathbf{M} \cdot \mathbf{rr}}{r^5} \right) dv \tag{2-4.14}$$

In Eq. (2-4.14), **M** represents a distribution of magnetic dipoles within a volume V bounded by a surface S. If $\mathbf{B}^0$ represents the magnetic field vector due to sources outside V, then the net force and moment on the material in V are given by integrals of force and moment densities:

$$F = \int_V (\mathbf{M} \cdot \nabla)\mathbf{B}^0 \, dv \tag{2-4.15}$$

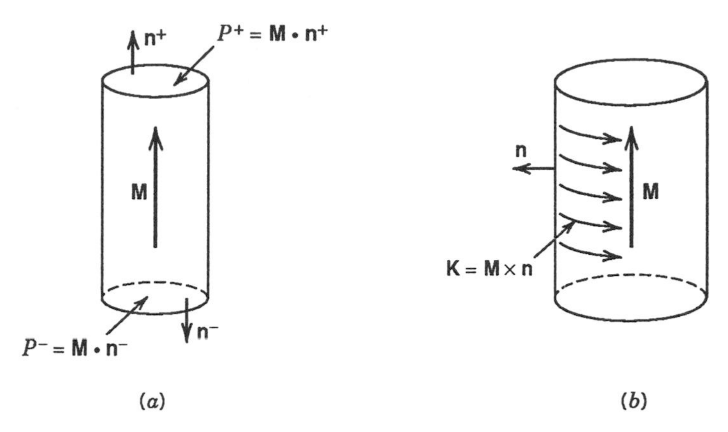

Figure 2-29 (*a*) Magnetic pole model. (*b*) Ampere-current model.

and

$$\mathbf{C} = \int \left[\mathbf{r} \times (\mathbf{M} \cdot \nabla)\mathbf{B}^0 + \mathbf{M} \times \mathbf{B}^0 \right] dv \qquad (2\text{-}4.16)$$

Equation (2-4.15) can be transformed to a different form if we use a vector identity (see Brown, 1966):

$$\mathbf{F} = \int_V \mathbf{M} \cdot \nabla \mathbf{B}^0 \, dv = \int_V (-\nabla \cdot \mathbf{M})\mathbf{B}^0 \, dv + \int_S \mathbf{n} \cdot \mathbf{M}\mathbf{B}^0 \, dS \quad (2\text{-}4.17)$$

This expression has a simple interpretation if we consider a slender rod uniformly magnetized along the axis of the rod. For this case, $\nabla \cdot \mathbf{M} = 0$ in V, and we obtain [see Figure 2-29*a*]

$$\mathbf{F} = MA\left(\mathbf{B}^0(\mathbf{r}^+) - \mathbf{B}^0(\mathbf{r}^-)\right) \qquad (2\text{-}4.18)$$

We imagine that positive and negative magnetic monopoles or magnetic charges of intensity $p_m = \pm MA$ are concentrated at the ends of the rod and that the force on each pole is analogous to the force on a charge $q\mathbf{E}$; that is,

$$\mathbf{F} = p_m \mathbf{B}_0, \qquad p_m = \mathbf{M} \cdot \mathbf{n} \qquad (2\text{-}4.19)$$

With this interpretation, $\mathbf{M} \cdot \mathbf{n}$ represents a surface distribution of magnetic "charge" or poles and $-\nabla \cdot \mathbf{M}$ represents a volume distribution of magnetic poles. The right-hand expression in Eq. (2-4.17) is sometimes called the *pole model*.

Thus, the representation of the magnetic body force is not unique and may even be replaced by surface tractions. This raises serious questions regarding the internal stress state in a magnetized body.

The total force can be further transformed to a form resembling the Lorenz force (see Brown, 1966); that is,

$$\mathbf{F} = \int_V (\nabla \times \mathbf{M}) \times \mathbf{B}^0 \, dv + \int_S (-\mathbf{n} \times \mathbf{M}) \times \mathbf{B}^0 \, dS \quad (2\text{-}4.20)$$

For a uniformly axially magnetized rod, the volume integral vanishes and the force resembles that on a solenoid with surface current density $\mathbf{K} = -n \times \mathbf{M}$ [see Figure 2-29*b*). This representation is called the *Ampere-current model*, and $\nabla \times \mathbf{M} \equiv \mathbf{J}_m$ represents the equivalent magnetization current density and $\mathbf{J}_m \times \mathbf{B}^0$ represents the body force density.

CHAPTER 3

SUPERCONDUCTING MATERIALS

Superconductivity may ultimately replace the wheel.
—Henry H. Kolm, Magnet Physicist, *Technology Review*, February 1973

3-1 THE PHENOMENA OF SUPERCONDUCTIVITY

This introduction will deal mostly with the phenomena of superconductivity and the new materials now available that could be useful for levitation applications. Of particular interest are the material processes that can be used to design a particular material for a specific levitation application. Readers interested in a deeper understanding of the theory are referred to Tinkham (1975). A nonphysicist-oriented introduction can be found in the book by Simon and Smith (1988), including a nice history of the recent discoveries of high-temperature superconducting materials.

The phenomenon of superconductivity was discovered by H. K. Onnes in 1911 in mercury ($T_c = 4$ K or $-269°$C) after he had also discovered how to liquefy helium (4.2 K) in 1908. He was awarded a Nobel prize for this work in 1913. The behavior known as *flux exclusion*, described below, was discovered by Meissner and Ochsenfeld (1933) and is now known as the *Meissner effect*. Nobel prizes for superconductivity were also awarded to John Bardeen, Leon Cooper, and Robert Schrieffer for their theory of the superconducting state in 1957, and in 1974 Brian Josephson and Ivan Giaever won the prize for discovery of a quantum tunneling in superconductors known as *Josephson junctions*, which are the basis of many electronic devices.

In 1987 K. Alex Mueller and J. Georg Bednorz of IBM in Zurich won the Nobel prize for their discovery of a superconducting component with a critical temperature above 30 K (barium–lanthanum copper oxide). However, the discovery that made this book possible was that of C. W. Paul Chu of the University of Houston and Maw-Kuen Wu and James Ashburn of the University of Alabama (Huntsville) in January 1987 in which they produced the first superconducting material (Yttrium–barium–copper oxide) with a critical temperature greater than that of liquid nitrogen (77 K) [see Simon and Smith (1988) for a more complete account of these discoveries].

Levitated devices were proposed in the late 1950s and early 1960s using niobium and lead [see, e.g., Harding (1965a, b)]. However, practical superconducting wire was not available until the early 1960s. Niobium–tin had a critical temperature of 18 K (−255°C), but is rather brittle and difficult to make into wires or cables. The other material is niobium–titanium ($T_c = 9.8$ K), which is the "workhorse" of superconducting wire and is now used in thousands of magnetic resonance imaging machines in hospitals as well as in thousands of research magnets in universities and research laboratories around the world.

The three salient features of superconducting materials that are relevant to levitation are:

- Zero resistance to steady current flow
- Exclusion of magnetic flux lines at low fields
- Flux trapping or pinning at higher magnetic fields

There are many other important properties of superconductors that distinguish them from normal conductors, but the above are important to both the principle and practice of creating levitation devices.

Zero resistance is crucial to the creation of large magnetic fields in multiturn coil magnets and is essential to the induction of supercurrents in passive bulk levitation magnetic bearings. Complete or partial flux exclusion is necessary for repulsive levitation bearings. Furthermore, flux pinning or trapping is important for producing materials with high J_c as well as for bulk levitation devices that create suspension forces.

Zero Resistance

The salient properties of superconductivity only occur below certain critical temperatures T_c, depending on the particular material. How-

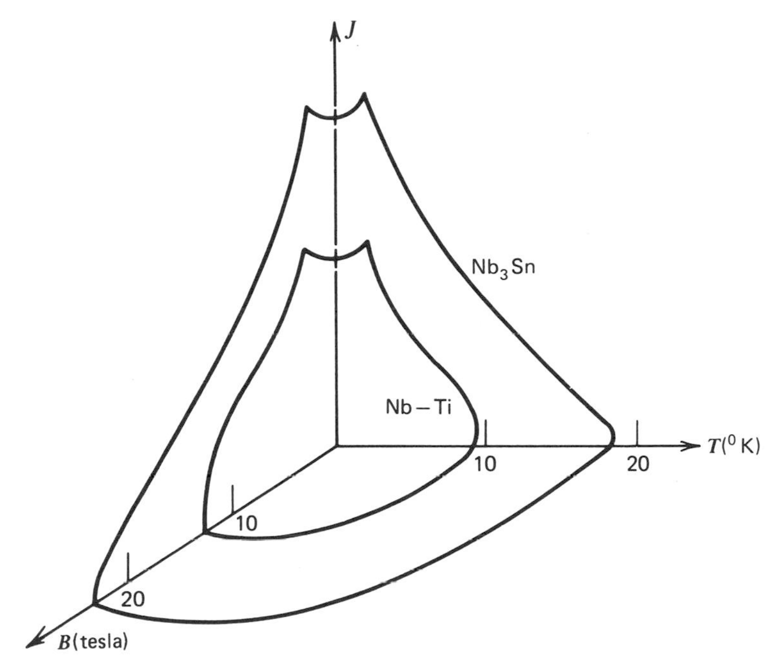

Figure 3-1 Superconducting-normal conducting transition surface for Nb_3Sn and Nb–Ti.

ever, the conducting state can become normal if the value of the transverse magnetic field becomes too high. Furthermore, transport current can create a transverse field. Thus, a limiting value of J exists, for a given temperature and field, above which the material becomes normal. These properties are illustrated in Figures 3-1 and 3-2.

For a superconducting wire-like conductor carrying current in a transverse magnetic field, the material will become normal if the values of T, B, and J do not lie in a corner of the space of (T, B, J), where $T > 0$. If any of the three variables put the state out of this corner, the material will become resistive. Typical values of T, B, and J on the critical surface are shown in Table 3-1 for a number of materials.

When the transport current is small, the dependence of the critical magnetic field on the temperature is given by the relation

$$B_c = B_{0c}\left(1 - \frac{T^2}{T_c^2}\right) \tag{3-1.1}$$

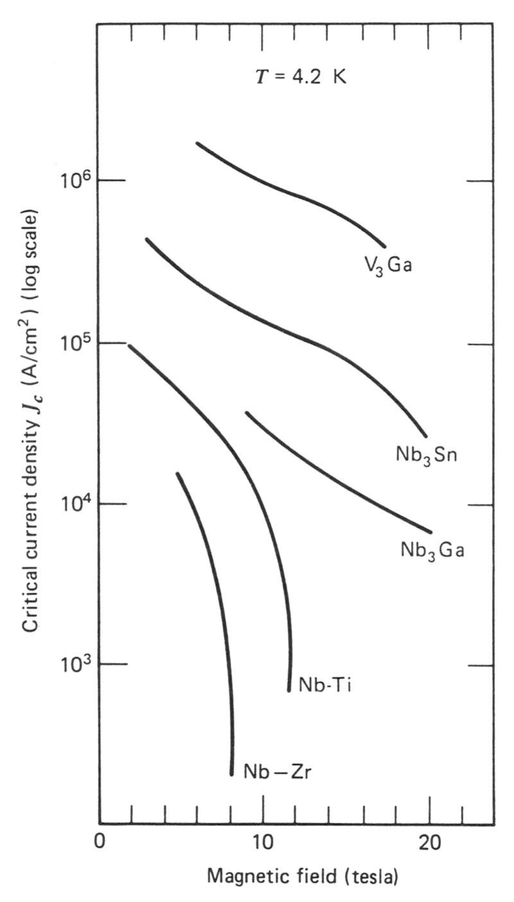

Figure 3-2 Critical current density versus magnetic field for several superconducting materials. [From Hein (1974), with permission from the American Association for the Advancement of Science, copyright 1974.]

TABLE 3-1 Low-Temperature Superconductors

Superconductor	T_c (K)
Nb–Ti	9
Nb_3Sn	18
Nb_3Al	19
Nb_3Ga	20
Nb_3Ge	22

For example, for Nb–Ti at T = 4.2 K (liquid helium) we have B_c/B_{0c} = 0.82. If conductor motion in the magnet produces friction and heating which raises the temperature by 2 K, then $B_c/B_{0c} = 0.62$. If this value were below the design field in the magnet, the conductor would become locally normal.

One of the paradoxes of superconductivity is the fact that these materials are generally more resistive in their normal state than are normal conductors, such as copper or aluminum, as illustrated in Figure 3-3.

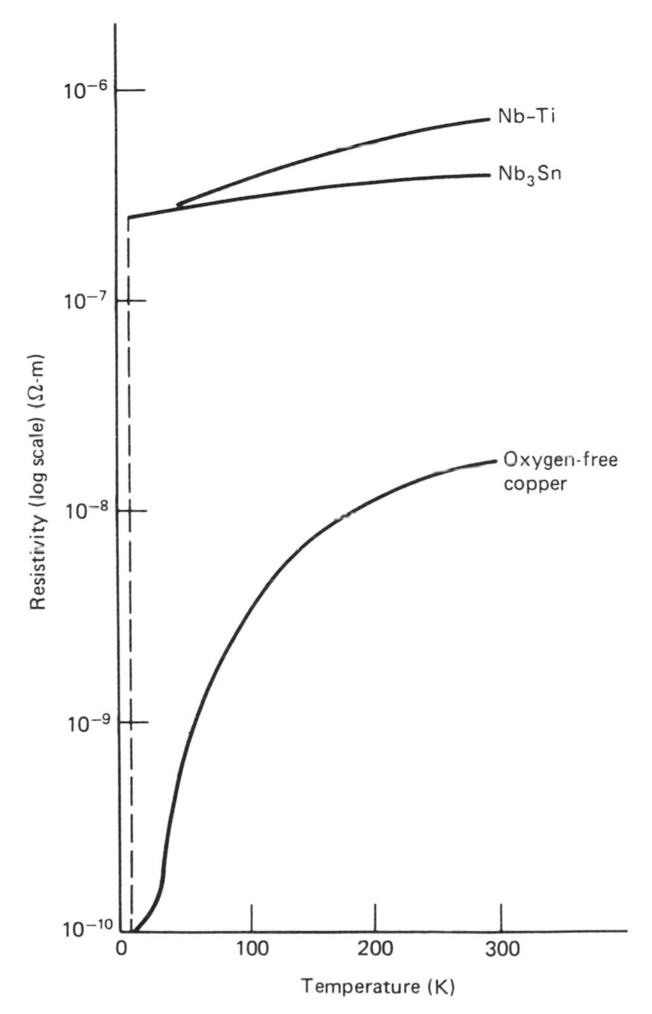

Figure 3-3 Electrical resistivity of superconducting wires and oxygen-free copper. [Adapted from Brechna (1973), with permission from Springer-Verlag, Heidelberg, copyright 1973.]

Flux Exclusion

One of the principal properties of superconductors is their ability to screen out magnetic flux from the interior of the conductor. This is illustrated in Figure 3-4 for a cylindrical conductor in a transverse magnetic field. Above the critical temperature T_c, the field penetrates the boundary of the conductor. Below T_c and for small values of the magnetic field, the flux will be excluded from all but a thin layer near the conductor surface. For low enough values of the magnetic field, this exclusion is complete and is called the *Meissner effect*. The screening is accomplished by persistent currents that circulate near the boundary of the superconductor. Exclusion of flux from a material is often called *diamagnetism*, as contrasted with paramagnetism or ferromagnetism in which flux lines are attracted to the material. In effect, we can think of the material as having a negative magnetic susceptibility. (Suppose $\mathbf{B} = \mu_0(\mathbf{H} + \mathbf{M}) = 0$ and $\mathbf{M} = \chi\mathbf{H}$, then $\chi = -1$.) In an external field $\mathbf{B}^0$, a diamagnetic body can experience a

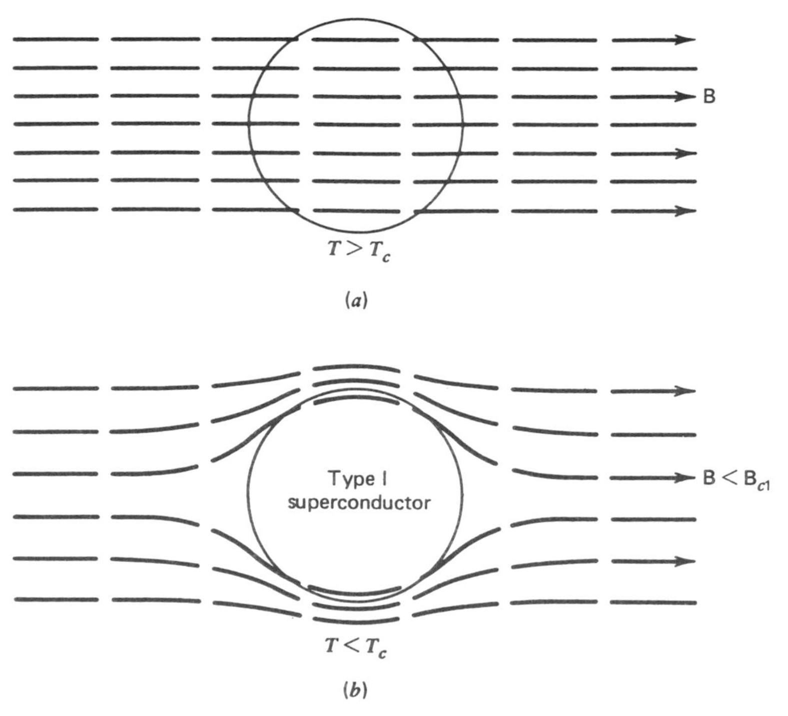

Figure 3-4 Meissner effect: (*a*) Superconductor in a magnetic field above the critical temperature and (*b*) superconductor in magnetic field below critical temperature.

force and couple given by

$$\mathbf{F} = \mathbf{m} \cdot \nabla \mathbf{B}^0 \quad \text{and} \quad \mathbf{C} = \mathbf{m} \times \mathbf{B}^0 \tag{3-1.2}$$

where for a uniformly magnetized body we have $\mathbf{m} = \mathbf{M}V$ (V is the volume of the diamagnetic body). Physically, the equivalent force on a superconducting body results from the Lorentz force $\mathbf{J} \times \mathbf{B}^0$ on the circulating screening currents.

Flux Vortex Structures — Type I and Type II Superconductors

Superconducting materials are classified as either Type I or Type II (sometimes referred to as either *soft* or *hard*). The Type I materials are often the pure metals and have low values of the critical magnetic field and critical current. Within the superconducting state, they exhibit the property of complete flux exclusion.

Type II materials are generally alloys or compounds, such as Nb–Ti or Nb_3Sn, or the high-temperature superconducting oxides such as YBCO or BSSCO, which are able to carry very high current density in high transverse magnetic fields without becoming normal. At low magnetic fields, Type II materials will exhibit perfect diamagnetism. However, another state exists in which the flux penetrates the material in clusters of flux lines. A schematic of this effect is shown in Figure 3-5 for a thin-film superconducting material. In this case the screening currents circulate around each flux bundle like small vortices. The center of each vortex is normal, whereas the region of zero to low field is superconducting.

There exist two values of the critical magnetic field $\mathbf{H}$. Below H_{c_1} the material behaves as a Type I material with complete flux exclusion. Above H_{c_1} and below H_{c_2} the flux can penetrate the material, thereby creating normal and superconducting regions. This is illustrated in Figure 3-6, and values of H_{c_2} are tabulated in Table 3-4 for a few materials. For example, the lower critical field for Nb_3Sn is 0.023 T (230 G), while the upper critical field is 23 T. Thus, for practical applications, flux penetrates a Type II superconductor in bundles of flux lines. (See e.g., Anderson and Kim, 1964.)

Critical Current

The critical transport current for a Type II material is governed by the interaction of the current with the magnetic fluxoid lattice. One of the

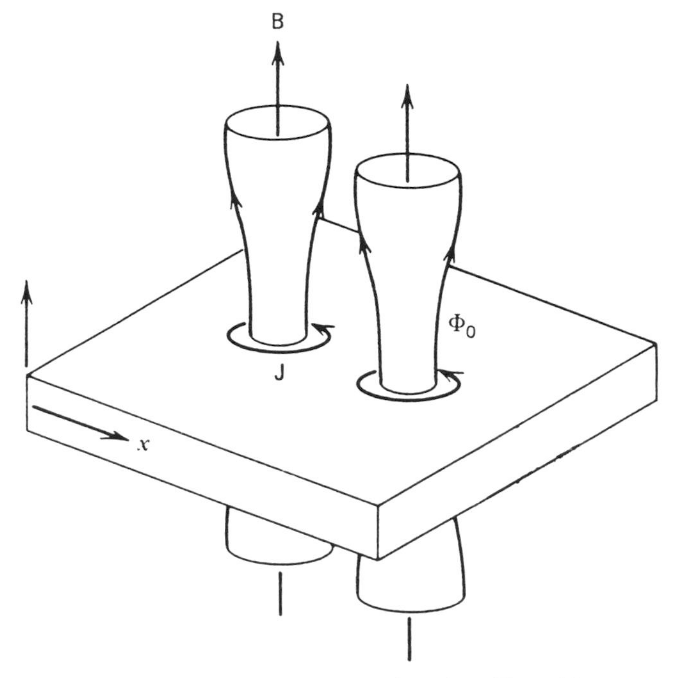

Figure 3-5 Flux bundles and current vortices in a Type II superconductor.

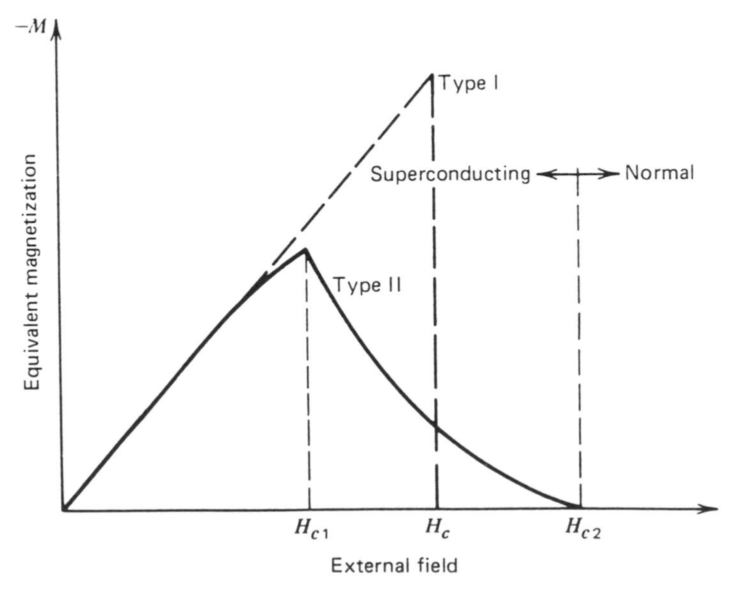

Figure 3-6 Equivalent diamagnetic magnetization versus magnetic field for Type I and Type II superconducting materials.

important concepts in the understanding of these effects is the force on the fluxoid due to the transport current.

To calculate the force on the circulating current surrounding a fluxoid, we assume that the flux is confined to a normal cylindrical region as shown in Figure 3-7. The circulating current acts as a magnetic dipole with pole strength Φ_0/μ_0, where Φ_0 is the total flux through the cylinder. If we denote the dipole strength per unit length by $\mathbf{M}$, the force on the dipole due to an external field $\mathbf{B}^T$ is given by

$$\mathbf{F} = \mathbf{M} \cdot \nabla \mathbf{B}^T \tag{3-1.3}$$

Suppose that a transport current flows transverse to the fluxoid axis with a density J_x^T. Using the axes shown in Figure 3-7, the magnetic field associated with J_x^T must satisfy the following equation:

$$\mu_0 J_x^T = \frac{\partial B_z^T}{\partial y} \tag{3-1.4}$$

For uniform current density J_x^T, B_z^T is a linear function of y, and the nonzero component of the force is given by

$$F_z = \frac{\Phi_0}{\mu_0} \frac{\partial B_z^T}{\partial y} = \Phi_0 J_x^T \tag{3-1.5}$$

If B_0 denotes the average fluxoid flux per unit area, then the force on the fluxoid per unit area is $J_x^T B_0$.

Thus, in the absence of resistive forces, flux lines in Type II superconductors would move freely through the conductor, leading to dissipation and eventually driving the material into the normal state.

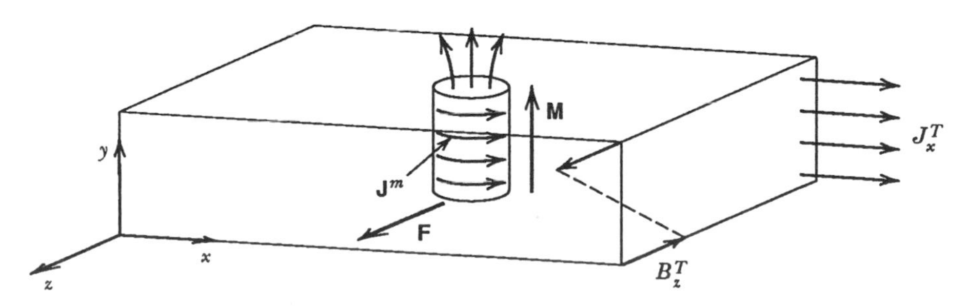

Figure 3-7 Sketch of the force on a fluxoid in a superconductor due to the flow of transverse current.

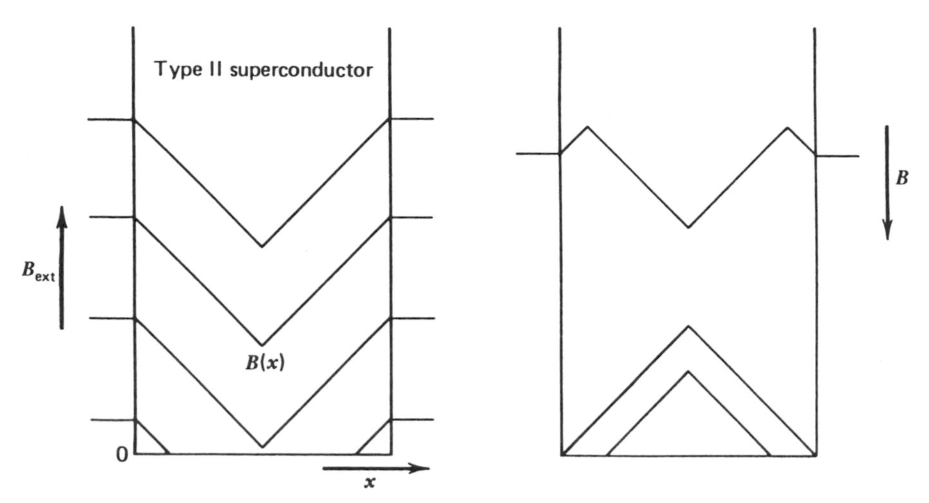

Figure 3-8 Penetration of magnetic flux into a Type II superconductor—Bean critical state model.

In practical materials, however, the flux-line motion is resisted by lattice defects. These defects act to "pin" the fluxoid at various points along the vortex, resulting in so-called *pinning forces*. The pinning forces depend on the magnetic field, the temperature, and the nature of the defects. [See, for example, Kramer (1975) and Huebener (1979) for a review of pinning forces in low temperature superconductors.]

In one theory for the critical transport current advanced by Bean (1964, 1972), it is assumed that all flux lines move to maximize the pinning force. For a one-dimensional problem, with the transport current J orthogonal to B one has

$$JB = \alpha_c(B) \tag{3-1.6}$$

where α_c is the maximum value that a pinning force can attain. This is called the *critical-state model*. Bean, for example, assumes that $\alpha_c \approx B$, so that J has a constant value whenever it penetrates the superconductor. (Critical current behavior in commercial superconductors is more complicated; see, for example, Figure 3-8.) The Bean model leads to a constant field gradient in the Type II superconductor as illustrated in Figure 3-8. This figure illustrates the hysteretic behavior of flux penetration in hard superconductors.

Flux Quantization

Another of the fundamental properties of the superconducting state is the fact that the flux contained in a closed circuit of current is an integral number of flux quanta. The smallest flux that can penetrate a circuit is $\Phi_0 = 2.07 \times 10^{-15}$ Wb. This value is given by the ratio of Planck's constant and the electron charge—that is, $\Phi_0 = h/2e$. This property is a macroscopic manifestation of quantum mechanics. One of the consequences of flux quanta is that magnetic flux can only penetrate a superconductor in discrete flux bundles called *fluxoids*. One important application of this property is a very sensitive magnetic-flux measuring device called a "SQUID," an acronym for "superconducting quantum interference device."

Derivation of flux quantization can be found in any number of reference texts on superconductivity. In the theory of electromagnetics, the momentum of a charged particle, q, has a part proportional to the mass and velocity, $m\mathbf{v}$, but also has a component proportional to the vector potential associated with the magnetic field, i.e., $\mathbf{B} = \nabla \times \mathbf{A}$; since $\nabla \cdot \mathbf{B} = 0$, and the momentum of the charged particle is given by $\mathbf{p} = m\mathbf{v} + q\mathbf{A}$.

Now imagine a closed superconducting circuit with current density $\mathbf{J} = nq\mathbf{v}$. The flux threading a circular circuit is given by

$$\Phi = \int \mathbf{A} \cdot d\mathbf{l} = 2\pi r A_0 \tag{3-1.7}$$

Thus the momentum takes the form

$$p_\theta = \frac{mJ}{qn} + q\frac{\Phi}{2\pi r} \tag{3-1.8}$$

From elementary quantum mechanics the momentum is related to a probability wave function which has wavelength λ; that is, $p_\theta = h/\lambda$, where h is Planck's constant. For a closed orbit, λ must be a fraction of the orbit circumference; that is, $\lambda = 2\pi r/N$, where N is an integer. To simplify the problem, we assume that all the current flows on the surface of the closed circuit. Thus, inside $J = 0$, and

$$\frac{hN}{2\pi r} = q\frac{\Phi}{2\pi r}$$

or

$$\Phi = \frac{Nh}{q} \tag{3-1.9}$$

It has been observed that superconducting electrons travel in pairs; thus $q = 2e$, or $\Phi = Nh/2e$. Therefore the flux can only take on integral values of the quantity $\Phi_0 = h/2e$, as mentioned earlier.

Penetration Depth

In the quasiclassical theory of electron motion in a superconducting solid, Newton's law for an electron of mass m and charge e is given by

$$\frac{d\mathbf{J}}{dt} = \frac{ne^2}{m}\mathbf{E} \tag{3-1.10}$$

This represents a constitutive law for $\mathbf{J}$ in superconducting materials and replaces Ohm's law $\mathbf{J} = \sigma\mathbf{E}$, for normal conductors. When Eq. (3-1.10) is combined with Maxwell's equations [Eqs. (2-1.10)–(2-1.13)] (where displacement currents are neglected), an equation for the magnetic field in a superconductor is obtained:

$$\nabla^2\dot{\mathbf{B}} = \lambda^{-2}\dot{\mathbf{B}} \tag{3-1.11}$$

where $\lambda^2 = m/ne^2\mu_0$. Equations (3-1.10) and (3-1.11) form the basis of one of the early theories of superconductivity proposed by the London brothers in 1935. They tried to explain the Meissner effect using a one-dimensional solution of Eq. (3-1.11), namely,

$$B = B_0 e^{-x/\lambda}$$

which predicts that flux will be confined to a thin layer near the surface of a superconductor.

3-2 REVIEW OF THEORY

The theory of superconductivity requires a knowledge of both classical electrodynamics and quantum mechanics. The theory attempts to describe the motion of electrons in a solid lattice of ions. In a solid, not all the electrons associated with an atom can participate in electric conduction. In metals, only the outer electrons of the atoms are free to move in the lattice. However, in normal electrical conductors, these electrons experience collisions. Only through the application of an applied electric field do these electrons become organized to create a transport current. The great mystery concerning superconductors is how the electrons can become organized so as not to suffer the loss of energy under collision. The collisions can occur due to the vibrating ions at finite temperature or due to defects in the lattice structure. Electrical resistance in normal conductors thus decreases as the ionic vibration approaches zero or as the absolute temperature approaches zero. Even then the defects in the lattice will result in a finite resistance at $T = 0$.

The key to understanding superconduction of electrons was the discovery that in some solids the motion of one electron could be coupled to another through the interaction of the lattice deformation. Electrons normally repel one another, but at sufficiently low temperatures, this electron-lattice coupling creates a weak attractive force between the electrons.

In the low-temperature metallic superconductors, this weak pairwise coupling of electrons, otherwise known as a *Cooper pair*, is responsible for the possibility of collective motion of electrons without loss due to collisions and without the need for an external electric field to produce an electric current. This theory developed in 1957 by Bardeen, Cooper, and Schrieffer won them a Nobel prize in Physics. This so-called BCS theory allows one to calculate the critical temperature from basic physical and atomic properties [Schrieffer (1964)].

One of the physical variables in this theory is the *coherence length* which determines the characteristic separation of electrons in a Cooper pair. One of the differences between low-temperature metallic superconductors and the new oxide superconductors is that the coherence length for the metallic materials is larger than that for the ceramic oxide superconductor.

It is believed that electron pairs are still the key to superconduction in the new, higher-temperature ceramics. However, it is not accepted whether the coupling mechanism is through the lattice deformation or some other mechanism. [See Phillips (1989), or Orlando and Delin (1991) for a more detailed review of the latest theory.]

The next building block of superconductor theory after the success of BCS in determining T_c was explaining the magnetic field effects on superconductors and the so-called "mixed state" or double critical fields H_{c1} and H_{c2} in Figure 3-6. In low fields the London theory predicted the Meissner screening effect of the surface supercurrents.

However, Russian scientists Landau, Ginzburg, and Abrikosov in the 1950s predicted that the complete exclusion of flux was not the lowest energy state. The lowest energy state was found to be a hexagonal array of flux lines penetrating the superconductor as shown in Figure 3-5. Each flux bundle was a quantum of magnetic flux 2×10^{-15} Wb which was surrounded by a superconducting vortex current. The core of each vortex was a normal conductor, and its size is proportional to the coherence length discovered in the BCS theory. The average separation of these flux quanta or supercurrent vortices decreases as the magnetic field increases; for example, $d \approx (\Phi_0/B)^{1/2}$. When transport current flows in the superconductor transverse to these flux bundles, the magnetic field of the transport current creates a lateral force on the vortices which tries to move these vortices. Without flux pinning mechanisms to fix these supercurrent vortices, the motion of these vortices, under transport current, can lead to a breakdown of the superconducting state.

In low-temperature superconductors, the vortex lattice due to an applied field appears to exhibit a regular pattern. However, there is now evidence that in high-temperature materials the vortex pattern can become disordered [see, e.g., Nelson (1988)]. Some physicists refer to this as flux lattice "melting." Some theorists believe that in a good high-temperature superconductor with high J_c, the flux line vortices are pinned by defects in a disordered pattern into what is referred to as a vortex "glass."

Thus, at the time of this writing (late 1993), there is still discussion as to the basic theory that predicts the correct T_c in the new materials. However, what may have more impact on practical applications is obtaining a proper understanding of the nature of flux line pinning in the Type II state and how to design a "dirty" enough material whose critical current density approaches the limits found in thin films.

3-3 LOW-TEMPERATURE SUPERCONDUCTING MATERIALS

Superconducting materials for wire wound magnet application have been available since the mid-1960s. The two principal materials are niobium–titanium (Nb–Ti) and niobium–tin (Nb_3Sn). The properties of these materials are shown in Table 3-1. These materials require operating temperatures near that of liquid helium (4.2 K). Nb_3Sn, with its higher-critical temperature (18 K) and higher critical current density, would be the preferred material if it were not for its extremely brittle properties which are shared with the new, higher-temperature superconducting materials. Thus, by far the most widely used material in the past 20 years has been Nb–Ti which is made in the form of multifilament wire. The principal applications for these materials are as follows: (a) for winding magnet coils in various shapes such as cylindrical magnets (solenoids) for medical imaging magnets (MRI); (b) flat racetrack shapes for dipole magnets for levitated vehicles, accelerators, motors, and electric generators; and (c) three-dimensional shapes in the form of baseball seams for mirror fusion machines and for toroidal arrays of magnets for tokamak fusion reactors. [See Montgomery (1980) and Wilson (1983) for superconducting magnet design issues.]

Conventional superconducting materials that can be manufactured in large quantities and that are used in practical devices include alloys

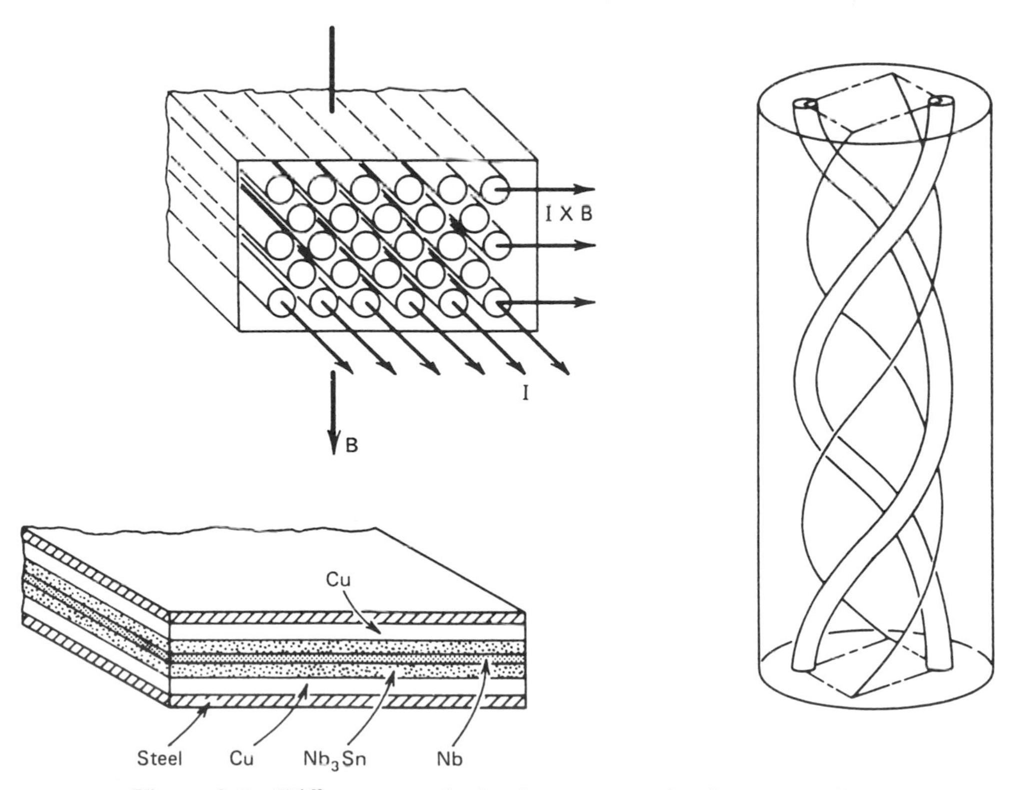

Figure 3-9 Different topologies for superconducting composites.

of niobium, such as Nb–Ti or Nb–Zr, or compounds of niobium and vanadium, such as Nb_3Sn, Nb_3Ge, and V_3Ga. These materials are made in long lengths for winding into magnets. The cross-sectional geometry has a variety of topologies, depending on the material and the application. Several topologies are illustrated in Figure 3-9.

An isolated superconducting filament can suffer a thermal instability. A small rise in temperature produces a normal resistive zone that grows under Joule heating, until the whole magnet becomes normal. This problem has been mitigated in practical materials by placing a good conductor, such as copper or aluminum, in parallel with the superconducting conductor. This has led to two basic types of conductors: the multifilament and the layered or flexible-tape superconducting, composite conductor.

In the continuous-filament-type conductor, such as Nb–Ti, the normal conducting matrix is often copper. The wire is made by stacking up arrays of copper and Nb–Ti rods inside a copper cylinder. This billet is then drawn down by passing the billet through successive dies until a certain cross-sectional shape and size are achieved (see Figure 3-10*a*). This process obviously involves large strains and creates problems for brittle materials. Nb–Ti has sufficient ductility to be drawn down to a 10-μm size without fracture, but Nb_3Sn is very brittle and requires different methods of manufacture.

One method of manufacturing a multifilament Nb_3Sn composite is to array niobium and bronze rods in a billet, draw the billet down to wire form, and then thermally react the wire to form Nb_3Sn. A cylindrical tantalum barrier is often used to prevent diffusion into the copper (Figure 3-10*b*).

The properties of various Type II superconductors are given in Table 3-1. The exact value of the critical current depends not only on the magnetic field, but on the degree of thermal stability required and the heat transfer properties. The maximum applied current density versus applied transverse magnetic field is shown in Figure 3-2 for a number of superconductors. Commercially available conductors of Nb–Ti range from 0.5-mm-diameter wire, with 400 filaments which can carry 200 A at low fields ($\sim$ 1.0 T), to a 1.2×1.2 cm^2 conductor designed for 6000 A for the mirror fusion yin–yang magnets used at Livermore, California. For applications which call for fields below 9 T, Nb–Ti/Cu composites are presently employed. For higher-field environments, Nb_3Sn composites are currently used. The critical current, critical temperature T_c, and field H_{c2} in these conductors are also strain-sensitive.

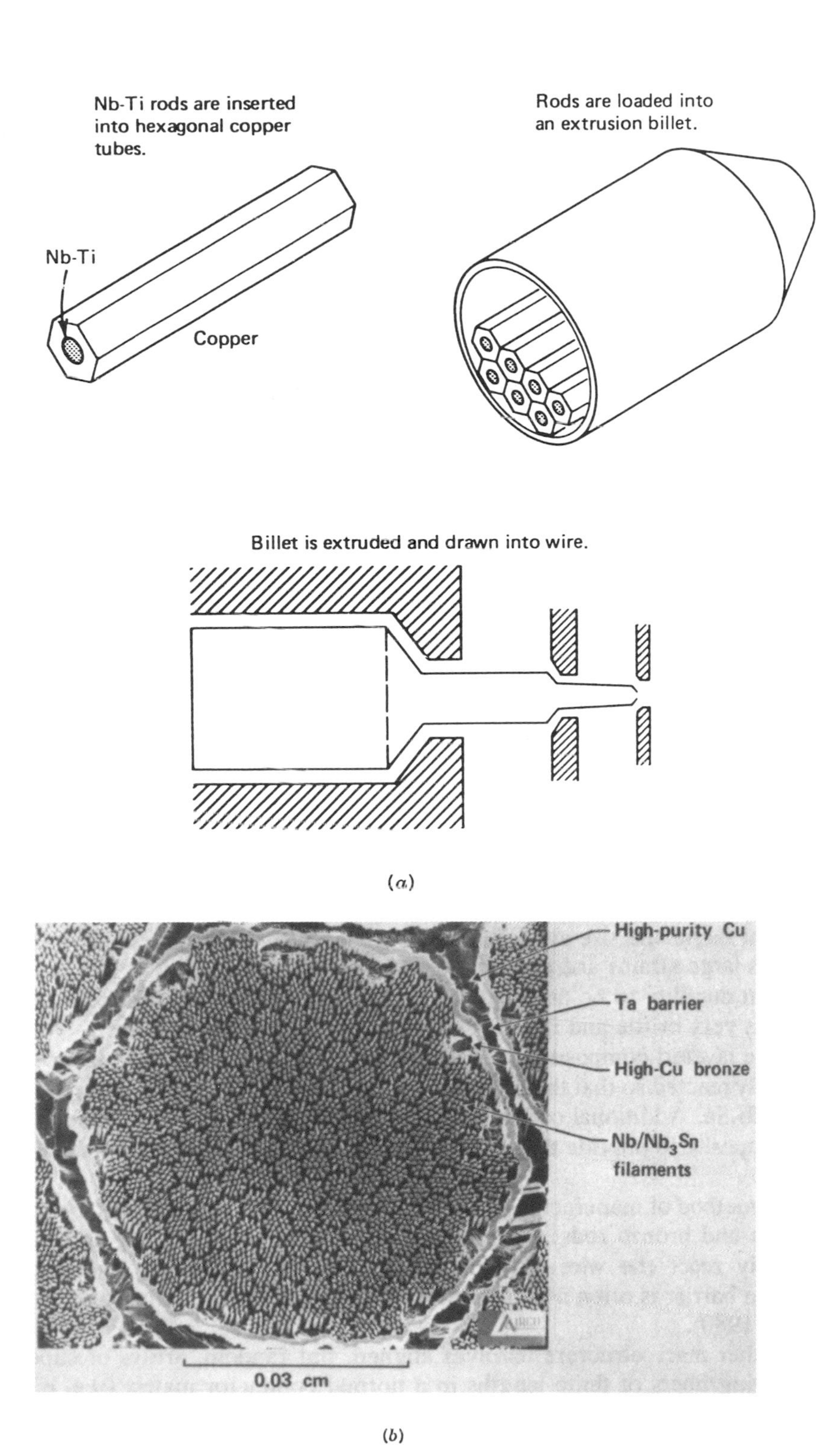

Figure 3-10 (*a*) Steps in the manufacture of low-temperature superconducting composites. [From Scanlan (1979), with permission.] (*b*) Photograph of the cross section of a multifilamentary superconducting composite. [From Hoard (1980), with permission.]

3-4 HIGH-TEMPERATURE SUPERCONDUCTING MATERIALS

The adjective "high-temperature" is a relative term and generally refers to those new classes of materials that have critical temperatures T_c above 30 K. This class includes the new ceramic materials based on copper and oxygen as well as the new "bucky ball" or fullerene materials based on carbon. The superconducting ceramic oxides are called *perovskites* because of their atomic structure (see Figure 3-11) and are similar to ferroelectric materials such as barium titanate ($BaTiO_3$), which is used for piezoelectric applications. Our discussion here will focus mainly on these ceramic superconductors which are listed in Table 3-2 along with their critical temperatures. These temperatures range from 35 K to 133 K. While much attention has been paid to those materials which have T_c above that of liquid nitrogen (78 K), there are applications for which a lower T_c would still

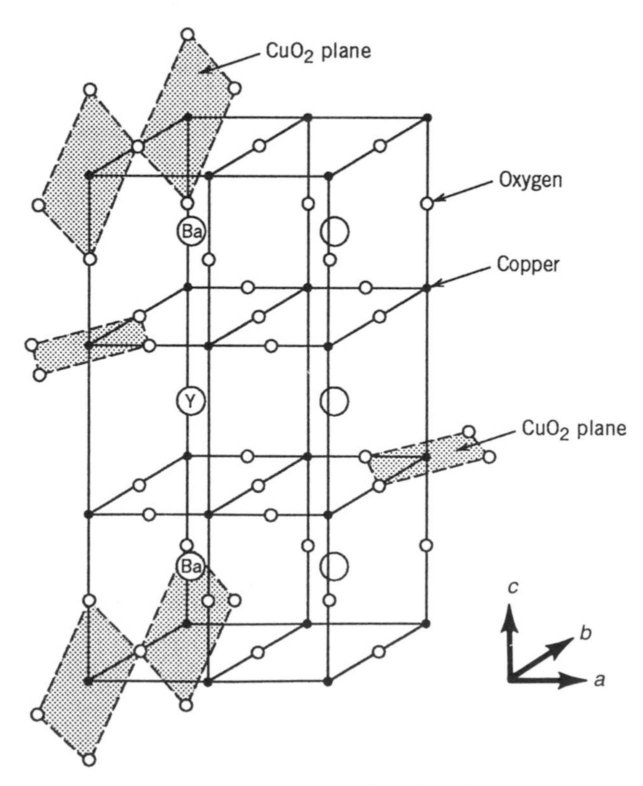

Figure 3-11 Atomic structure of $YBa_2Cu_3O_7$ high-temperature superconducting material. [After Phillips (1989).]

TABLE 3-2 High-Temperature Superconductors[a]

Superconductor	T_c (K)
$La_{1.85}Ba_{0.15}CuO_4$	35
$La_{1.85}Sr_{0.5}CuO_4$	40
$YBa_2Cu_3O_7$	95
$Bi_2Sr_2CaCu_2O_8$	85
$Bi_2Sr_2Ca_2Cu_3O_{10}$	110
$Tl_2Ba_2CaCu_2O_8$	108
$Tl_2Ba_2Ca_2Cu_2O_{10}$	125
$HgBa_2Ca_2Cu_3O_{6+\delta}$	133.5

[a]Based on Orlando and Delin (1991).

be useful, including (a) cryofluid-based devices such as liquid hydrogen pumps and (b) magnet-based physics experiments.

These new materials are processed in three different forms: bulk, thin film, and wire or tape. All three forms are useful for magnetic levitation application, so we will review the properties of each. It is recognized that new classes of superconducting materials may be discovered with completely different properties than these and perhaps with much higher T_c. At present, however, the cuprate ceramic oxides provide a benchmark to design new materials.

The first material to be discovered with $T_c > 30$ K was a barium–lanthanum–copper oxide system (Ba–La–CuO) which was announced by Bednorz and Müller in September 1986. Chu and Wu followed quickly (early in 1987) with the first liquid nitrogen superconductor, yttrium–barium–copper oxide (YBCO), which has attracted worldwide interest ever since. This material is known as the 1–2–3 conductor because of its chemical subscripts—that is, $Y_1Ba_2Cu_3O_7$. However, this representation assumes an ideal atomic ordering as in Figure 3-11. However, Phillips (1989) points out in his book that real materials can have significant deviations from ideal geometric arrangements due to defects in the lattice arrangements. Thus, many authors write the chemical formula with variable subscripts—for example, $YBa_2Cu_3O_{7-x}$ $(0 < x < 1)$ or $La_{2-x}Ba_xCu)_4$; in the case of Bednorz and Müller, $x = 0.15$ ($T_c = 35$ K). In this book we have not used this more precise notation because our emphasis is on levitation and not processing. Also, in most cases, we use an acryonym such as YBCO or LBCO to denote the materials. Those interested in the more detailed chemistry and materials processing should consult other sources such as Phillips (1989), Poole et al. (1988) or other modern references on superconductivity.

TABLE 3-3 Comparison Between Niobium-Based Superconductors and CuO-Based Superconductors

Niobium-Based Superconductors	CuO-Based Superconductors
$T_c < 23$ K	30 K $< T_c <$ 133 K
Metallic	Ceramic
Relative ductility	Brittle
Normal electrical conductor	Normal electrical insulator
$H_{c2} <$ 20–30 T	30 T $< H_{c2} <$ 200 T

As of late 1993, the highest temperature superconductor was a mercury-based cuprate Hg-Ba-Ca-CuO, with a critical temperature of 134 K. Scientists at the University of Houston however have been able to increase the temperature to 164 K by placing a small sample under high pressure (456 MPa). Also a French group has claimed evidence for a 250 K multilayer Bi–Sr–Ca–Cu–O (BSCCO) superconductor. The film thickness was 30 nm on a $SrTiO_3$ substrate. However, this claim has yet to be confirmed by other researchers at the time of publication. If it holds up, it would be a most exciting development.

The first question one can ask is, How do the new ceramic oxide superconductors compare with the widely used niobium-based, low-temperature materials such as Nb_3Sn or Nb–Ti? This comparison is summarized in Table 3-3.

Cuprates — Crystal Structure

The structures of these new materials are similar to those of naturally occurring minerals called *perovskites* [such as calcium titanate ($CaTiO_3$)] or to those of the ferroelectric material barium titanate ($BaTiO_3$) used in piezoelectric applications. These materials are highly anisotropic as illustrated by the structure shown in Figure 3-11 for YBCO. Here the *c*-axis dimension is nearly three times the *a–b* dimensions. These materials also have planar structures (CuO_2) and chainlike structures (CuO) and have been referred to by some theorists as *two-dimensional superconductors*. It is believed that the superconducting currents flow along these planes and chains. The planelike structures are also found in the bismuth and thallium compounds (see Table 3-2). This anisotropy means that the supercurrent capacity can be significantly greater for transport along the planes than perpendicular to the planes. This implies that in multicrystalline materials, the currents will have difficulty passing from one crystal grain to another unless they are aligned. Thus the highest current-carrying densities have been found in single-crystal and thin-film specimens. Improvements in current-carrying capacity for bulk multicrystal materials have been made by making larger crystal structures and using so-called *texturing*. The latter is achieved by process methods that lead to alignment of the planelike structures. The mechanical behavior of these materials is brittle and presents problems for fabrication into wire or tape. [See, e.g., Bobrov (1989).]

Cuprates — Superconducting Properties

Critical temperatures of the ceramic CuO based superconductors are listed in Table 3-2. Besides having higher critical temperatures than the niobium-based superconductor (Table 3-1), the new superconductors have higher upper critical magnetic fields. Niobium-based materials are superconducting below 12 T. However, YBCO materials are believed to have critical fields anywhere from 30 T to 200 T. This means that very strong magnetic field devices could be built with these new materials if we could design the material and its host structure to withstand the very high stress fields at 30 T or higher.

Critical Currents

For levitation applications, the current-carrying capacity is paramount in these materials. There are two measures of supercurrent capacity: magnetization currents and transport currents. Magnetization currents circulate within one or a few crystal grains. Magnetization currents are important in bulk superconductor magnetic bearing applications. Transport currents must flow across the entire macroscopic conductor and are generally much lower than magnetization currents. The maximum current density for which the material remains superconducting is denoted by J_c and is usually measured in amperes per square centimeter (A/cm^2). J_c is very strongly magnetic-field-dependent. The dependence of J_c on magnetic field for YBCO is shown in Figure 3-12. Also, because of the anisotropic nature of the material, an important factor is whether the applied field acts parallel to or perpendicular to the path of the current (see Salama et al., 1992). As mentioned in Section 3-3 on the low-temperature materials, J_c is strongly temperature-dependent and dramatic increases can be obtained the further below T_c one designs the applications.

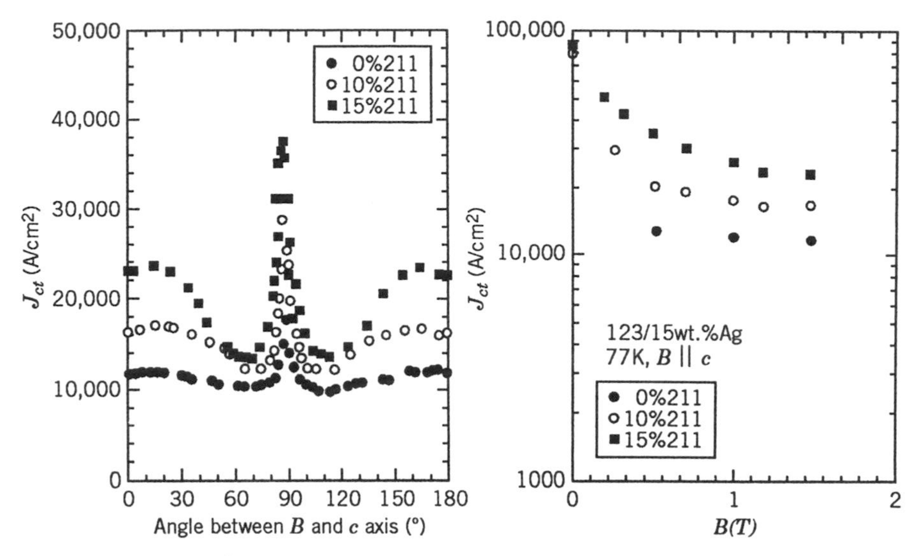

Figure 3-12 (left) Critical current of YBCO versus direction of magnetic field for three samples with different amounts of nonsuperconducting phase of YBCO (211). (right) Critical current versus magnetic field for high-temperature material (YBCO). [From Lee et al., (1992) with permission © 1992 World Scientific Publishing Company Ltd.]

In addition to larger grain sizes, critical currents have been increased in these materials by clever processing methods that create so-called "flux-pinning" centers in the materials. As discussed in the theory section, magnetic flux penetrates the superconductor in the Type II regime in the form of supercurrent vortices. These vortices experience magnetic forces that would tend to move them in the material which could lead to a breakdown of the superconducting state. *Flux pinning* refers to microstructures in the materials that tend to prevent these supercurrent vortices from moving under magnetic sources. In low-temperature materials, these mechanisms take the form of point defects, dislocations, interstitial atoms, and other topologies that depart from the ideal crystal structure. In YBCO materials, dramatic increases in J_c have been obtained by creating a network of nonsuperconducting phases of the material. For example, Murakami (1989) has developed a melt-quench process that disperses a so-called (211) nonsuperconducting phase of YBCO (Y_2BaCuO_7) in the $YBa_2Cu_3O_7$ superconducting phase (see Section 3-5). This quest for flux-pinning mechanisms suggests the paradox that the ideal superconducting material is one that has an imperfect atomic lattice arrangement. This is one of the reasons why pure superconductors are

TABLE 3-4 Properties of Low- and High-Temperature Superconductors

Material	Critical Temperature, T_c (K)	Magnetic Field, B_{c2} (T)	Current Density, J_c (A/cm^2)
Nb–Ti	9.6	8	2×10^5
Nb_3Sn	18	23	3×10^5
$YBa_2Cu_3O_7$			
Bulk-melt-quenched	95	18	3×10^4
Thin film	95	35–200	6×10^7

poor conductors and "dirty" superconductors become the workhorses of magnetic technology.

Although the job of creating high-current superconducting materials, wires, and tapes has proven difficult, the good news is that the upper limit as measured in thin-film specimens has been found to be very high: ~ 10^7 A/cm^2 (Table 3-4). At the present time (late 1993), bulk YBCO materials in 1-T fields can be processed with current densities of 3×10^4 A/cm^2 or higher at 78 K. Wire superconductors have now been fabricated in lengths greater than 10 m with current densities greater than 10^4 A/cm^2. Current densities around 2×10^4 A/cm^2 can provide useful application (e.g., magnets for levitated vehicles). For safety margins, however, the goal is to produce material with $J_c \approx 10^5$ A/cm^2, still well below thin-film capabilities.

Thin Films

There are many methods to create thin-film structures such as molecular beam epitaxy (MBE), pulsed laser deposition, magnetron sputtering, chemical vapor deposition (CVD), and metal oxide CVD (MOCVD). The thin film superconductor is deposited on a substrate.

The anisotropy in crystalline structure of CuO-based ceramic superconductors creates preferred planes. In the case of YBCO (Figure 3-11), this means that the *c* axis is aligned normal to the substrate. Material scientists use a number of substrate materials such as MgO, $SrTiO_3$, or $LaSrGaO_4$. The host substrate is chosen to be compatible with the superconducting layer, including atomic lattice dimensions.

The principal applications of thin-film superconductors are electronic devices. However, we have made levitation force measurements on thin-film YBCO specimens and find the promise of magnetic bearing application. There is also the potential for levitation of micromachine components using thin-film superconducting bearings.

Thin-film thicknesses range from 200 Å to 10^4 Å. We recall that 1 Å equals 10^{-10} m (10^4 Å = 1 μm). Measurements of J_c in thin-film specimens range from 10^5 to 10^7 A/cm^2 at liquid nitrogen temperatures (78 K). The collective experience seems to be that as the film thickness increases beyond 1 μm, significant decrease in J_c occurs. The potential of superconducting thin-film bearings lies in creating thick-film specimens (5 μm) with J_c greater than 10^5 A/cm^2.

Wires and Tapes

Both YBCO and BSCCO superconductors have been made into single and multifiliment wire. In the first couple of years after the discovery of the liquid nitrogen superconductor, YBCO, there was a general consensus that it would be a decade or more before a commercial, high-T_c superconducting wire would be available. However, at the time of this writing (late 1993) there is great excitement at the progress in high-T_c wire development (especially in the BSCCO material), and it may not be long before long lengths of this material ($>$ 100 m) become available.

Low-temperature superconducting wire has revolutionized magnetic technology. Its applications include (a) Mag-Lev trains, (b) magnetic medical imaging, and (c) high-field magnets for high-energy and plasma physics. The availability of high-temperature wire has the potential to revolutionize the power industry with applications to electrical generators, power transmission lines, energy storage, and so on. At this time, companies in Europe, Japan (e.g., Sumitomo Electric Industries), and the United States [e.g., Intermagnetics General Corp. (IGC)] are racing to produce the first high-quality, high-T_c wire.

One of the leading groups in this field is that of Ken-ichi Sato of Sumitomo Electric Industries, Ltd. in Osaku, Japan. Their greatest success is with bismuth-based high-T_e material (e.g., Bi–Sr–Ca–CuO). There are two superconducting phases: a 2223 phase with T_c = 110 K and a 2212 phase with T_c = 80 K (Sato et al., 1991). BSCCO wire is usually made by sheathing the material in a silver tube (sometimes called the "powder-in-tube" method). Early results of a short test sample of this wire at liquid nitrogen temperatures are shown in Figure 3-13. In 1991, current densities of 5×10^4 A/cm^2 in a zero filed were achieved, and measurements of 12,000 A/cm^2 in a 1-T field were also achieved.

In Japan a group at Toshiba and Showa Electric have built a small superconducting solenoid using BSCCO tape (Kitamura et al., 1991).

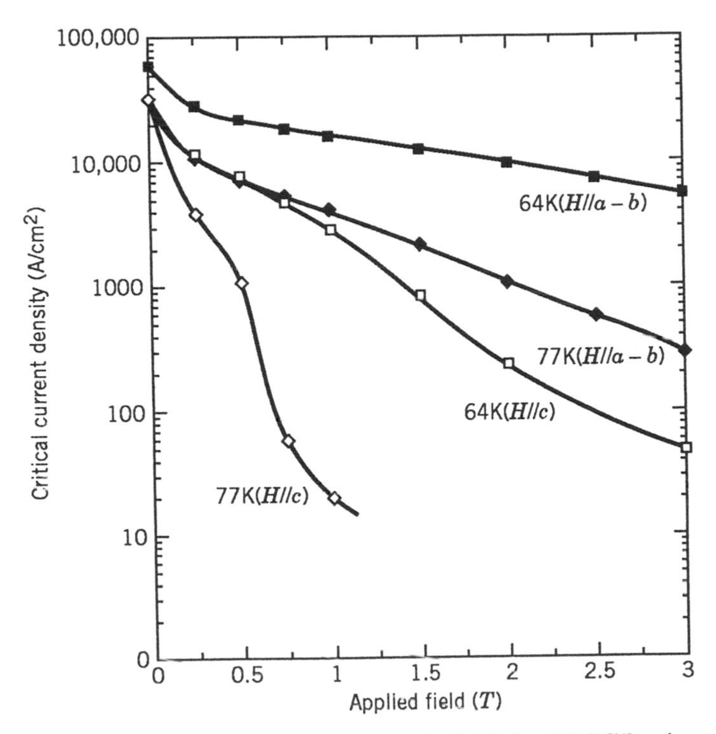

Figure 3-13 Critical current versus magnetic field for BSCCO wire using the "powder-in-tube" method. [From Sato et al. (1991), with permission.]

The powder-in-tube method was used, and 0.15-mm-thick by 3.5-mm-wide tape in lengths of 3 m were produced. The magnet was comprised of 8 coils, each with 28 turns per coil. The tape had a J_c of 3.9×10^4 A/cm^2 at 4.2 K and a J_c of 6.3×10^3 A/cm^2 at 77 K, both measured in a zero magnetic field. The magnet produced a 1.15-T field at 4.2 K.

In the United States a group at IGC (in Guilderland, New York) has produced a multifilament wire using BSCCO [see Haldar and Motowidlo (1992)]. The process is schematically illustrated in Figure 3-14, and a cross section of the wire is shown in Figure 3-15. The measured critical current at 77 K was $J_c = 3 \times 10^4$ A/cm^2 ($B = 0$). The process involved mixing the chemicals BiO, PbO, and CaO (or CaCO), SrCO, and CuO in a cation ratio of 1.8, 0.4, and 2.0 2.2, and 3.0). The mixture was heat-treated at 840°C and reground. The powder was then packed into a 4-mm-inner-diameter and 6-mm-outer-diameter silver tube 15 cm long. The tube was swaged and drawn to 0.1–0.2 mm thick and annealed at around 830–870°C for 24–150 h. (See also Neumüller et al., 1991.)

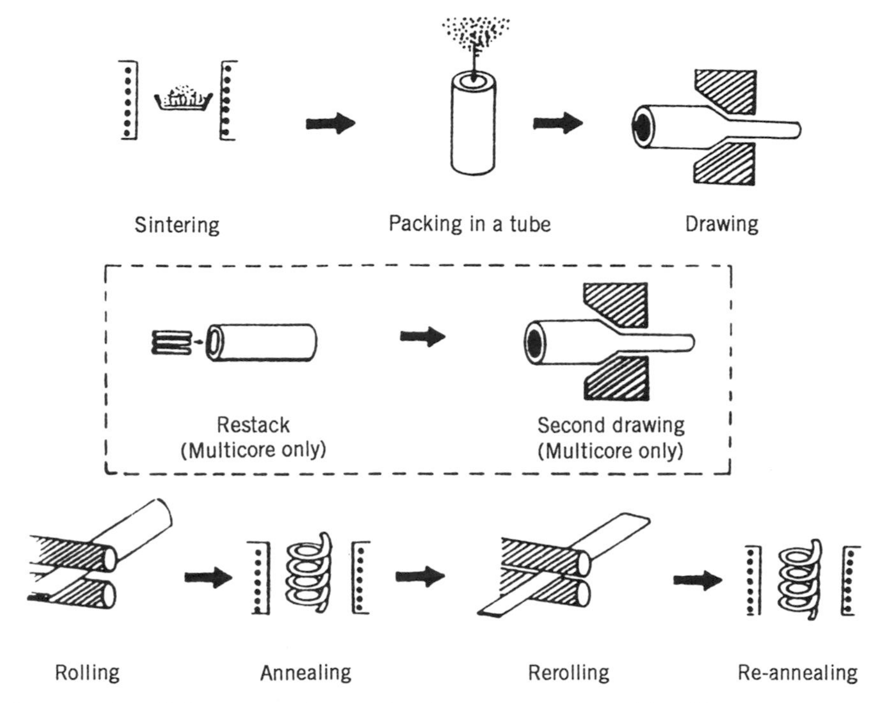

Figure 3-14 Schematic of steps in the process of making high-temperature bismuth-based BSCCO, Ag-clad wire using the powder-in-tube method. [From Haldar and Motowidlo (1992), with permission.]

Fullerenes

Recently discovered molecular structures of carbon known popularly as "bucky balls" or as "C_{60}, and C_{70}" have added a new class of materials to the race for high-temperature superconductors. This class of materials have structures that look like geodisic domes and have been called *fullerenes* after the architect and inventor Buckminster Fuller. Recently, alkali-metal doping of C_{60} and C_{70} have produced superconductors with $T_c = 33$ K (Tanigaki et al., 1991).

C_{60} is produced from pure carbon rod from the soot generated in an arc discharge. The C_{60} powder was mixed with a small amount of Tb and Cs is quartz tubes and reacted at 400°C for 24 h. The system $Cs_2Rb_1\ C_{60}$ was found to have a superconducting transition temperature at 33 K.

As of late 1993, it is not clear if this class of materials will prove useful for superconducting applications. However, it does show that one should expect new classes of superconducting materials to be

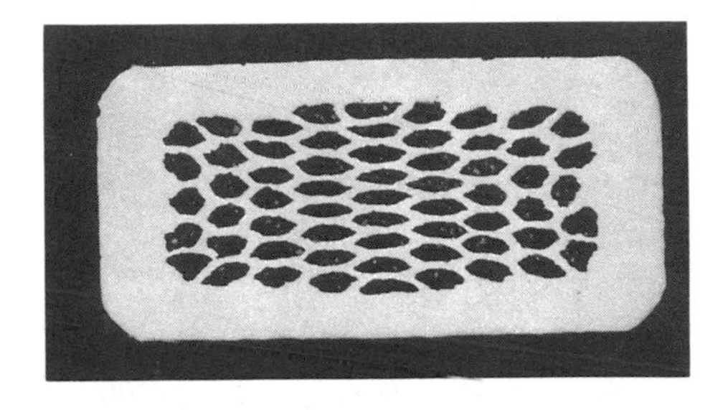

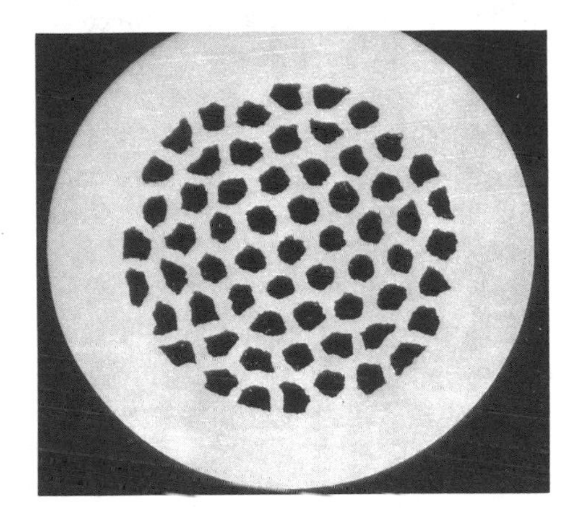

Figure 3-15 Photographs of the cross section of multifilament BSCCO (2223) wire. Bottom: Diameter, 2.26 mm, Length 11 m. (Courtesy of Intermagnetics General Corp., Guilderland, New York.)

discovered in the next decade which may prove much more advantageous than the current crop of CuO-based materials.

3-5 PROCESSING OF BULK SUPERCONDUCTORS

Knowledge of the chemical constituents of a superconductor such as $YBa_2Cu_3O_7$ is not sufficient to produce good material with optimal levitation properties for bearings. The history of materials development for the A15 metallic superconductors such as niobium–tin showed that one or two decades of engineering and scientific work are necessary to produce a reliable, commercial-quality superconductor at a reasonable price. This process for the oxide high-temperature superconductors is, as of late 1993, only one half a decade into development. However, during this short period, dramatic improvements in J_c and levitation properties have been made in bulk YBCO and in BSCCO wire through optimization of the processing methods. In bulk YBCO, for example, J_c has gone from around 10^2–10^3 A/cm^2 at 77 K to over 3×10^4 A/cm^2 in a 1-T field. In addition, levitation magnetic pressures on small rare earth magnets have risen from 0.1 N/cm^2 to over 20 N/cm^2.

The original processing method involved using co-precipitated nitrate precursors, sintering, cold pressing the powders into a pellet, and oxygen annealing. This process created a porous material with very small grain sizes and poor interconnection between grains.

The recent methods for processing bulk YBCO have involved raising the temperature of the process in order to produce a partial melt. These methods have been called *melt-texturing* or *quench–melt–growth* (QMG)—or, more recently, the *melt–powder–melt–growth* (MPMG) technique (developed in Japan). These methods were developed in several laboratories such as Bell Labs (Jin et al. 1988, 1990), University of Houston (Salama et al., 1989), Catholic University of America (Washington, D.C.) (Hojaji et al., 1989, 1990), and ISTEC Labs (Tokyo) (Murakami et al., 1989a). We describe one of these methods, MPMG, developed by M. Murakami of ISTEC, as presented in several of his published papers. This technique has produced very good bulk material with both high transport and magnetization J_c. The method is also one of the few which has been published in sufficient detail that it has been replicated by many university and industrial laboratories in Japan.

MPMG Process for YBCO

1. Appropriate amounts of powders of Y_2O_3, $BaCO_3$, and CuO (99.9% pure) are mixed.
2. Calcinate at 900°C for 24 h in flowing oxygen.
3. Melt powders at 1300–1400°C in platinum crucibles.
4. Rapidly cool between cold copper hammers into a plate.
5. Pulverize the plate into a powder and mix.
6. Press into a pellet of desired shape.
7. Reheat sample to 1100°C for 20 min.
8. Cool slowly to room temperature in platinum crucible under 1 atm air at a rate of 1°C/hr to 20°C/hr in the range 950–1000°C followed by furnace cooling.
9. Anneal sample at 600°C for 1 hr.
10. Cool slowly in flowing oxygen.

A schematic of the process is shown in Figure 3-16.

Murakami describes the process as an attempt to solve two problems. The first is to produce large grains on the order of 1 mm or

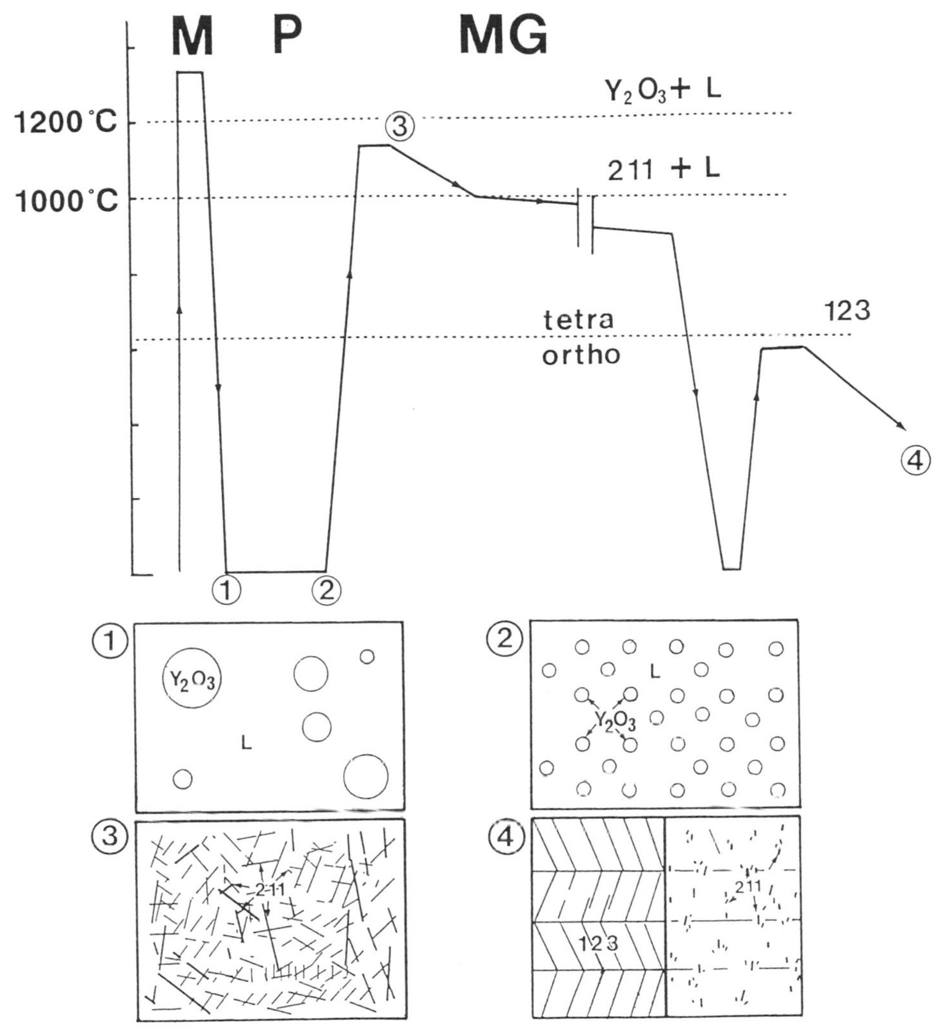

Figure 3-16 Schematic of "melt–powder–melt–growth" method of M. Murakami of ISTEC, Tokyo, Japan.

larger. The second is to produce good flux pinning in order to increase J_c. He attributes the high J_c in his samples to the dispersion of small nonsuperconducting inclusions of Y_2BaCuO_5 or the so-called (211) phase. He starts with a composition whose stoichiometry is changed from the 1 : 2 : 3 ratios toward the (211)-rich region. He ends up with a volume fraction of (211) ranging from 25% to 30%. The (211) particles have diameters of 1–3 μm. Murakami claims that these (211) particles dispersed in the (123) matrix provide both increased flux pinning and

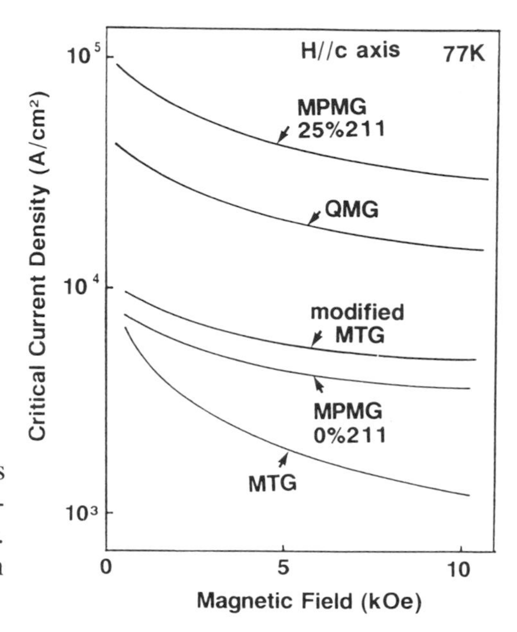

Figure 3-17 Critical current versus field for YBCO bulk superconductor for various processing methods. [From Murakami et al. (1991), with permission.]

increased fracture toughness. Measurement of transport J_c exceed 30×10^3 A/cm^2 (see Figure 3-17).

A similar process by Salama et al. (1989) at the University of Houston has also produced large grain samples with very high J_c values.

No doubt further progress will be made which will improve the performance and decrease the cost of these materials. The key point here is that the superconductor properties not only can be optimized, but can be tailored to meet specific application needs such as high levitation force properties. For example, if one wishes to use the flux drag force in bulk superconductors to create a vibration damper, perhaps a different processing scheme would be required to shape the force–displacement–velocity behavior of the superconducting material.

3-6 MAGNETIZATION AND LEVITATION FORCES

Levitation forces between permanent magnets and bulk superconductors such as YBCO are believed to be the result of induced supercurrents in the material due to the field of the magnetic field source (e.g., the permanent magnet). These currents flow within grains as well as in closed loops that span several or many grains in the superconductor.

In free-sintered processed YBCO, it is believed that intergrain currents are small and that the induced currents flow in small loops in the small crystalline grains. The flow of current in a small circuit can be viewed as a dipole. Thus, the collection of induced dipoles looks like an equivalent magnetization **m**. If the applied induction field **H** applied is uniform, we can then assign an average magnetization per unit volume $\mathbf{M} = \mathbf{m}/V$. Thus, the magnetic properties of the superconductor can be treated like a ferromagnetic or diamagnetic material. In particular, the relation **M** (**H** applied) is of importance. Examples of such measurements are shown in Figure 3-18 for both free-sintered and quench–melt–growth processed YBCO samples [see, e.g., Moon et al. (1990)] showing the effective magnetization as the applied field is taken through a cycle. One can see that in the case of the free-sintered material there is little remnant magnetization when the field is removed. However, in the case of the quench–melt–growth sample, significant remnant magnetization remains after the applied field is removed. In general, it has been observed that the larger the peak magnetization, the larger the levitation force [see, e.g., Murakami et al. (1990) and Figure 4-27].

The large magnetization is attributed to two effects. First, larger grain sizes and better intergrain current paths in melt–quench processed material allow for larger current path diameters, which increases the effective magnetization. Second, the greater flux-pinning strengths will increase the critical current J_c within grains and will also increase the transport J_c, which again increases the magnetization.

A large ΔM effect at $\mathbf{H} = 0$ also produces a suspension force effect as will be discussed in Chapter 4.

3-7 SUPERCONDUCTING PERMANENT MAGNETS

One of the promising developments in these new materials is persistent, high-field superconducting magnets. In this phenomena a bulk rectangular or cylindrically shaped YBCO specimen is cooled in a high magnetic field of 1–10 T. When the applied field is removed, circulating persistent currents in the materials trap magnetic flux so that the sample acts as a permanent magnet. A group at the University of Houston under R. Weinstein has produced samples with remnant fields of 1–4 T (see Weinstein et al., 1992) (Figure 3-19). This development is significant in that it adds another way of producing a primary field source.

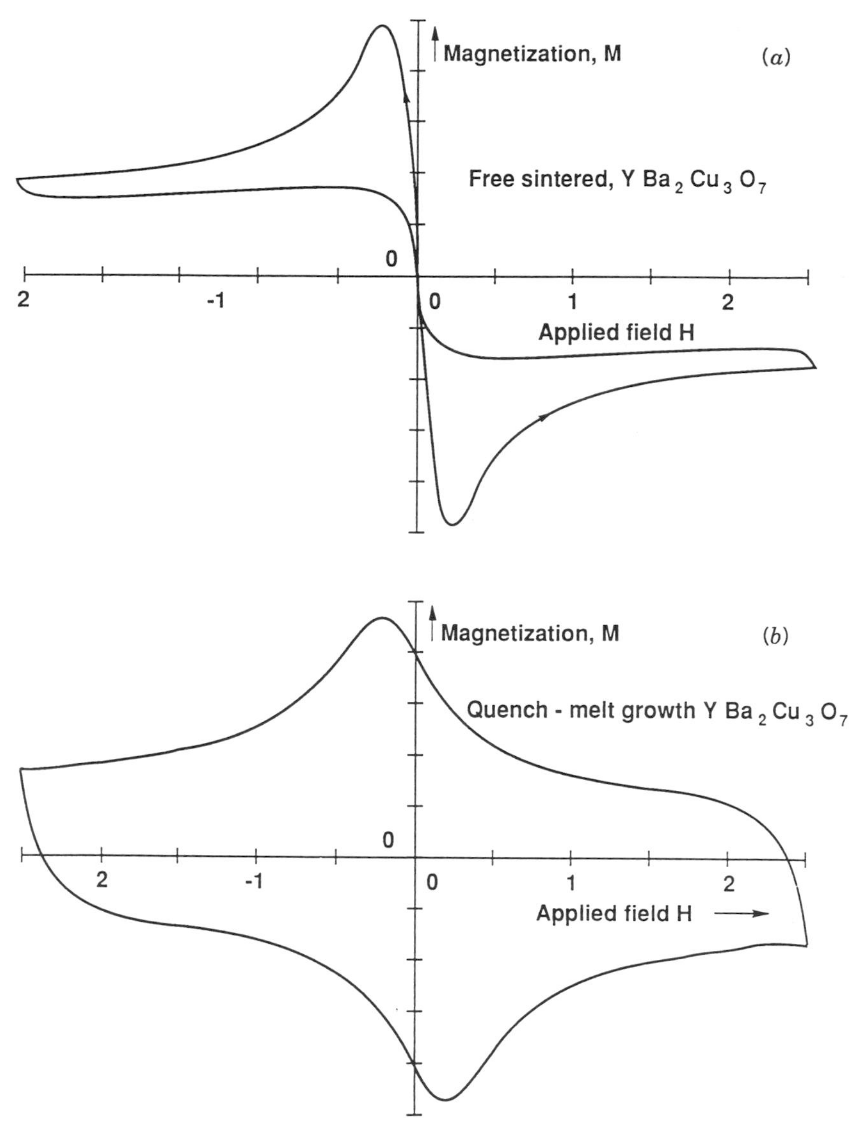

Figure 3-18 Magnetization curves for YBCO in an applied magnetic field for free-sintered (*a*) and quench–melt–growth (*b*) processed material. [From Moon et al. (1990), with permission.]

Conventional rare earth permanent magnets can retain flux densities of 1–1.5 T in a closed ferromagnetic circuit. However, in air the high reluctance will limit the maximum field at the poles to around 0.5 T (see Chapter 2 for a discussion of reluctance). To create higher flux densities, one can use an electromagnet (i.e., a current-carrying coil surrounding a ferromagnetic core, which can produce fields of

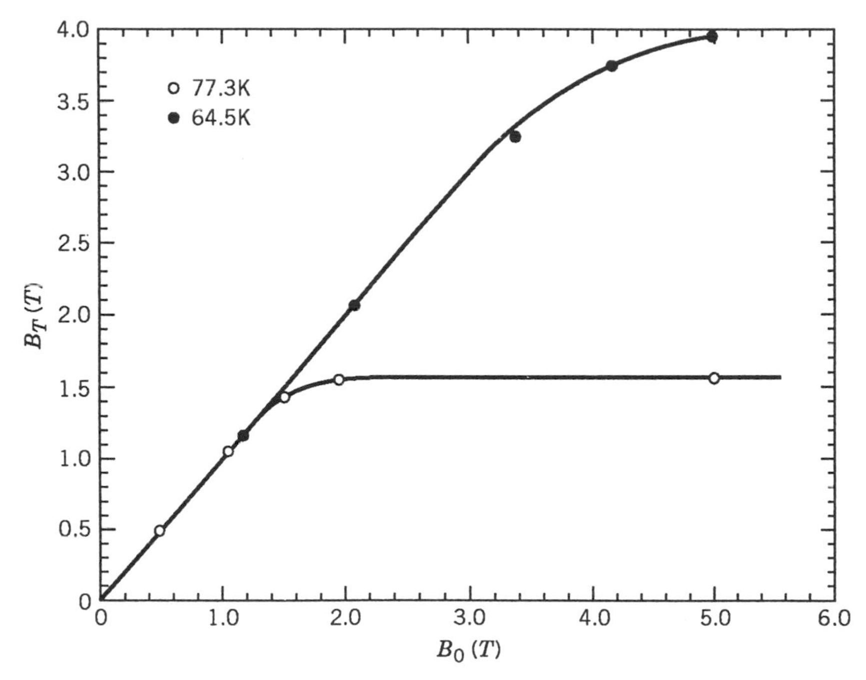

Figure 3-19 Residual field for YBCO superconducting permanent magnets. [From Weinstein et al. (1992), with permission.]

1–2 T) or one can build a multiturn magnet out of superconducting wire. Superconducting magnets can create fields greater than 6 T, but are very expensive. The creation of bulk superconducting permanent magnets raises the possibility of producing a portable field source of 1–5 T at low cost. At the time of this writing, very little work had been done on the levitation properties of these superpermanent magnets. However, they certainly present new options for the levitation or magnetic bearing engineer. (See Chapter 7 for a discussion of potential Mag-Lev application.)

The Houston group has used proton and helium particle irradiation to increase the J_c of the melt-quenched samples from 13,000 A/cm^2 to 45,000 A/cm^2 at liquid nitrogen temperatures. They have also used an excess stoichiometry of yttrium in producing a specimen that traps 4 T at 64 K.

Another group at Nippon Steel in Japan has trapped 1.7 T magnetic field in a quench-melt growth sample of YBCO at 63 K.

Flux Creep

One of the issues for the application of superconducting permanent magnets is flux creep. This phenomenon involves the loss of field with time according to the following equation:

$$\frac{B(t)}{B(0)} = 1 - \beta \log t$$

In the Houston tests, one sample lost 13% of its field in 1 week. However, because of the logarithmic relation, the next 13% would be lost in 19 years. Thus, these magnets are not so permanent, but are still relatively stable after the first drop in field intensity.

CHAPTER 4

PRINCIPLES OF SUPERCONDUCTING BEARINGS

> Thus it is settled by nature, not without reason, that the parts nigher the pole shall have the greatest attractive force; and that the pole itself shall be thc seat, the throne as it were, of a high and splendid power.
>
> —W. Gilbert, *DeMagnete* (1600)

4-1 INTRODUCTION

Conventional Bearings

Bearings are so pervasive in modern electric and mechanical machines that they are often taken for granted; they are usually the last thing a designer thinks about, until there is a failure. Conventional bearing systems involve roller, ball, hydrodynamic, and gas support systems. Nonconventional methods include active ferromagnetic bearings, active electric field systems, and superconducting bearings. Bearing systems come in linear and rotary types. The principal function of a bearing is to allow the relative motion of two machine parts with a minimum of resistance, wear, noise, friction, and heat generation. In roller and ball bearing systems these functions are realized through the kinematic mechanism of rolling of two hard surfaces. In hydrodynamic and gas bearings the two machine parts are separated by a fluid or gas film. In some sense this can be called hydrodynamic or aerodynamic levitation. In bearing technology the distinction is often made between hydrostatic and hydrodynamic bearings as well as between aerostatic and aerodynamic systems. In the aerostatic bearing, pressurized gas must be fed to the bearing to obtain a lift force, whereas in the aerodynamic and hydrodynamic cases the lift is self-generating due to the relative motion of the moving parts.

The macroscale application of an aerostatic bearing was in the air-cushion vehicles built in the 1970s in the United States and Europe as passenger-carrying vehicles. This transportation technology was superseded by magnetic levitation vehicles which were developed in the 1970s in Japan and Germany. The air-cushion vehicles required large motors to pump pressurized air under the vehicle. A self-lift generating concept called the *ram air cushion* was studied by the United States Department of Transportation and Princeton University in the 1970s, but it never reached the prototype stage. Of course, many so-called hover-craft air-cushion boats and ferries are employed around the world. To data there is no magnetic levitation alternative to air-cushion levitation of boats or ships.

Magnetic Bearing Systems

The separation of one moving part from another using magnetic fields in most systems requires a material source of magnetic field and a field-shaping or field-trapping material.

Field source systems include:

- Current-carrying coils (normal or superconducting)
- Permanent magnetic materials
- Superconducting permanent magnets

Field-shaping or *field-trapping materials* include:

- Soft ferromagnetic materials (e.g., silicon steel)
- Passive superconductors
- Normal conductors (e.g., eddy current systems)

With various combinations of the above, one can conceive of a variety of magnetic suspension systems as illustrated in Figure 4-1. One can, of course, produce levitation forces with two field sources such as two magnets or two current-carrying coils. In most practical designs, however, it is often more efficient to put the active field source on only one of the machine parts.

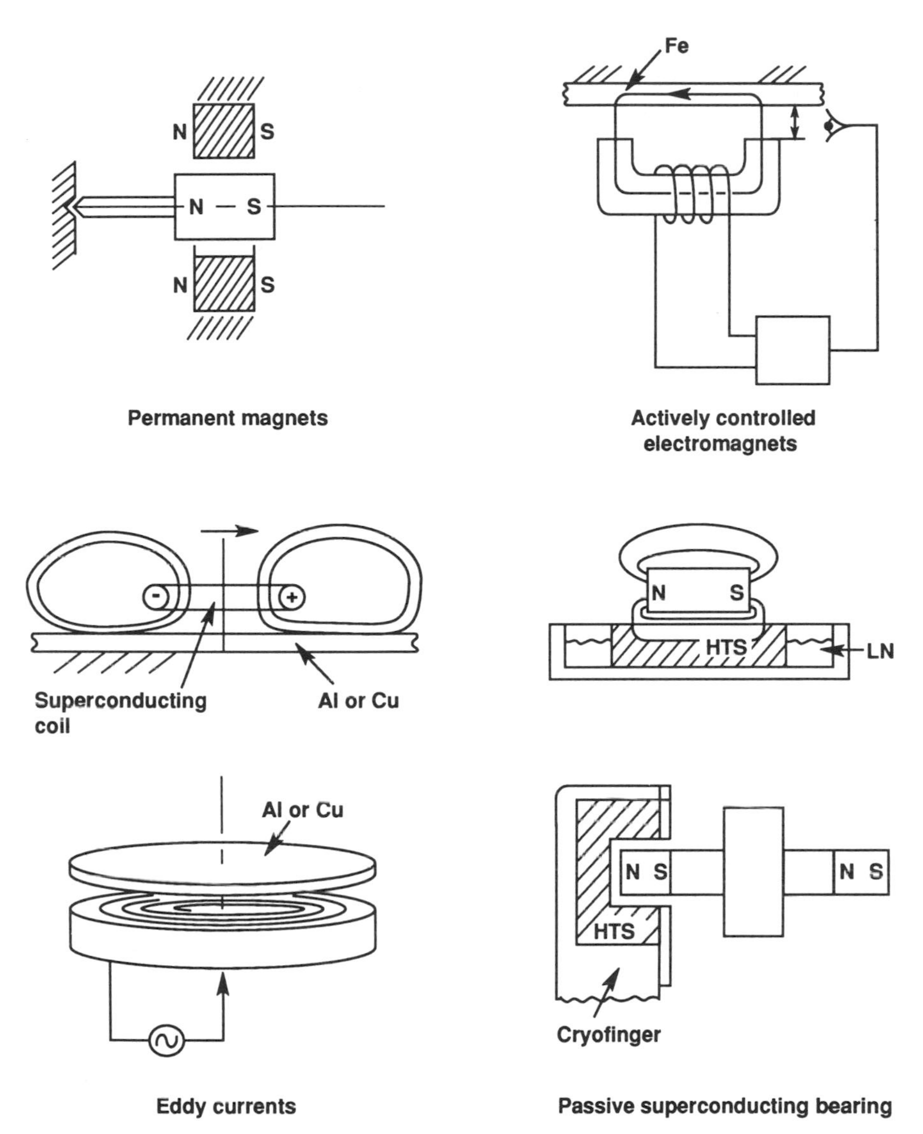

Figure 4-1 Six different magnetic bearing systems.

A couple of general remarks can be made about magnetic bearings independent of the source of the field or the material.

1. Exclusion of magnetic flux is required for repulsive levitation.
2. Penetration of flux in the magnetic material is required for attractive levitation or suspension.

These two statements follow from the expression for the magnetic force in terms of magnetic pressure and magnetic tension [Eq. (2-2.21)].

The focus of this book is on those levitation systems that use superconducting material as either a field source or a field-shaping function. However, we will make a few observations concerning non-superconducting magnetic levitation schemes—in particular, permanent magnet and active electromagnetic levitation systems. [See Conner and Tichy (1988) and Tichy and Conner (1989) for a discussion of eddy current bearings.]

Levitation with Permanent Magnets

As noted in Chapter 1, Earnshaw's theorem on stability leads one to the conclusion that a permanent magnet cannot be in stable equilibrium in the field of a set of other permanent magnets. This means that at least one degree of freedom of the levitated body is unstable. If one is willing to stabilize the unstable modes using mechanical means, then permanent magnets can serve as useful bearings for the other degrees of freedom. One widespread application of permanent magnet bearings is in utility watt-hour meters (see Figure 4-2). In this application, the magnetic force equilibrates the gravity force but

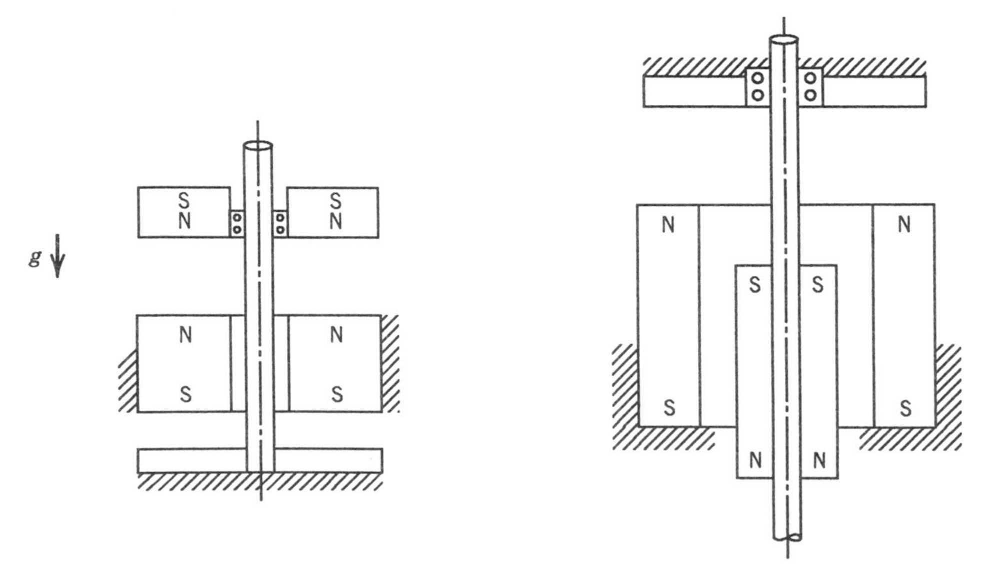

Figure 4-2 Two permanent-magnet bearing configurations with mechanical constraints. [After McCaig and Clegg (1987).]

mechanical bearings are required for the lateral stability (see McCaig and Clegg, 1987; also see Polgreen, 1966). In Figure 4-2, two magnet configurations are shown. In one the levitation force is repulsive, whereas in the other the force is attractive.

The bearings in Figure 4-2 can be classified as thrust bearings. Permanent magnets can also be used for radial bearings in which the gravity load is transverse to the axis of rotation as shown in Figure 4-1 (top left). In this case the axial degree of freedom is unstable and must be constrained with a mechanical contact.

A large-scale use of permanent magnets was in the German M-Bahn levitated vehicle system installed in Berlin (see also Chapter 7). In this system, rare earth magnets on the vehicle were attracted to a ferromagnetic rail. The inherent magnetic instability was stabilized using mechanical springs (Dreimann, 1989). Lateral stability was achieved using guide wheels.

Besides using mechanical contact or guidance wheels to stabilize permanent magnet levitators, one can use active electromagnets. In the example shown in Figure 1-3 (top right), the permanent magnet is used to equilibrate the gravity load while the active electromagnet is used to provide positive magnetic stiffness using feedback control (see also Weh, 1989).

4-2 ACTIVE ELECTROMAGNETIC BEARINGS

Early successes with actively controlled electromagnetic suspension were reported by Beams and co-workers at the University of Virginia in the 1950s. Beams, a physicist and president of the American Physical Society, was able to levitate small submillimeter ferromagnetic spheres and spin them to speeds of over 10^6 rpm (Allaire et al., 1992 and Beams et al., 1962). Even by today's standards this is a remarkable achievement. Beams was able to develop a small centrifuge using his high-speed spheres to test coatings. An early history of magnetic levitation devices may be found in Geary (1964).

Active suspension was further developed in the 1970s to suspend small models in wind tunnels (Johnson and Dress, 1989). But perhaps the most dramatic application began in the early 1970s with prototype vehicles suspended by actively controlled magnets from steel rails [see, e.g., Jayawant (1981)] at speeds of up to 500 km/h. This work was developed in Germany, Japan, and North America. Today the Germans have begun to market a full-scale, high-speed revenue vehicle, called Transrapid, and Japan Airlines has developed a commuter vehicle called HSST (see also Chapters 1 and 7).

Active magnetic bearings have made great progress due in part to the development of microprocessors and small position and velocity sensors so that rotors weighing more than 1 tonne can be suspended. Reliability of these systems has increased dramatically to where active magnetic bearings are used in underground gas pipeline pumps in remote parts of North America (see also Chapter 6). This field is becoming a mature technology area as evidenced by the number of companies in Europe, Japan, and, to a lesser extent, the United States. Some companies are looking at levitated machine tool spindles for high-speed machining. Also, some aircraft engine manufacturers are even looking into the possibility of active magnetic bearings for replacement of fluid film bearings for jet engine rotors.

The major drawbacks to active magnetic bearings are the obvious: high cost, increased complexity (hence, potential reliability problems), and the size of control, sensing, and power electronics. Some of these problems will see solutions in the near future, such as the downsizing of the ancillary electronics systems. The integration of intelligence, actuation, and machine in these types of systems has been called *Mechatronics* in Europe and Japan. In the United States the term *Smart machines* has been used. In any case, in the next decade the use of active magnetic bearings will likely expand to applications where cost is not a major factor. Active magnetic bearings using superconducting coils or windings have been proposed by Eyssa and Huong (1990) and by the Grumman Aerospace Corp. [see Chapter 7 and U.S. DOT (1993a)].

Stiffness of Active Bearings

The magnetic stiffness of active magnetic bearings depends on the specific system and the control system. However, a review of some published data shown in Table 4-1 shows values on the order of 10^5–10^6 N/m. In many systems the stiffness depends on whether the rotor is stationary or moving. In either case the stiffness of active magnetic bearings is greater than current passive superconducting bearings because of two factors. First, the use of soft ferromagnetic material in active bearings and current-carrying coils as field sources creates magnetic flux densities over 1 T. This means that levitation bearing pressures of over 40 N/cm^2 are possible. The second factor is the use of feedback control which allows flexibility in designing in magnetic stiffness.

TABLE 4-1 Active Magnetic Bearings: Mass vs. Stiffness

Application	Rotor Mass (kgm)	Magnetic Stiffness (N/m)
Flywheel	6.5	1.78 10^6 (Axial) 2.12 10^5 (Radial)
	5.7	2.46×10^5
	2.4	7.6×10^5
Particle Beam choppers	12	2.1×10^5
	8	1.0×10^5
	1	1.6×10^5
Centrifuge	9	8×10^4

Source: Schweitzer (1988).

In superconducting passive magnetic bearings using permanent magnets and bulk yttrium–barium–copper oxide (YBCO), the fields are usually less than 0.5 N/m. The use of ferromagnetic material to concentrate the flux would likely tend to destabilize the levitation, although this topic has not received much attention. One way to increase magnetic stiffness in passive bearings is to increase the magnetic flux density.

In the remainder of this chapter we will focus the discussion on superconducting magnetic bearing systems. Three systems have received attention in the research literature in recent years:

- Passive superconductors and permanent magnets
- Passive superconductors and direct-current superconductors
- Superconducting permanent magnet systems (see Chapter 3)

In the next section we discuss the simplest superconducting levitation system, namely, a bulk or thin-film superconductor interacting with the field of a permanent magnet.

4-3 PASSIVE SUPERCONDUCTING BEARINGS

The levitation of small magnets over $YBa_2Cu_3O_7$ (YBCO) samples at liquid nitrogen temperatures has become more of a symbol of the new age of high-temperature superconductivity than even the loss of resistance (Hellman et al., 1988). One could even consider the name "superlevitators." However, in spite of the fact that levitation had become a symbol of the new superconductors in the public eye, most scientists initially did not view the property of stable levitation as having serious practical applications. One reason for this was the low levitation pressures measured at that time, namely, $p \approx 0.2\ \mathrm{N/cm^2}$. In spite of this skepticism, several laboratories around the world began to investigate levitation applications using YBCO. First, small rotors (< 10 g) were spun to speeds of over 10^5 rpm [Cornell University; see, e.g., Moon and Chang (1990)]; recently, they were spun to speeds of over 500,000 rpm by the Allied Signal Corporation. Then material scientists began improving the processing to where pressures of over 5 $\mathrm{N/cm^2}$ and higher were measured—a 25-fold increase. Soon laboratories in Japan and the United States were levitating much higher loads. At the ISTEC lab in Tokyo, they achieved 120 kg without rotation. Meanwhile, industrial laboratories built heavier rotary devices of 2–4 kg at speeds of up to 30,000 rpm. At the time of this writing (late 1993), there are plans for 100-kg levitated rotors at speeds of over 10,000 rpm for use as energy storage flywheels.

Thus, we enter a new age of magnetic bearings using the levitation properties of new materials such as YBCO and bismuth–strontium–calcium—copper oxide (BSCCO). The following sections will give a primer on the properties of these new materials and how they may be used in bearing applications. It is likely that in the next half decade, new superconducting materials will be discovered. Already mercury-based cuprates have demonstrated critical temperatures above 130 K. However, the issues raised below and the characterization tools are likely to remain valid.

Magnetic Forces

Because both force and magnetization are vector quantities, one has to define the direction of the magnetic forces relative to the superconductor–magnet geometry. There are two basic configurations for bearings for rotary machines (see Figure 4-3): (1) trust bearing and (2) journal bearing. As is demonstrated below, it is desirable to have a magnetic field source which has symmetry about the axis of rotation. This leads to the use of cylindrically shaped permanent magnets (usually of the rare earth kind) for the rotor element. These magnets can be a solid cylinder or ring or shell-type geometry as shown in Figure 4-4, with the magnetization aligned with the cylinder axis of rotation or in the radial direction. For this magnet geometry the principal load (e.g., gravity) is aligned with the axis of rotation in the thrust bearing, and the load is transverse to the symmetry axis in the journal bearing.

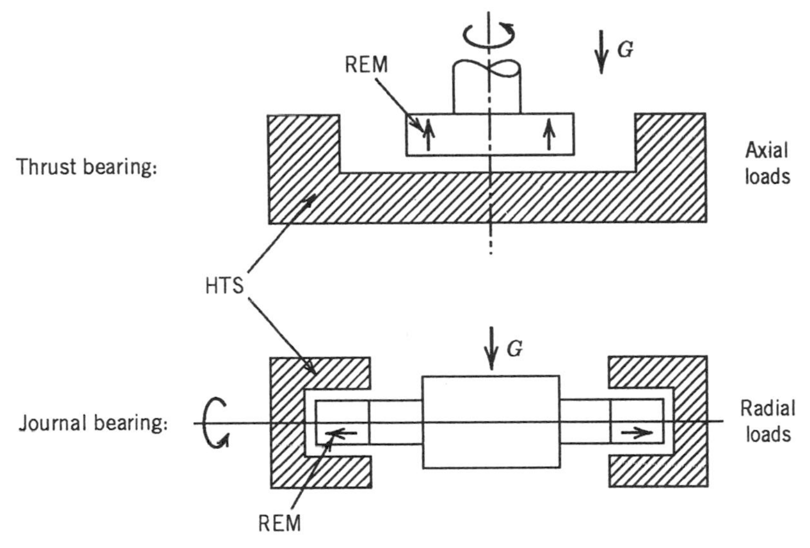

Figure 4-3 Passive superconducting bearing concepts. REM, rare earth magnet; HTS, high-temperature superconductor.

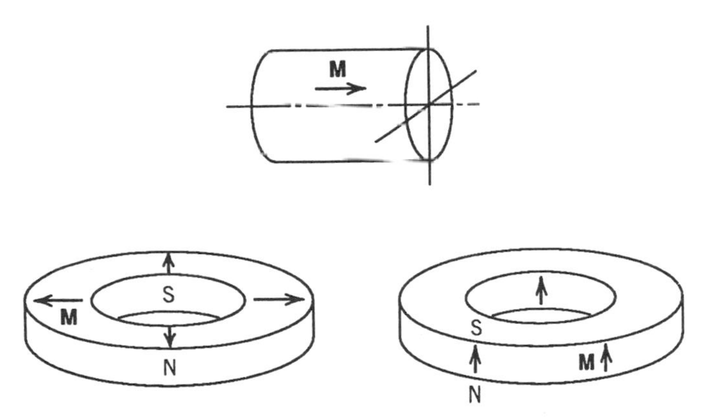

Figure 4-4 Three permanent-magnet shapes with symmetric magnetic fields suitable for high-temperature superconducting bearings.

Of course, one can design a hybrid bearing to take both axial and transverse loads. In any case, the bearing stator, which is the superconductor, need not be symmetric and can be fabricated in either a monolithic or discrete element configuration as shown in Figure 4-5*a*.

We begin our discussion with the thrust bearing configuration. In order to compare different materials, a standard test was developed at Cornell University (see Moon et al., 1988) where the test magnet was

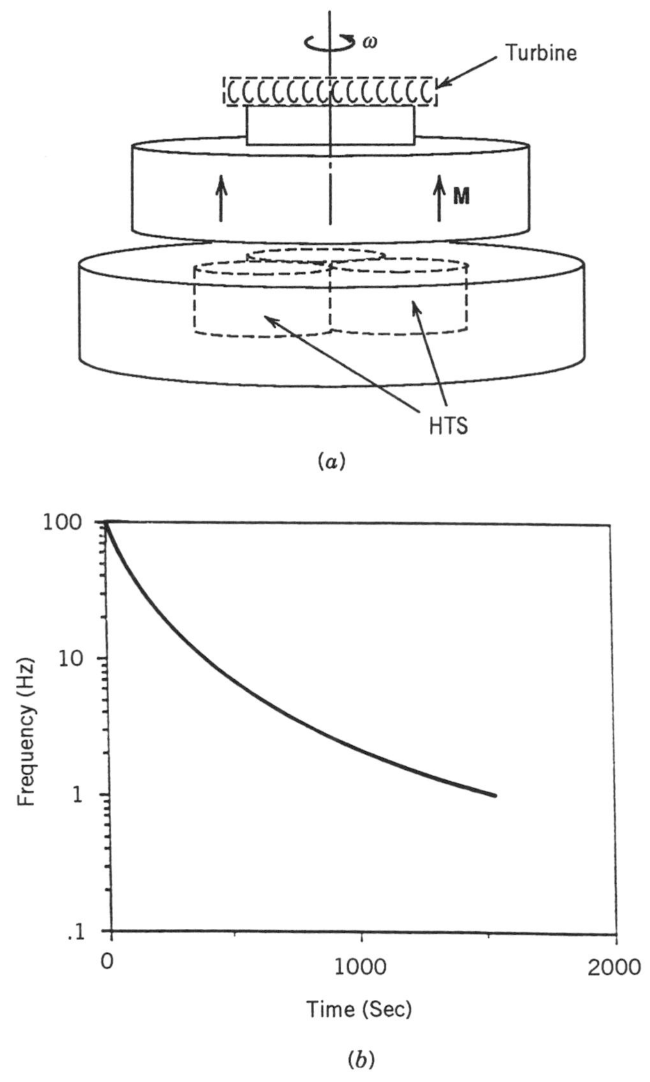

Figure 4-5 (*a*) Permanent-magnet rotor with a three-element, high-temperature superconducting bearing. HTS, high-temperature superconductor. (*b*), Spin-down time history in atmospheric pressure. [From Moon et al. (1993), with permission.]

much smaller than the superconductor. In this test the important geometric ratios were the aspect ratio of the magnets, the gap to magnet face diameter ratio, and the superconductor thickness to magnet length ratio. In this configuration one can then measure (a) the normal force as a function of distance and (b) the lateral force versus lateral displacement.

Another control in these measurements is whether the superconductor was cooled below its critical temperature in the field of the permanent magnet (field-cooled) or zero-field-cooled. This distinction may be important in some superbearing designs because in operation the magnet may not be physically removed from the housing during the cooldown process.

Bearing Pressure

One of the measures for comparison of the levitation properties of different superconductors is to use the average bearing pressure. This is the levitation force divided by the projected magnet area in the direction of the force. To get an idea of the limits to bearing pressure we can look at the ideal case of complete flux exclusion from the superconductor. This is equivalent to the interaction of two magnetic dipoles of opposite polarity as shown in Figure 2-8*a* with $m_1 = -m_2$. The force of repulsion is given by the integral of magnetic pressure on a plane midway between the two magnets:

$$F = 2\pi \int_0^\infty \frac{B_r^2}{2\mu_0} r\,dr$$

Neglecting demagnetization effects (see Chapter 2), we would obtain the maximum force when the magnets are touching pole face to pole face. Replacing each magnet by a point dipole at the center of each mass (separated by the magnet length L) we obtain an estimate of the magnetic moment

$$m = \frac{B_s A L}{\mu_0}$$

where A is the magnet pole face and B_s is the average surface field on the magnet face. The force between two opposing dipoles ($m_1 = -m_2 = m$) is given by Eq. (2-2.6) (see Chapter 2):

$$F = \frac{3}{2}\frac{\mu_0 m^2}{\pi r^4}$$

Setting $r = L$ and $A = \pi D^2/4$, we find a rough estimate of the repulsive force:

$$\frac{F}{A} \approx \frac{3}{4}\left(\frac{B_s^2}{2\mu_0}\right)\left(\frac{D}{L}\right)^2$$

This clearly indicates that the measured bearing pressure F/A will be proportional to the magnetic tension stress on the face of the magnet pole. Thus, for a maximum field strength of $B_s = 0.5$ T (which is typical for good rare earth magnets), one would expect bearing pressures on the order of 10 N/m^2 (14 psi) depending on the geometry of test magnet.

This also indicates that beyond a certain J_c all superconductors will yield similar bearing pressures. Thus, the focus shifts from improving material processing to optimizing the magnetic field configuration. This is a lesson learned by earlier generations of electrical engineers, namely, that improved machines are linked to careful shaping of magnetic flux paths and geometry.

4-4 CHARACTERIZATION OF LEVITATION FORCES IN HIGH-T_c MATERIALS

The characterization of superconducting materials for levitation applications depends on the magnetic field intensity and distribution as well as on the local material properties and geometry of the superconductor. Physicists and material scientists tend to characterize materials in terms of local properties such as critical temperature T_c, magnetization, or critical current J_c. However, the levitation force between a field source and superconductor is an integrated effect. In order to specify material properties for candidate superconductors for bearings, it is necessary to be able to measure integrated properties such as magnetic force–distance relations, magnetic stiffness, and damping. To this end, a number of different measurement schemes have been developed. One such system is shown in Figure 4-6 which utilizes a cantilevered beam force sensor (Moon, 1990). The cantilever is employed so that the force sensor can be located some distance away from the cryogenic temperatures. In this system a test magnet is attached to the end of the beam and strain gages are secured to the beam at the clamped end to measure the strain produced by the magnetic force at the tip.

The distance between the test magnet and the superconductor is measured by an optical sensing device. Both the strain signal and the position signal can be digitally stored for later analysis and can be displayed on an x–y plot showing magnetic levitation force versus distance between the magnet and the superconductor. To provide flexibility in the measurements, the clamped end of the beam is set on

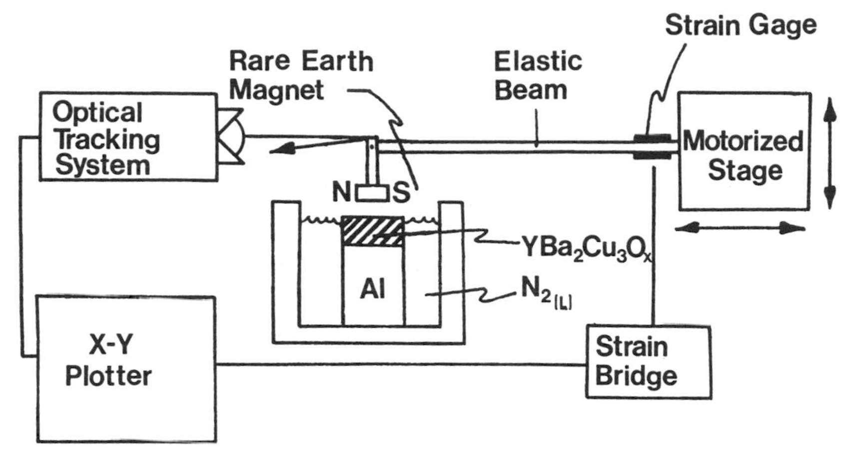

Figure 4-6 Schematic of system for measuring levitation force behavior of superconducting materials. [From Moon et al. (1988) and Moon (1990b), with permission.]

a two-degree-of-freedom motorized stage which can move the test magnet normal or tangential to the test specimen surface.

When the levitation force tests are used to evaluate superconducting materials, the test magnet is chosen to be smaller than the test superconductor. In this case the forces developed might be very small and the beam is designed to be flexible for sensitivity. However, for prototype testing, often the test magnet and the superconducting bearing are of comparable sizes and the forces could be relatively large (e.g., 1–10 N). In this case a stiffer elastic beam sensor can be used. When the beam is very stiff, the displacement of the clamped end of the force sensor can be used as a measure of the test magnet–superconductor separation.

Other systems employed to measure levitation force characteristics have also been used [see, e.g., Ishigaki et al. (1990), Johansen et al. (1990), Marinelli et al. (1989), Weeks (1989) and Weinberger et al. (1990b).]

Levitation Force Hysteresis

It has been observed in high-temperature superconducting ceramics that the induced current or effective magnetization behaves differently for increasing and decreasing applied fields (Moon et al., 1988)

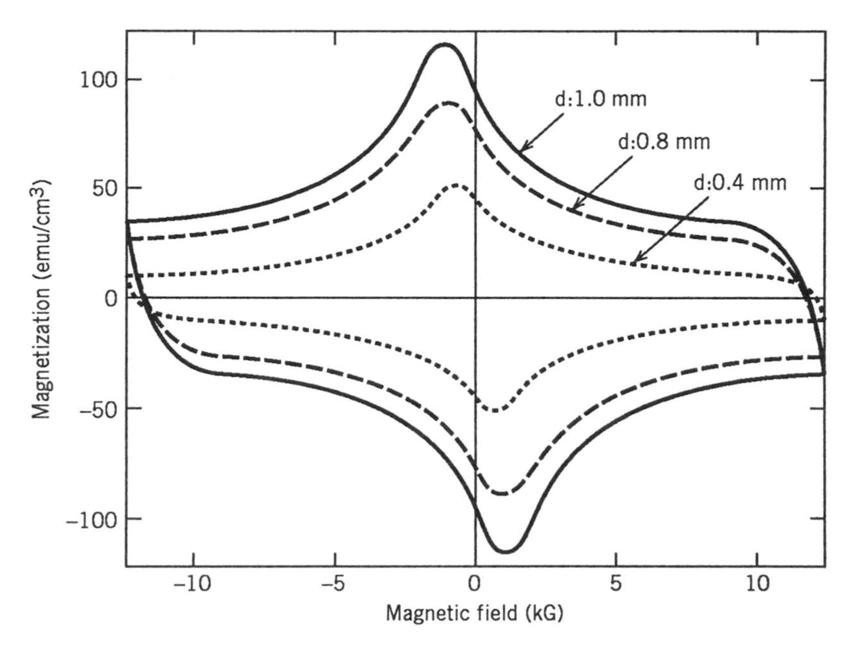

Figure 4-7 Magnetization diagrams for Quench–melt–growth YBCO for different thickness samples. [From Murakami (1989d) with permission.]

(Figure 4-7). Thus it is no surprise that the levitation force behavior should depend on the applied field history. If the applied field is produced by a permanent magnet, then the magnetic force–distance relation should be hysteretic as can be seen in Figures 4-8. In Figure 4-8 the hysteresis implies that levitation against a fixed gravity force might be possible in a range of heights $h_1 \leq h \leq h_2$ depending on the magnet–superconductor separation history.

Also, for certain ceramic superconductors, the levitation force can be either repulsive or attractive, depending on the separation history. If the applied field is produced by a direct-current coil, then hysteresis can result as the applied current or field is increased or decreased even when the coil–superconductor geometry is fixed.

It has been noted in experiments that for certain melt–quench processed materials, the hysteresis effect decreases as the operating temperature is decreased below 20 K (see Chang, 1991).

Levitation Force – Distance Relation

It is interesting to note an apparent paradox in the observed force–distance relationship for a magnet levitated over a flat ceramic

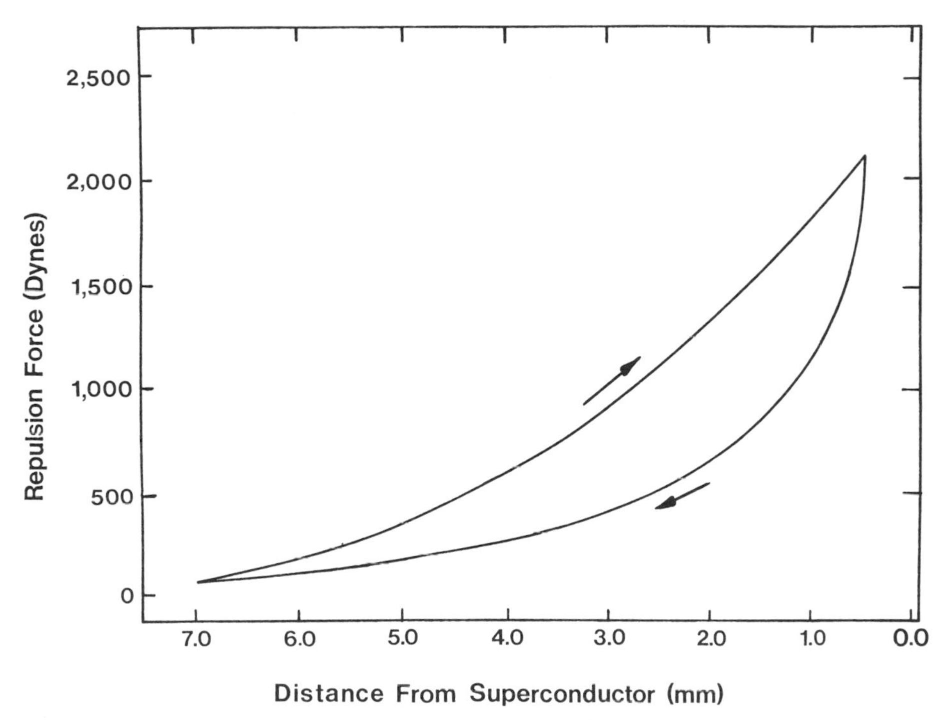

Figure 4-8 Levitation force versus distance from the surface of a free sintered YBCO specimen using a rare earth test magnet. (See Figure 4-6.)

superconductor. If the force were modeled as an interaction between two dipole magnets, then one should expect an inverse power law relation. However, experiments with small test magnets levitated over large ceramic superconducting show an exponential relation:

$$F = F_0 e^{-\alpha z}$$

where z is the distance from the center of the test magnet to the superconductor surface. This is illustrated in a series of experiments shown in Figure 4-9 (Chang et al., 1990).

Magnetic Stiffness

As discussed in Chapter 1, stable levitation requires that the levitation force change proportionally as the separation distance increases or decreases. This change in levitation force is called the *magnetic stiffness*. For a cylindrical permanent magnet, there are five magnetic stiffnesses corresponding to the five degrees of freedom; heave, pitch,

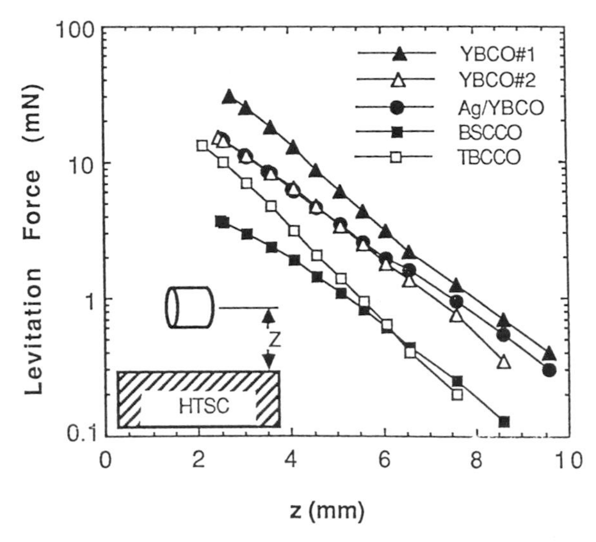

Figure 4-9 Levitation force versus vertical distance for free sintered YBCO superconductors as well as Bismuth and Thallium Cuprates (Chang et al., 1990, with permission).

yaw, lateral, and axial displacements. It is possible for one or more of these magnetic stiffnesses to be negative, which implies that the levitated body is unstable in that degree of freedom.

The magnetic stiffness can be measured in two ways: a quasistatic test and a dynamic test.

Static Measurement In this method a test magnet is fixed to a force transducer as in Figure 4-6, and a reference force–distance curve is first taken as in Figure 4-8. This curve represents a magnet displacement cycle. To find the magnetic stiffness, small changes in the magnet position are made at different points around the cycle. When there is a large hysteresis in the force–displacement relation, the stiffness is *not* given by the derivative of the force–displacement curve, as can be seen in Figure 4-10. However, when the hysteresis is small, the slope of the force–displacement curve is a good measure of the stiffness.

It should be noted that unlike the elastic stiffness in mechanical structures, the magnetic stiffness is a very nonlinear function of magnet–superconductor distance. An important curve that has been found experimentally is a relation between the magnetic stiffness κ and the levitation force F: (Chang et al., 1990):

$$\kappa = cF^a$$

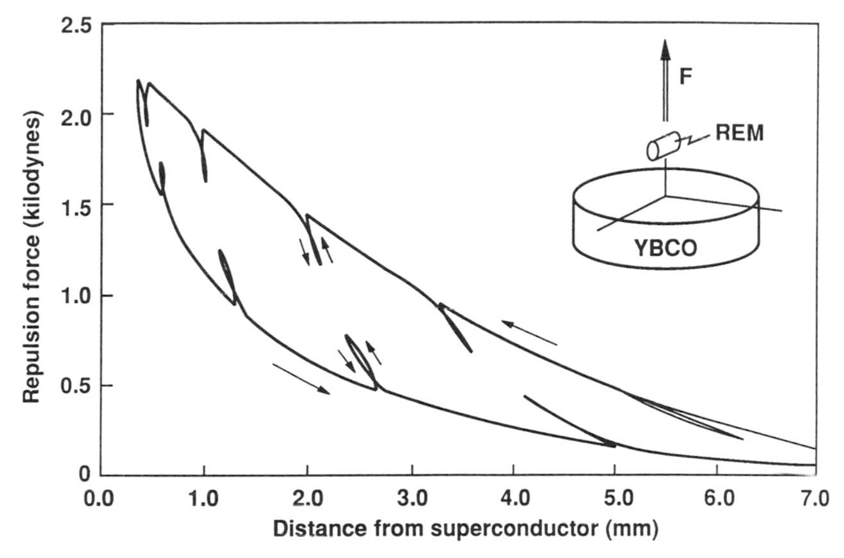

Figure 4-10 Levitation force–distance properties for a sintered YBCO specimen showing magnetic stiffness loops. REM, rare earth metal. [From Moon et al. (1988), with permission.]

where a has typical values of $1.0 < a < 1.6$, depending on the material and relative orientation of the test magnet and superconductor. Several examples are shown in Figure 4-11 for YBCO.

Dynamic Measurement This technique draws on the analogy with a mass on elastic spring. Here the elastic stiffness is related to the natural vibration frequency ω_0 and the mass m:

$$k_e = \omega_0^2 m$$

When the test magnet is supported by an elastic structure of stiffness k_e, we suppose that the magnetic and elastic stiffnesses act in parallel and thus we obtain

$$k_e + \kappa = \omega^2 m$$

If we measure the elastic stiffness when the superconductor is normal, then $k_e = \omega_0^2 m$ and we obtain

$$\kappa = \left(\omega^2 - \omega_0^2\right) m$$

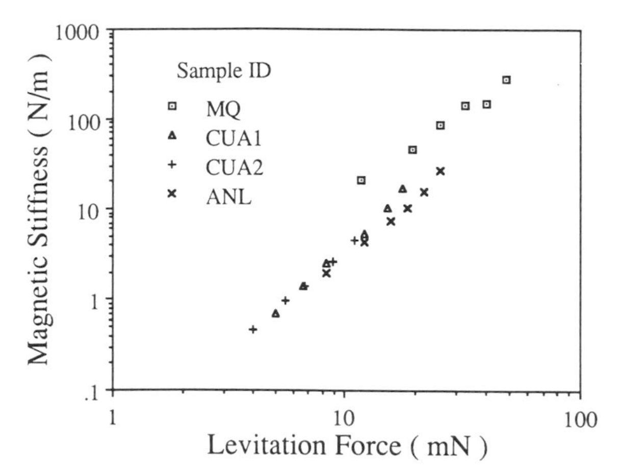

Figure 4-11 Magnetic stiffness versus levitation force for YBCO materials. MQ: melt–quench process, Catholic University of America. CUA1, CUA2: free-sintered process, Catholic University of America. ANL: free-sintered process, Argonne National Laboratory.

This technique has been used in a number of experimental studies [see, e.g., Moon et al. (1989)]. In the study just cited, a comparison of both static and dynamic methods were made and a good agreement was found. Another study of the dynamic stiffness is that of Williams and Matey (1988).

Amplitude-Dependence of Magnetic Stiffness The nonlinear nature of the magnetic stiffness manifests itself in the fact that the measurements described above may be amplitude-dependent. Two such studies have shown, however, that the stiffness increases as the vibration amplitude decreases. A study of a sintered YBCO superconductor at 78 K conducted at Argonne National Laboratory (Hull et al., (1990) and Basinger et al. (1990) showed a dramatic increase in stiffness (up to 100%) as the amplitude decreased from around 10^{-3} m to 10^{-6} m. However, in another study by Yang and Moon (1992) conducted at Cornell University, using a melt–quench processed material, only a modest 10–20% increase was found over the same vibration amplitude range. These results are summarized in Figure 4-12*a*, *b*.

Suspension or Attractive Forces

In magnetic transportation systems, one can have either an attractive or repulsive levitation force. In stationary superconducting levitation,

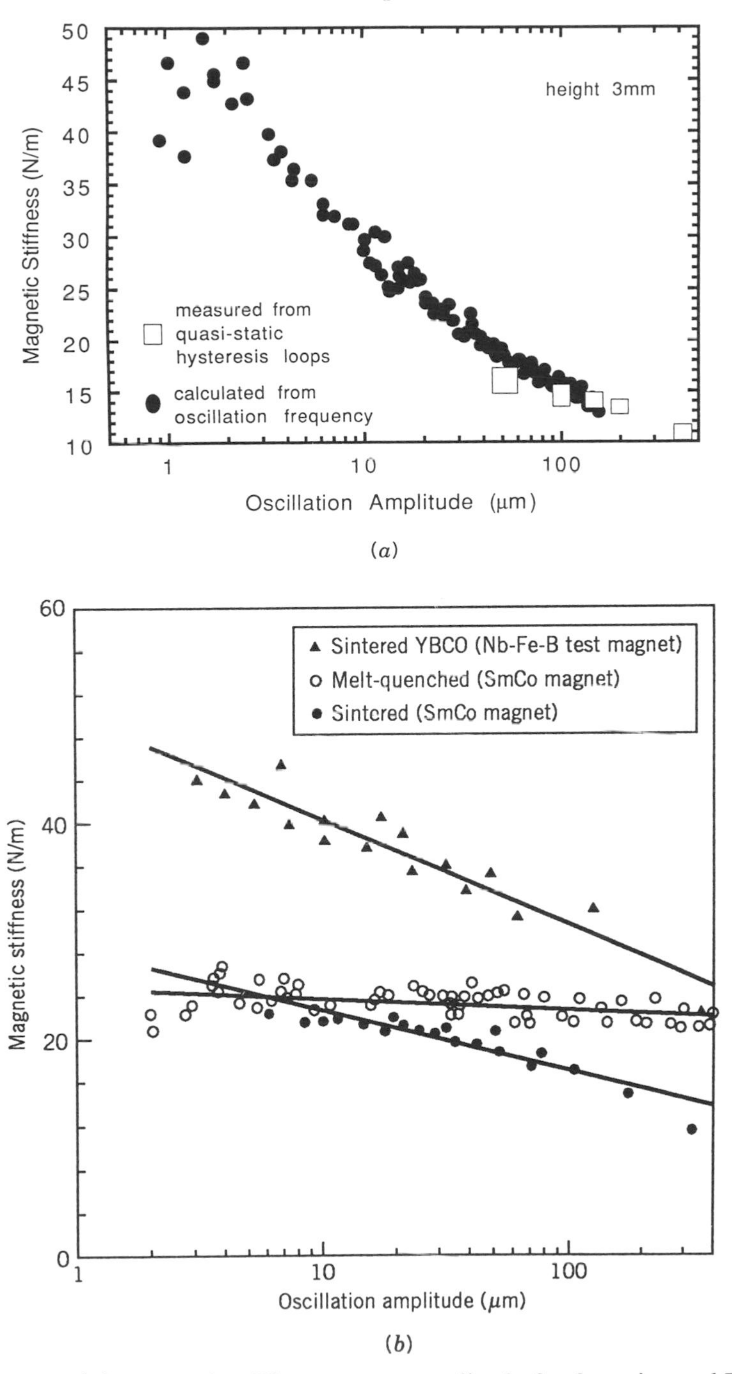

Figure 4-12 (*a*) Magnetic stiffness versus amplitude for free-sintered YBCO. [From Basinger et al. (1990), with permission.] (*b*) Stiffness vs. amplitude for melt–quenched YBCO. [Yang and Moon (1992) reproduced by permission of the American Institute of Chemical Engineers © 1990 AIChE all rights reserved.]

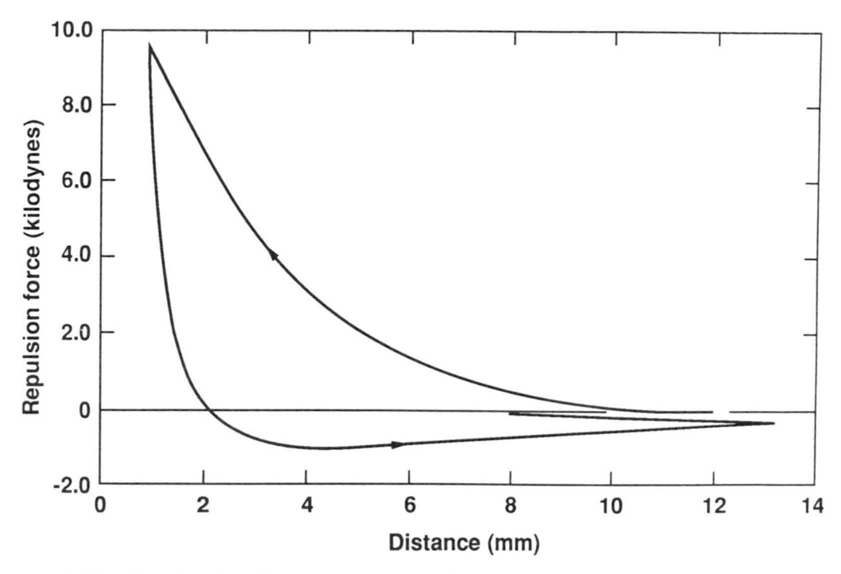

Figure 4-13 Levitation force versus vertical distance for a YBCO superconductor (melt–quench processed material) showing suspension force effect.

both repulsive and attractive or suspension-type forces are also possible. We recall here the basic expression for the magnetic force normal to a superconducting surface in terms of Faraday–Maxwell stresses:

$$F_n = \frac{1}{2\mu_0} \int \left(B_n^2 - B_t^2 \right) dA$$

where n indicates a component normal to the surface and t indicates a component tangential to the surface. A flux-repelling surface ($B_n = 0$) produces a magnetic pressure or repulsive force, whereas a flux-attracting surface ($B_t = 0$) results in a magnetic tension or attractive force. Superconductors in a Type I state (the Meissner regime) exclude flux, and therefore they can only act as a repulsive force generator. However, most applications involve the Type II state, $B > B_{c1}$, and some flux penetrates the superconductor. Thus, some ceramic superconductors can exhibit suspension due to attractive forces as well as repulsion effects as seen in Figure 4-13. This is especially true for melt–quench processed materials which have a large magnetization loop with the ability to trap flux when the applied field is removed. This may have important consequences for the design of practical levitators. On the one hand the designer can take advantage of both effects to design both attractive and repulsive forces. On the other hand, one must be careful of the flux history; otherwise a repelling device might become an attracting one. A few papers that discuss magnetic suspension include Adler and Anderson (1988), Kitaguchi et al. (1989), and Politis and Stubhan (1988).

In closing, one should point out that ordinary permanent magnets near ferromagnetic materials are attracting force devices. However, they cannot be used for stable levitation because of the instability expressed in Earnshaw's theorem.

Magnetic Damping

The observation that a displacement cycle of a test magnet near a ceramic superconductor produces a hysteretic magnetic force suggests that the cycle is not reversible in the sense of thermodynamics and that energy is lost in each cycle. What holds for large displacements of the test magnet will also hold for small displacements during vibrations of a test magnet. This nonreversible process manifests itself as magnetic damping. Measurements of magnetic damping have been reported by Moon et al. (1989). Magnetic damping is produced regardless of whether the magnet vibrations are normal to or parallel to the superconductor surface. Typical data are shown in Figure 4-14. The damping is seen to increase as the flux density increases or as the test magnet is moved closer to the surface. A damping measure can be found by assuming that the decay is similar to a linear viscous damper:

$$m\ddot{z} + c(B)\dot{z} + (k_e + \kappa)z = 0$$

where the explicit dependence of c on the field is acknowledged. The nondimensional damping is defined as

$$\gamma = \frac{c}{2m\omega_0}, \qquad \omega_0^2 = \frac{k_e + \kappa}{m}$$

and is sometimes represented as a "percent of critical damping." Critical damping results when oscillation motion changes to simple exponential decay ($\gamma = 1$). Measurements at low frequencies (2–20 Hz) suggest a relative damping of up to 10% critical damping for $B \approx 0.4$ T for YBCO superconductors.

There is some debate as to whether the damping is actually proportional to the velocity—that is, whether the viscous analog is correct.

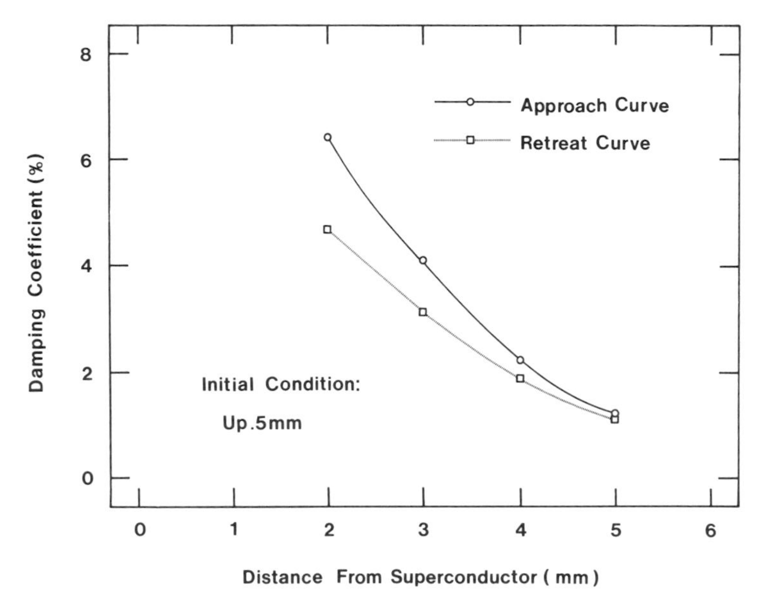

Figure 4-14 Damping versus distance for a vibrating rare earth magnet oscillating above a YBCO superconductor. [From Moon et al. (1989), with permission.]

From lateral magnetic drag measurements (see below) there is evidence that the damping may be velocity-independent—that is, more like Coulomb dry friction damping. Theoretical studies of magnetic damping were still under development as of late 1993.

Lateral Magnetic Drag Force

Magnetic forces normal to the surface of a superconductor generally involve either compression or stretching of magnetic flux lines as represented by the Faraday–Maxwell model [Eq. (2-2.21)]. However, if a test magnet is moved parallel to the surface of a ceramic superconductor, the flux lines are sheared, resulting in a lateral magnetic drag force. This force has two parts: a reversible part represented by a lateral magnetic stiffness and a hysteretic or a magnetic drag force. An example of test measurements is shown in Figure 4-15. The large loop represents a hysteretic effect, whereas the small loops represent a stiffness effect. What is interesting about these measurements is that the lateral force seems to be asymptotic to a constant value which suggests a similarity to a Coulomb dry friction force. The asymptotic value seems to be independent of low values of the velocity.

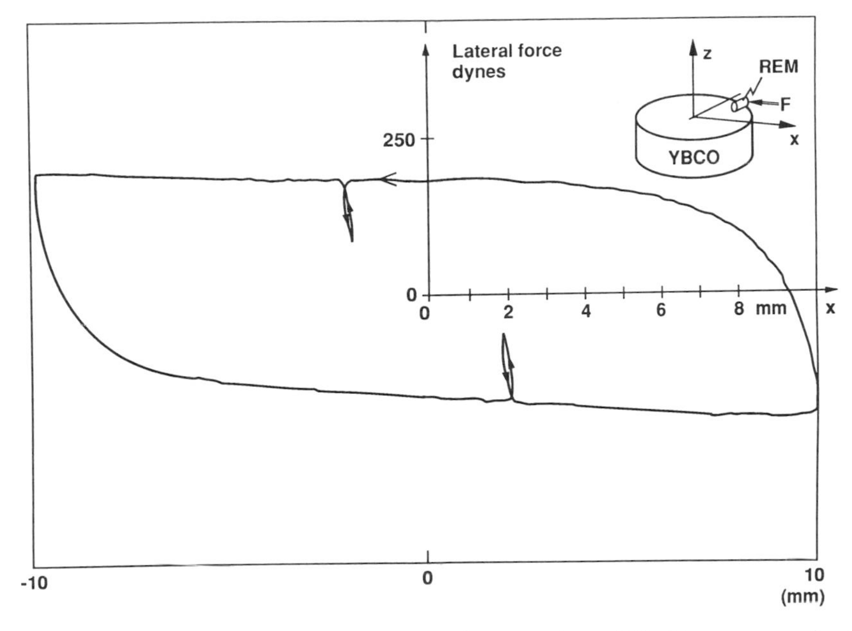

Figure 4-15 Lateral drag force versus distance for a YBCO superconductor. REM, rare earth magnet. [From Chang et al. (1990), with permission.]

An interesting result of these tests is that the lateral stiffness appears to be related to the drag force as shown in Figure 4-16. That the lateral stiffness should depend on the field intensity is shown in Figure 4-17, which shows a monotonic relation between lateral stiffness and the levitation force. Experiments on lateral magnetic forces depend on whether the superconductor is zero-field-cooled or field-cooled. The above results are for the zero-field-cooled case. Further discussion of the magnetic drag or friction force may be found in Brandt (1988) and Johansen et al. (1991).

Rotary Drag Torque

Imagine a cylindrical magnetic field generator (e.g., a permanent magnet or superconducting coil) levitated above or below a superconducting surface. Let us further assume that the magnetic axis of symmetry coincides with the spin axis of the levitated body. Then one can ask, "How does the superconductor react to the spinning magnet?" The answer in theory is that there should be no reaction torques on the spinning magnet if the magnetic field is perfectly symmetric. That is, torques are developed in reaction to a change in flux at the surface

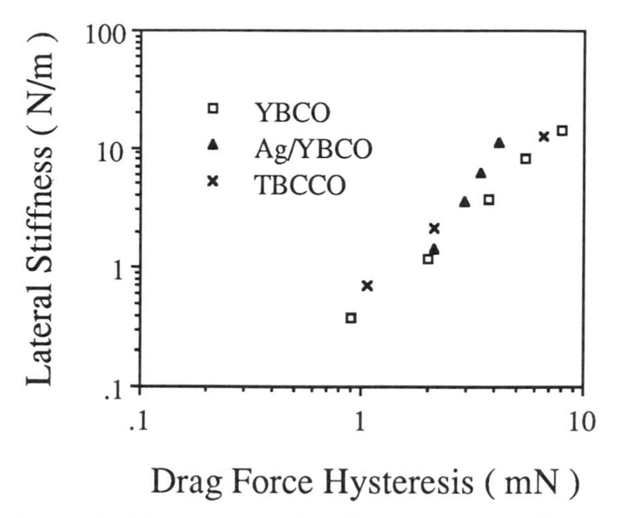

Figure 4-16 Lateral stiffness versus drag force hysteresis. [From Chang et al. (1989), with permission.]

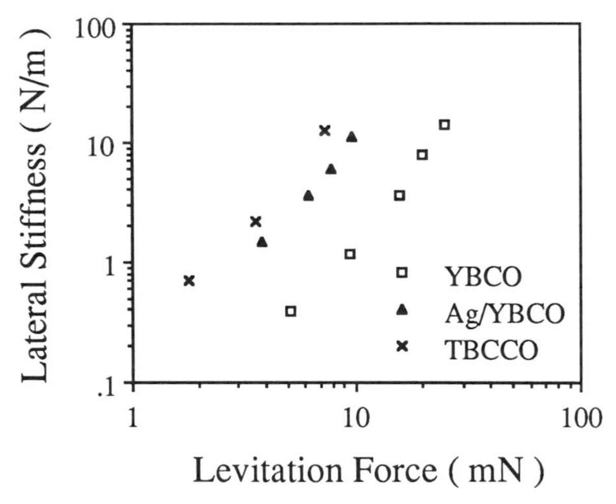

Figure 4-17 Lateral stiffness versus levitation force for YBCO and thallium-based superconductors. [From Chang et al. (1989), with permission.]

of the superconductor. Note that unlike the spinning magnetic body, the superconductor does not need to be symmetric in any sense.

Of course in laboratory experiments, a rotating object would have aerodynamic or viscous drag torques. Also, the field generated by the levitated body might have some small asymmetry that will produce reaction or damping torques.

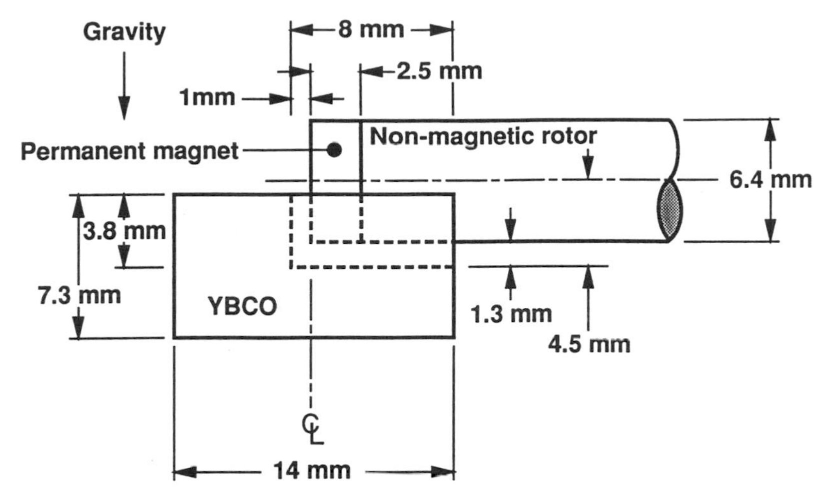

Figure 4-18 Dimensions of a YBCO bearing system which achieved 120,000 rpm. [From Moon and Chang (1990), with permission.]

Experiments on the torque developed on a spinning levitated magnet have been carried out both in air and in vacuum (see Moon and Chang, 1990). The geometry is shown in Figure 4-18. At 1 atm pressure, the decay in angular frequency exhibits an exponential time history characteristic of a viscous torque. However, in a vacuum the decay in frequency shows constant deceleration which is characteristic of a force that is independent of velocity (Figure 4-19). Measurement

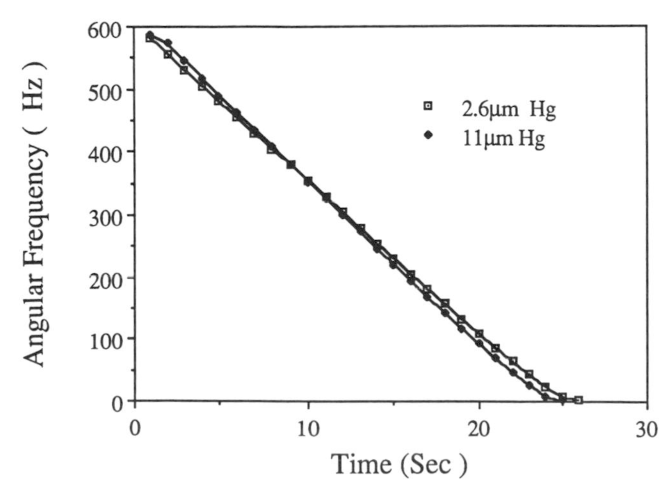

Figure 4-19 Rotor spin frequency versus time in vacuum. [From Moon and Chang (1990), with permission.]

of the field symmetry in the rotating rare earth magnet showed that the field had about a 5% asymmetry which is thought to be the source of the flux drag torque.

Another source of torque on a spinning magnetic body could be small mutation or precession motions. When the rotor is unbalanced, or its initial spin is not aligned with its symmetric spin axis, the spin axis might translate or wobble, producing a small change in flux at the superconductor surface. When the spin rate is close to one of the lateral or rigid body natural frequencies, the rotary motion could couple into one or more of the other rigid body modes, thereby producing damping. This has been observed by Weinberger et al. (1991) at the United Technologies Research Laboratories in Connecticut. (See also Chapter 5.)

Low-Temperature Levitation of High-T_c Superconductors

As observed in Chapter 3, the critical current capacity depends strongly on temperature, decreasing to zero as one approaches T_c and increasing as one lowers the temperature toward that of liquid helium (4.2 K) (see Figure 3-1). It is not surprising that one should observe similar behavior in the levitation force between a permanent magnet and a ceramic superconductor. Yet few measurements of the levitation force in the range 4.2–78 K have been reported. One of the first appeared in the Ph.D. dissertation of P.-Z. Chang of Cornell University.

Using a glass, low-temperature helium dewar, experiments were performed on two different melt–quench processed YBCO materials 10–12 mm in diameter and 2–5 mm thick (Figure 4-20). The test magnet was an $SmCO_5$ cylinder 6.4 mm long and 3.2 mm in diameter with a peak field of 0.35 T.

Two results are noteworthy. First, the relative hysteresis in the force–distance curve was markedly reduced at lower temperatures. Second, the magnitude of the levitation force increased from 78 K to 4.2 K (Figure 4-21). The increase was a factor of 12 for a sintered YBCO specimen and a factor of 7 for a melt–quench processed sample. For this sample, the magnetic pressure was on the order of 10 N/cm^2 on the projected magnet area.

Levitation Force Versus Magnetic Field

In most stationary experiments of levitation of magnets above ceramic superconductors, the test magnet has been a rare earth permanent

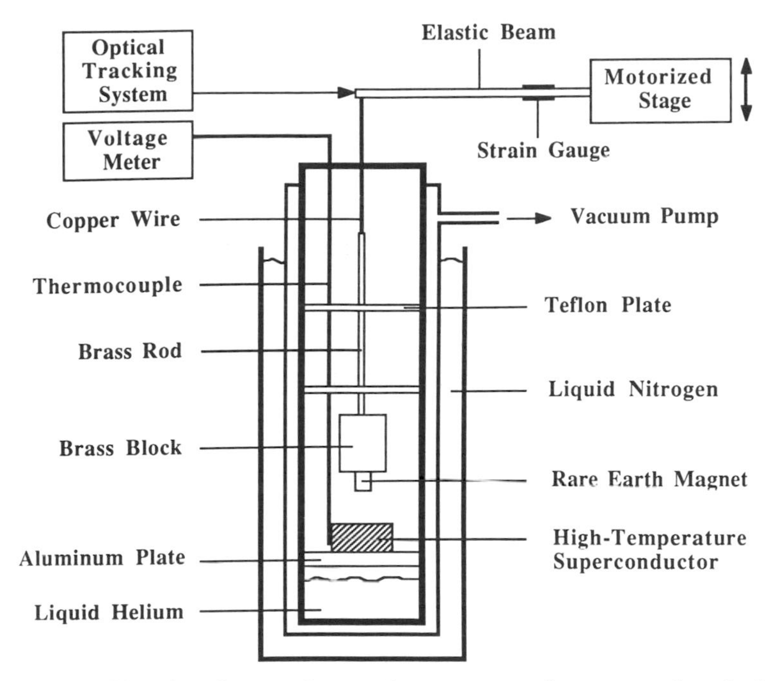

Figure 4-20 Sketch of experimental apparatus for measuring levitation forces at low temperatures (4.2–77 K). [From Chang (1991), with permission.]

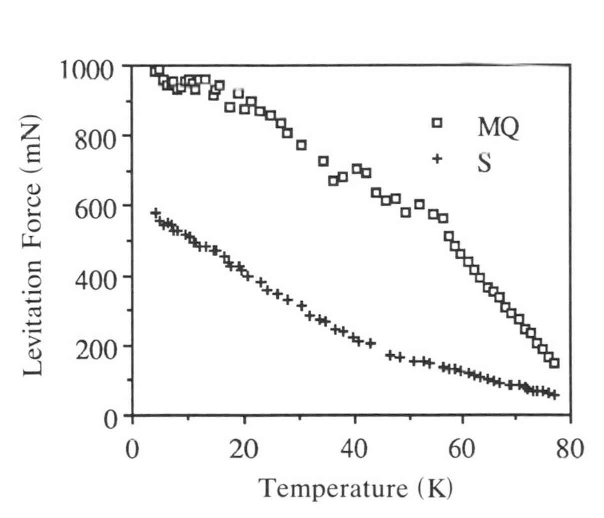

Figure 4-21 Levitation force versus temperature for sintered (S) and melt–quench (MQ) processed YBCO. [From Chang (1991), with permission.]

magnet. Thus, the only way to vary the field at the surface of the superconductor is to also vary the geometric distance between the two.

However, one can also create levitation forces between a current-carrying coil and a ceramic superconductor [see Moon (1992), Golkowski and Moon (1993).] In this case, the coil is also a superconductor, albeit at low temperatures. But the current, and hence the field, can be changed without changing the geometric configuration. Magnetic fields of up to 2 T can be created. Two experiments have been done at Cornell University: one with an Nb–Ti wire wound coil at 4.2 K, and the other with an Nb_3Sn wire wound coil at 15 K (Figure 4-22*a*).

The melt–quench processed YBCO superconductor had a diameter of 35 mm and a thickness of around 13 mm. The wire wound Nb–Ti superconductor cylindrical coil has 450 turns in 28 layers, with an outside diameter of 24 mm.

The force–distance relation for a fixed current was found to be of an exponential form:

$$F = ae^{-bz}$$

where z is the distance of the face of the coil from the superconductor. This is the same form as for a rare earth magnet (Figure 4-9).

The dependence of the levitation force on the current (and, hence, on the applied field) was found to be close to quadratic as shown in Figure 4-22*b*. This relation is to be expected, given the quadratic nature of the magnetic pressure as a function of magnetic field. The maximum field was on the order of 2 T in this experiment, and the maximum force was on the order of 30 kg or 300 N.

A numerical calculation was made of the magnetic force based on a flux exclusion assumption. In this experiment, the superconducting ceramic was zero-field-cooled. Because at these low temperatures the critical currents are believed to be greater than 10^4 A/cm^2 and much higher than at 78 K, the superconductor is believed to act as a good flux repeller. Based on this assumption, the flux distribution should be as shown in Figure 4-23*a*. A calculation of the magnetic force versus stiffness shows excellent agreement with the experiment as shown in Figure 4-24. The distribution of repulsive magnetic pressure based on the flux repeller model shown in Figure 4-23*a* shows a concentration of magnetic pressure at the edge of the coil (see Figure 4-23*b*).

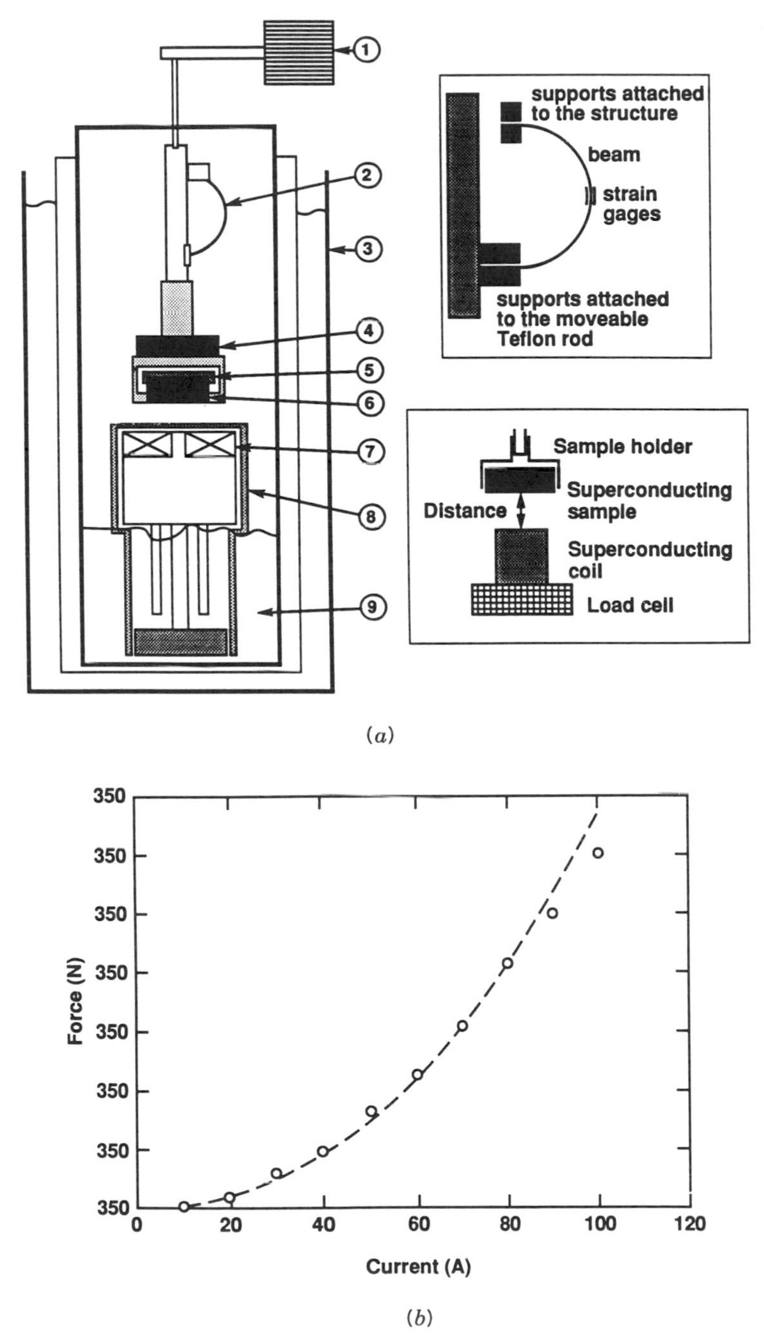

Figure 4-22 (*a*) Experimental apparatus for measuring levitation force between a bulk superconducting specimen and a superconducting coil. (*b*) Levitation force on a bulk YBCO superconductor vs. current in a superconducting coil. [From Golkowski and Moon (1993), with permission.]

Force Creep

One of the consequences of Type II superconductivity is the notion of flux creep [see, e.g., Tinkham (1975)]. In this theory, magnetic flux can diffuse into a superconductor over time, thus weakening the screening effect which is necessary to develop repulsive magnetic pressure or

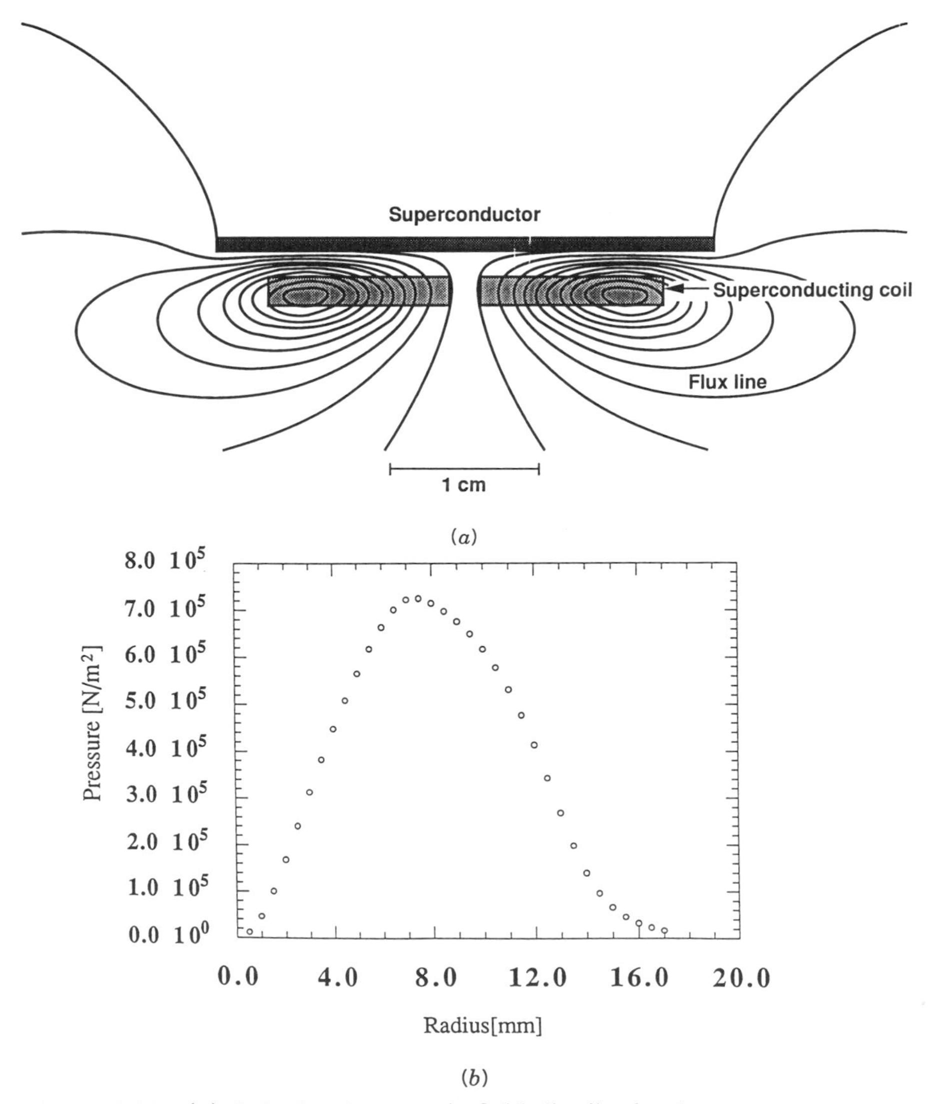

Figure 4-23 (*a*) Calculated magnetic field distribution between a superconducting coil and bulk superconductor. (*b*) Distribution of repulsive magnetic pressure based on the flux repeller model shown in part *a*.

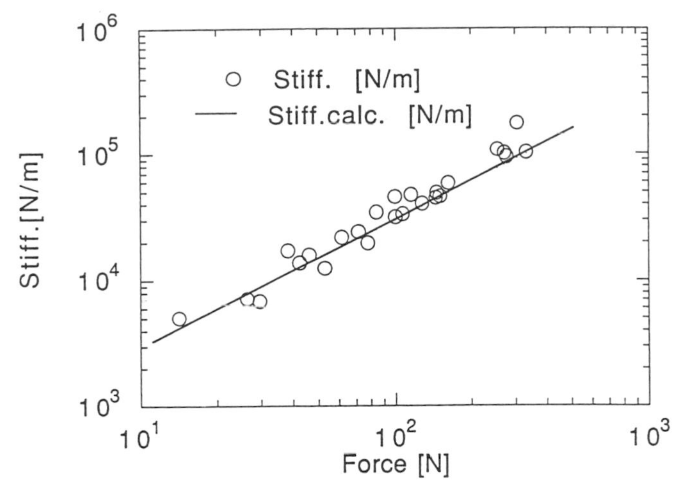

Figure 4-24 Magnetic stiffness versus levitation force for YBCO with a superconducting coil field source. [From Golkowski and Moon (1993), with permission.]

levitation force. Macroscopically, this effect can be observed in the decrease of the levitation forces if a magnet is suddenly placed in the vicinity of a superconductor. An example of this effect is shown in Figure 4-25. [See also Moon and Hull (1990) and Moon et al. (1990).] The data are for a melt–quench processed YBCO ceramic superconductor at 78 K. A small test magnet attached to a force transducer is

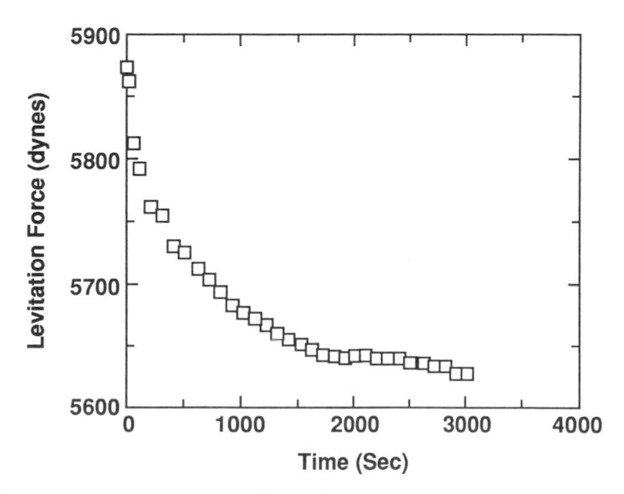

Figure 4-25 Levitation force relaxation versus time for a melt–quench processed YBCO specimen. [From Moon et al. (1990), with permission.]

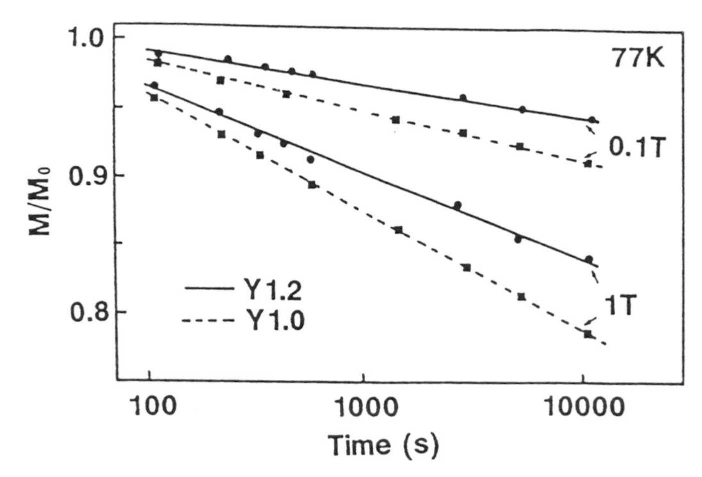

Figure 4-26 Flux creep versus time for two MPMG processed YBCO samples with 211 phase (Y1.2) and without the 211 phase (Y1.0). [From Murakami et al. (1990), with permission.]

suddenly placed a few millimeters above the superconductor surface, and the time decay of the force is measured. The data in Figure 4-25 reveal a 5% loss in force after 3000 sec. If this decay is exponential, then the time required for next 5% loss to occur will be much greater than 3000 sec. This result was for the vertical levitation force. However, one can also see a similar effect for a change in lateral force due to a sudden movement of the test magnet in a direction parallel to the superconductor surface.

The fundamental principle governing these force creep phenomena are believed to be related to flux-pinning theories. Also, magnetization relaxation in YBCO has been measured by Murakami et al. (1989b) in a quench–melt–growth YBCO sample at 78 K (Figure 4-26). Because the levitation force is proportional to the induced magnetization in a superconductor, it is not surprising to see these similar decay phenomena in both magnetization and levitation forces.

Material Processing and Levitation

Whereas the critical temperature (T_c) in ceramic superconductors seems to depend on some fundamental structure of the material (e.g., the chains and planes arrangement of atoms discussed in Chapter 3), other properties such as critical current (J_c), magnetization, and levitation forces all seem to be process-sensitive. In a way, this has been fortunate because great progress has been made in increasing the magnetic force capability by systematically changing the material processing variables.

One of the first studies of the effects of processing on levitation is the work of Wang et al. (1989), who varied the temperature and pressure in a sintered YBCO superconductor to optimize the levitation force. A 50% improvement was obtained.

A later work is the study by Murakami et al. (1991) of the Japanese ISTEC Superconductivity Research Laboratory. They looked at five different processes for YBCO: sintered, melt–textured growth, quench–melt–growth, and two melt–powder–melt–growth processes. The results are shown in Figure 4-27. These processes were described in Chapter 3. The dramatic improvement of the MPMG processes specimens is attributed to the inclusion of a 211 nonsuperconducting phase (Y_2BaCuO_x) interspersed in the superconducting 123 phase. Murakami et al. claim that these particles act as pinning centers which, in turn, improve the intragranular J_c. [See also Oyama et al. (1990).]

The magnetic force depends on the effective magnetization. The magnetization depends on both J_c and the size of the superconducting grains. Another study that attempted to show the importance of grain size was a joint project of Bell Labs and Cornell University (Chang et al., 1992). In this work, S. Jin of Bell Labs prepared a set of disk-like YBCO specimens with different average grain sizes ranging from 4 μm to 410 μm. Measurements of the magnetization showed a direct correlation with grain size. Also, the levitation force showed a monotonic increase in value with grain size as shown in Figure 4-28.

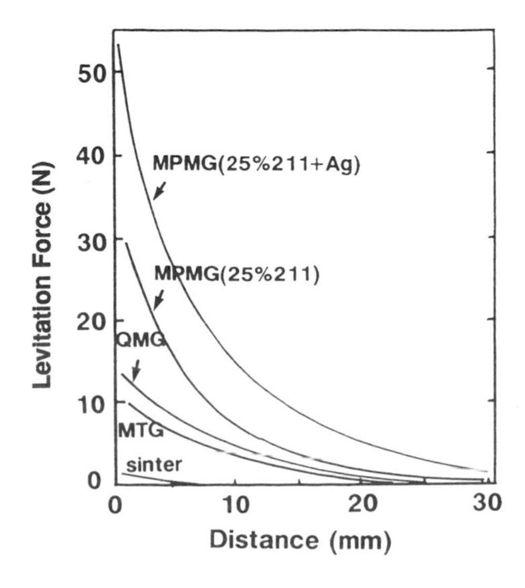

Figure 4-27 Levitation force versus distance of test magnet from YBCO samples. Effect of 211 phase and different processing methods. [From Murakami et al. (1991), with permission.]

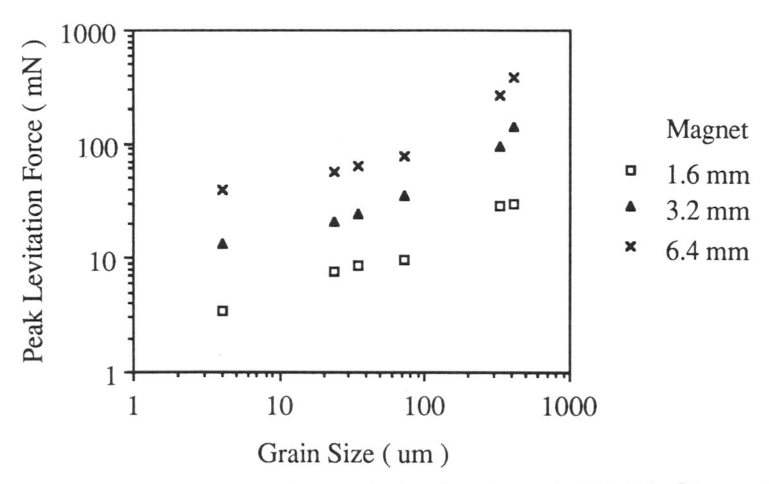

Figure 4-28 Effect of grain size on levitation force in YBCO. [From Chang et al. (1992), with permission.]

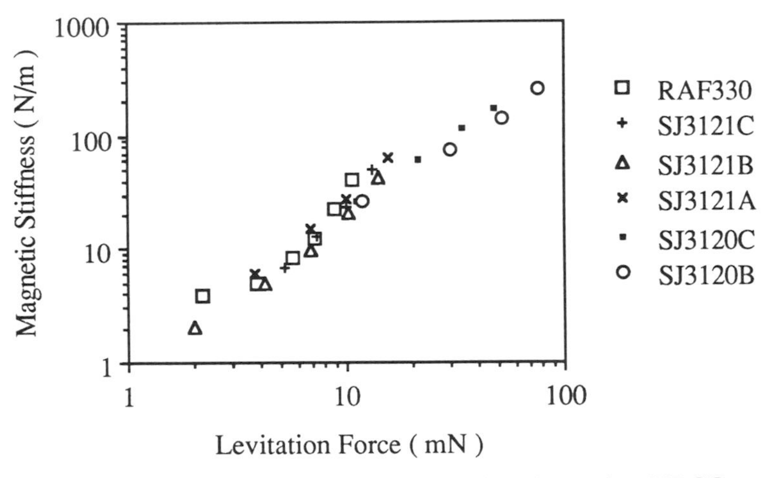

Figure 4-29 Magnetic stiffness versus levitation force for YBCO samples with different grain sizes. [From Chang et al. (1992), with permission.]

Roughly, two orders of magnitude increase in grain size resulted in an order-of-magnitude increase in levitation force at 78 K. What is interesting, however, is that the magnetic stiffness–force relationship seemed to be independent of the grain size as shown in Figure 4-29.

Levitation Forces — Thickness Effect

With levitation applications as a motivation, it is natural to ask the following question: How much bulk ceramic superconductor is necessary to achieve the highest possible levitation force for a given size magnetic field source? If the superconductor did indeed behave as a theoretical Type I Meissner material, one would only need a thin layer of the material under the magnet because the flux would be screened out from the interior. However, in a bulk ceramic material it is known that there is considerable flux penetration, and it is natural to ask how much material is needed or desired.

This question has been addressed in two studies for the case when the test magnet is small compared with the superconductor. In these studies a small test magnet approximately 6 mm in diameter is brought to the surface of a cylindrical sample of YBCO of diameter D and thickness Δ, where D is on the order of 12–20 mm. In the first study by Wang et al. (1989) a sintered sample of YBCO was used. The superconductor thickness was systematically reduced by machining

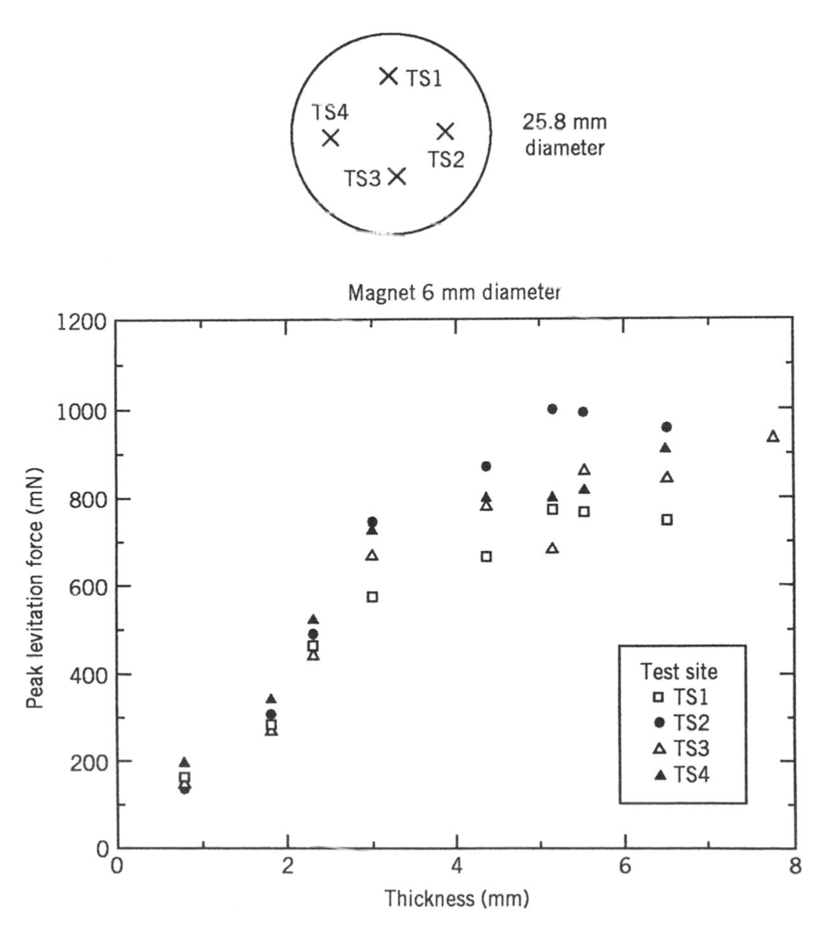

Figure 4-30 Effect of thickness of YBCO sample on levitation force.

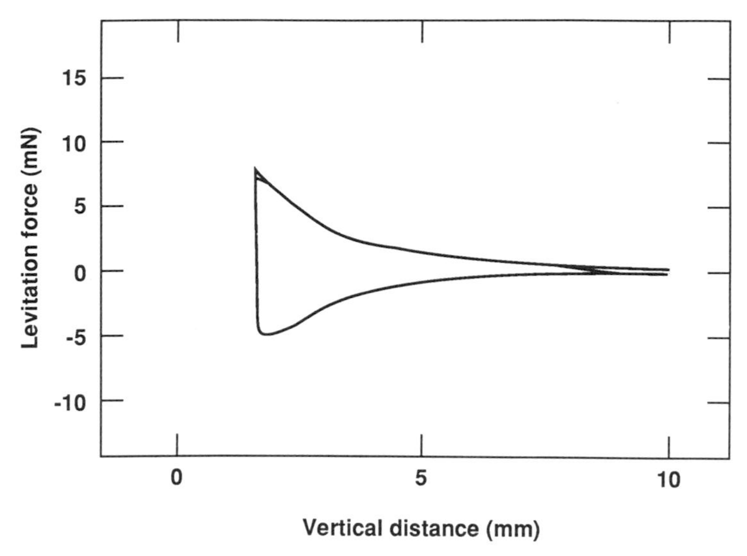

Figure 4-31 Levitation force versus distance for a YBCO thin-film sample.

from 12 mm to 1 or 2 mm. These results indicated that the force was constant until $\Delta = 5$ mm but then dropped linearly with decrease in Δ.

A second study by Moon and Chang et al. (1992) was performed on a sample of melt–quench processed YBCO at the Catholic University of America. The results again show a drop-off in force measurement below 5 mm (Figure 4-30). Whether this thickness of 3–5 mm is characteristic of YBCO or depends on the magnet size is not known. Also, studies in which the test magnet is much larger than the superconductor have not been reported. A theoretical study of the thickness effect on magnetic levitation force has been published by Johansen et al. (1990) and Yang (1992).

Levitation Forces in Thin-Film Superconductors

In contrast to bulk ceramic superconductors, which can be made many centimeters thick, thin-film superconductors are grown on a crystalline substrate to a thickness of a micron or less. In the case of YBCO the principal axis is aligned normal to the substrate. This alignment produces very high critical currents on the order of 10^7 A/cm^2. This high value has been taken as the ultimate goal of bulk superconductors which at present have J_c values in the range of 10^4 A/cm^2 at 78 K. Thus, for thin films a very small amount of superconductor

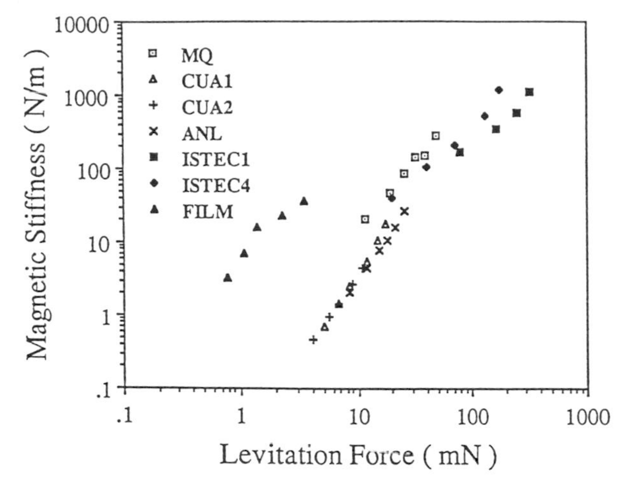

Figure 4-32 Comparison of magnetic stiffness versus levitation force for thin-film and bulk YBCO specimens.

should produce levitation forces comparable to a bulk conductor 10^2 times its thickness.

Measurements of levitation forces in thin films have been conducted in our laboratory at Cornell using YBCO thin films with thicknesses of 0.2 and 1 μm. A typical force–displacement curve shown in Figure 4-31 shows a suspension effect almost as large as the repulsion effect. Also, the magnetic–stiffness relation for the thin film seems to be much higher than that for the bulk material (Figure 4-32). The apparently higher flux pinning in these thin films also seems to lead to higher magnetic damping. These results suggest that if these films could be made an order of magnitude thicker (10–20 μm), they could generate magnetic pressures comparable to those of bulk superconductors 1 cm thick. Also, thin-film levitation may play a role in overcoming the friction problems inherent in micron-sized micromachine devices.

CHAPTER 5

DYNAMICS OF MAGNETICALLY LEVITATED SYSTEMS

They simply ride on a magnetic field. The logistics of it work just like a railway system. . . . But the tracks are magnetic, nothing at all like railway lines. The great thing about it is it's all silent and computer controlled.

—Fred Hoyle, *October The First Is Too Late*

5-1 INTRODUCTION

There are three types of dynamics problems in magnetically levitated systems which are application-specific: (i) anti-gravity levitation of a body with no motion, (ii) levitation of a spinning body with no translation of the center of mass, and (iii) levitation of a body in translation relative to a fixed guideway. The first two problems are typical of magnetic bearings for rotating machinery, whereas the third is attendant in problems of magnetically levitated transportation vehicles or projectiles. The second two problems have a similar attribute. Kinetic energy stored in rotational or translational motion can sometimes couple into the lateral, rigid body and elastic modes of the levitated body, which can lead to large oscillations and catastrophic failure. In designing a successful and safe levitation system, at least four specifications must be met:

1. The levitation system must equilibrate gravitational forces.
2. The system must be stable under small dynamic perturbations.
3. The system must be stable under occasional large perturbations.

4. The system must provide sufficient damping to ensure “ride quality” in the face of relative motion of the bearing platform or guideway.

The first requirement of equilibrating gravity is obvious, but for a levitation inventor the design process cannot stop there. As we have seen in Chapter 2, the nemesis of Earnshaw’s theorem in electromagnetic systems requires one to look at the stability about the equilibrium point. For a rigid body this generally requires a positive magnetic stiffness for at least five of six rigid body modes. (In rotating systems we allow rotation about an axis, and in transportation systems we allow translation along the guideway without restoring forces.) We have seen earlier that a ferromagnetic body in the field of permanent magnets will have at least one negative magnetic stiffness. This means that although magnetic systems can be designed to counter the force of gravity, the body will move away from equilibrium if perturbed ever so slightly. This fact of nature regarding ferromagnetic forces has spawned the new field of actively controlled magnetic levitation which has had dramatic successes in creating stable levitation systems using feedback control to produce positive magnetic stiffness [see, e.g., Schweitzer (1988) and Allaire (1992)]. In superconducting magnetic systems, positive magnetic stiffness can be achieved without feedback, but it is not always guaranteed.

What is unique about the dynamics of magnetically levitated dynamics vis-à-vis other suspension or bearing systems such as air-cushion, gas, or hydrodynamic bearings? The main features of magnetic forces are as follows:

- Long-range forces
- Hysteresis
- Nonlinear forces
- Active control (in feedback-based bearings)
- Self-feedback effects

These characteristics have important implications for design of magnetic levitation devices. These properties relate to both ferromagnetically based and superconducting Mag-Lev systems. The unique characteristics of superconducting levitation include:

- Flux-exclusion and flux-pinning forces
- Flux feedback in persistent current wire wound magnets

- Thermal stability phenomena (especially in wire wound magnets)
- The ability to produce both repulsion and attraction forces in bulk high-temperature superconducting (HTSC) materials
- Damping forces due to flux drag

Our description of the dynamics of levitated bodies follows similar treatments of rotor dynamics and vehicle dynamics, especially aircraft stability problems. We will examine two specific classes of problems: (i) the rigid body with a cylindrical symmetry axis and (ii) the vehicle in a guideway with a symmetric mass distribution on either side of the vertical plane.

The following terminology is used:

Heave—motion normal to gravity

Lateral or sway—motion in a plane normal to gravity and to either the magnet axis or direction of motion

Pitch—rotation about the lateral axis

Roll—rotation about the long axis of a vehicle or about the symmetry axis of a magnet

Yaw—rotation about an axis aligned with gravity

In both rotor dynamics and vehicle levitation systems, rigid body as well as flexible elastic degrees of freedom are possible. Thus, the number of degrees of freedom is, in general, greater than six. In this chapter we will only discuss the rigid-body modes, although in some important cases (such as flexible rotors) the elastic modes may indeed be important.

Literature Review

It is not our intention here to cite all references on the subject of dynamics and stability of levitated objects. Some work before 1964 has been reviewed by Geary (1964). Also, the subject of actively controlled magnetic bearings has a large literature which will not be reviewed here. The reader should see Schweitzer (1988) or the recent proceedings of the Magnetic Bearings Conference held in Virginia (Allaire, 1992) The book by Frazier et al. (1974) discusses tuned circuit magnetic bearings developed at the Draper Laboratory in Cambridge, Massachusetts. A recent literature review on Mag-Lev transportation has been prepared by the staff at Argonne National Laboratory and contains many references to Mag-Lev vehicle dynamics (see He et al., 1991).

Stability Of course the central study on the nature of magnetic and electric levitation is the paper of Earnshaw (1842) discussed in Chapter 1. This paper has been discussed in many books [see, e.g. Jeans (1925) and Scott (1959)]. However, another influential paper is that by Braunbeck (1939a) (in German). He attempted to prove that although Earnshaw's conclusion about the impossibility of stable levitation is correct for ferromagnetic or paramagnetic materials, stable levitation may be possible in the presence of diamagnetic or superconducting materials. This was a very important statement for its time. However, the theoretical proof rests on the assumption of a small test body (or one with symmetry, so that it acts as a point body). Levitated bodies of technical interest are extended masses (e.g., vehicles) with inhomogeneous distributions of currents or magnetization. Very little work about the stability of rigid bodies in magnetic fields has been published, except for a few special cases to be discussed below [see, e.g., Tenney (1969), Homer et al. (1977)].

In the 1950s and 1960s a number of works were published on the stability of diamagnetic and superconducting bodies in static magnetic fields. In a few cases, magnetic forces were calculated and positive magnetic stiffnesses were shown to exist for certain geometries and configurations of levitated body and magnetic fields.

Diamagnetic Levitation In a companion paper, Braunbeck (1939b) demonstrated experimentally that small diamagnetic bodies made of bismuth and carbon could be levitated in a static magnetic field. Waldron (1966) demonstrated diamagnetic levitation using a 3.8-g pyrolytic graphite sample and an electromagnet. An early paper on diamagnetic levitation is that by Boerdijk (1956).

Superconducting Spheres and Cylinders An early demonstration of stable levitation with superconductivity material is that by Arkadiev (1945, 1947). The dawn of the space age in the late 1950s and 1960s spawned a series of papers on magnetically levitated superconducting spheres for gyroscope applications. Geary (1964) lists two patents of Buchhold of the General Electric Company in 1958. In one, Buchhold describes the levitation of a niobium (Nb) sphere levitated between ten Nb wire coils, each carrying persistent currents. He describes experiments in which the sphere is rotated to speeds of 40,000–50,000 rpm. An earlier report by Culver and Davis (1957) of the Rand Corporation (Santa Monica, California) describes a superconducting gyroscope having the same design as that of Buchhold.

An extensive study of levitation of a superconducting sphere for a gyroscope application was carried out by John Harding and co-workers at the Jet Propulsion Laboratory, Pasadena, California, from 1960 to 1965. [See, e.g., Harding and Tuffias (1960) and Harding (1965a).] In these experiments both solid Nb and hollow spheres coated with Nb were levitated in the field of a set of superconducting rings. Harding (1965b) also calculated the perturbed magnetic forces on a superconducting sphere and established positive stiffnesses and hence stability. Another calculation of magnetic forces for a similar problem is given by Beloozerov (1966) in a Soviet physics journal.

In another work of the same genre, Bourke (1964) described the levitation of a cylindrically shaped aluminum body plated with lead which superconducted at 4.2 K. The cylinder was levitated and rotated in the field created by 200 turns of superconducting Nb wire operated in the persistent current mode. Bourke also calculated magnetic stiffnesses from which one can calculate the natural frequencies (see below).

Levitation of Superconducting Circuits Superconductor technology received a boost in the late 1960s and 1970s with efforts to develop magnetic fusion confinement systems. In one experiment at Princeton University, a superconducting ring was levitated in an axisymmetric field (File et al., 1968). This problem has been analyzed by Tenney (1969), who examined the magnetic forces and stability. An earlier study was done by Rebhan and Salat (1967) in Germany. Their conclusion was that a constant current loop cannot be stably levitated but that a persistent current superconductor may be stable under the proper geometry and field configuration. Another paper dealing with the magnetic energy of a set of current loops is that of Kozorez et al. (1976) from Ukraine. The magnetic stiffness of a circular superconducting coil in a toroidal magnetic field was calculated and measured by the author (Moon, 1979). Although the coil was supported by elastic springs, the experiment demonstrated that under constant current at least one magnetic stiffness is negative using the change in natural frequencies (see also Moon, 1980, 1984). Theoretical studies of levitated superconducting rings include Woods et al. (1970) and Marek (1990).

Finally, in another Ukrainian paper, Mikhalevich et al. (1991) show that a rectangular superconducting loop can be stably suspended below a pair of direct-current (dc) filaments (Figure 5-1) when the loop operates in the persistent mode. Experimental confirmation of

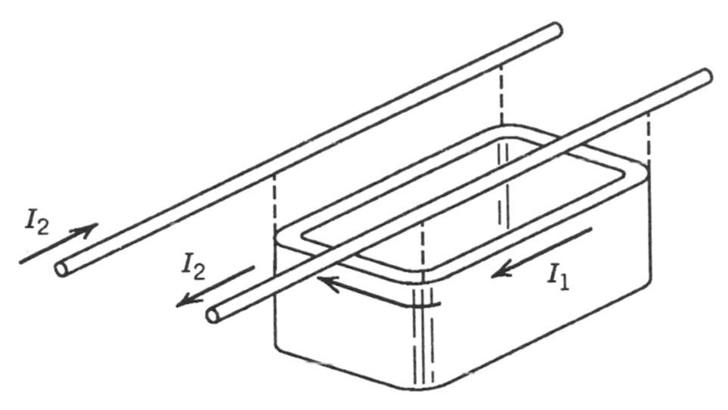

Figure 5-1 Sketch of the suspension of a superconducting coil under a pair of parallel dc-current-carrying wires. [After Mikhalevich et al. (1991).]

this levitation scheme is claimed. The authors propose this geometry for a Mag-Lev transportation system.

High-T_c Superconducting Levitation In most experiments and applications of bulk high-T_c superconducting levitation, the flux density is larger than H_{c1} and flux pinning and flux drag play a major role in the magnetic stiffness and other vibration properties. Thus, the dynamics of high-T_c materials differ from either diamagnetic or pure Meissner ($H < H_{c1}$) levitation. Soon after the discovery of YBCO in 1987, measurements were made of the natural frequency and damping as a function of magnet–superconductor gap (Moon et al., 1988). However, further vibration experiments have shown that the magnetic stiffness depends on the amplitude of the vibration. Basinger et al. (1990) of Argonne National Laboratory have measured an order-of-magnitude increase in magnetic stiffness of sintered YBCO when the vibration amplitude decreased from 1 mm to 1 μm. The amplitude effect appears to be not as great for melt–quench processed YBCO according to experiments reported by Yang and Moon (1992). This amplitude-dependent vibration frequency demonstrates the nonlinear nature of superconducting levitation.

Theoretical studies on the stability of levitation dynamics for high-T_c superconductors have been given by Davis et al. (1988), Davis (1990) and Brandt (1990c), as well as Nemoshkalenko et al. (1990b). Theoretical calculation of the natural frequencies of a levitated permanent magnet over a high-T_c superconducting surface have been made by a group in Norway (Yang et al., 1989). Another group in Poland (Braun et al., 1990) have made corrections to the Norwegian calculations and

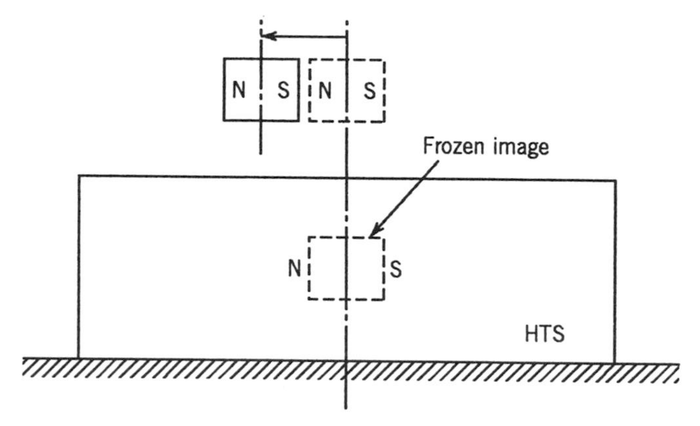

Figure 5-2 Frozen-image model to determine the lateral magnetic stiffness of a permanent magnet levitated over a flat, high-T_c superconducting (HTS) surface. [After Braun et al. (1990).]

have compared the results with experiments. For example, for a cylindrical magnet magnetized along its axis, they calculate the ratio of vibration frequency along the axis $f_{\parallel}$ to the frequency transverse to the axis and parallel to the plane of the superconductor $f_{\perp}$ to be $f_{\parallel}/f_{\perp} = \sqrt{3}$. Measurements on different-sized magnets show a ratio range of 1.3 to 1.9. The theoretical model used is sometimes called a "frozen-image magnet" model (Figure 5-2). Thus when the magnet moves, the supercurrents are assumed to remain fixed in place, thereby creating a restoring force or magnetic stiffness.

Another important effect in dynamics is magnetic drag or what some authors call *magnetic friction* (see Chapter 4). Measurements of magnetic friction and force hysteresis have been measured by several groups [see, e.g., Moon et al. (1989)]. However, a group from the former Soviet Union at the Institute of Chemical Physics in Moscow has made some interesting observations of the effect of alternating-current (ac) fields on levitation. By placing a levitated superconductor (YBCO) first in an inhomogeneous static field and then turning on an alternating field of less than 10^{-2} T, they claim to eliminate the hysteresis and produce "a unique position of stable levitation" (Terentiev, 1990).

In another work they investigate the effect on the rotation of a levitated superconducting disk in a static field as one places a rotating magnetic field of different frequencies and intensities. The levitated body is observed to change its orientation (Terentiev and Kuznetsov, 1990). In a third paper (Terentiev and Kuznetzov, 1992), the Moscow group investigates the important problem of levitation height drift or sinking. In classical superconductor theory, one can calculate so-called flux creep due to thermal fluctuations of the superconducting vortex lines in the materials. Measurements by the Cornell group have shown an initial levitation sinking in the first seconds after levitation (Moon et al., 1990; see Chapter 4). However, the Moscow group speculates that in rotating magnetic bearings, small variation in the field will produce an ac field fluctuation. They expose a levitated YBCO sintered specimen to a small ac field (50–100 Hz) and demonstrate a drift or sinking of the levitation height. This has important consequences for magnetic bearing applications and deserves further study. However recent experiments at Cornell University using YBCO materials, processed in the melt–textured or melt–quenched process, show that long term flux creep is not a problem.

Another important property for dynamics of rotating magnet–superconductor pairs is rotational torques. Several groups [such as Moon and Chang (1990) and a group at Koyo Seiko Co., a bearing company in Japan (Takahata et al., 1992)] have measured spin-down torques.

An interesting magnetic levitation phenomenon in YBCO is the self-oscillation of a magnet levitated over a YBCO surface (Figure 5-3). This was first observed by a graduate student at Cornell (J. D. Wang) (see Moon et al., 1987) and has been demonstrated at many lectures given by the author. A disk-type magnet begins to oscillate slowly, and it eventually turns over and accumulates a net angular momentum. Several groups have tried to explain this phenomena [Martini et al. (1990), Ma et al. (1991), and an unpublished work of John Hull at Argonne National Laboratory]. The assumption in these models is the interaction between the thermal gradient and the temperature properties of the magnetization of the magnet.

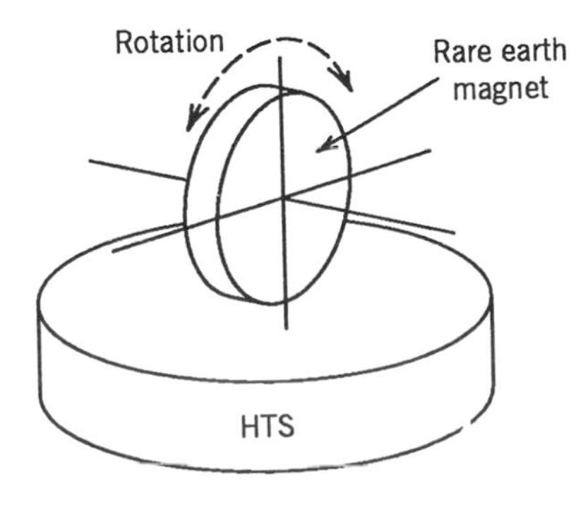

Figure 5-3 Sketch of the geometric arrangement for the self-oscillation of a rare earth magnet levitated above a YBCO superconductor. HTS, high-temperature superconductor.

Another nonlinear effect in the dynamics of superconducting levitation is observation of period doubling and chaotic dynamics of a magnet levitated over a YBCO disk (see below; also see Moon, 1989).

Dynamics of Superconducting Mag-Lev Vehicles

Instabilities. In EML Mag-Lev systems, dynamics and stability are first-order problems of control design. We shall not review the large amount of literature on this problem [see e.g., Popp (1982a)]. In superconducting EDL systems the key dynamics questions are stability and ride quality. Although dynamic vehicle instabilities under magnetic forces have been predicted theoretically and have been observed experimentally in models, there seems to have been less concern or studies done on the stability of full-scale systems. For example, early experimental work of the author reported static and dynamic instabilities of a levitated model in a "V"-shaped rotating aluminum guideway at Princeton University (Moon, 1974, 1977, 1978). Similar instabilities were observed at Cornell University in rotating wheel experiments with a discrete back-to-back "L"-shaped aluminum guideway (Chu, 1982; Chu and Moon, 1983). In both cases the models used permanent magnets for the magnetic field source. Several types of instabilities were observed, including heave, pitch and heave, and a lateral–yaw or snaking instability (see Figure 5-4). The latter problem was analyzed in depth in a M.S. dissertation at Cornell (Chu, 1982). This analysis showed that the instability was related to the nonconservative magnetic drag force which produced a yaw moment when the vehicle moved laterally. Recently these experiments have been reproduced at Argonne National Laboratory by Cai et al. (1992) using the old Cornell rotating wheel apparatus. Also, analyses of films of the tests of the MIT Magneplane dynamics (Kolm and Thornton, 1972, 1973) seem to show similar lateral–yaw motions as well as pitching instabilities.

Theoretical studies of EDL Mag-Lev instabilities can be found in the work of the Ford Motor Co. group (Davis and Wilke, 1971), who reported on the possibility of a growing vertical oscillation. Another similar study was by Fink and Hobrecht (1971). In the United Kingdom a group at the University of Warwick designed and built a split-sheet, EDL-levitated guideway model with a superconducting coil. They also studied the possibility of lateral instability in a theoretical study, though no experimental evidence was given (see Wong et al., 1976, Sakamoto and Eastham, 1991). A later theoretical study

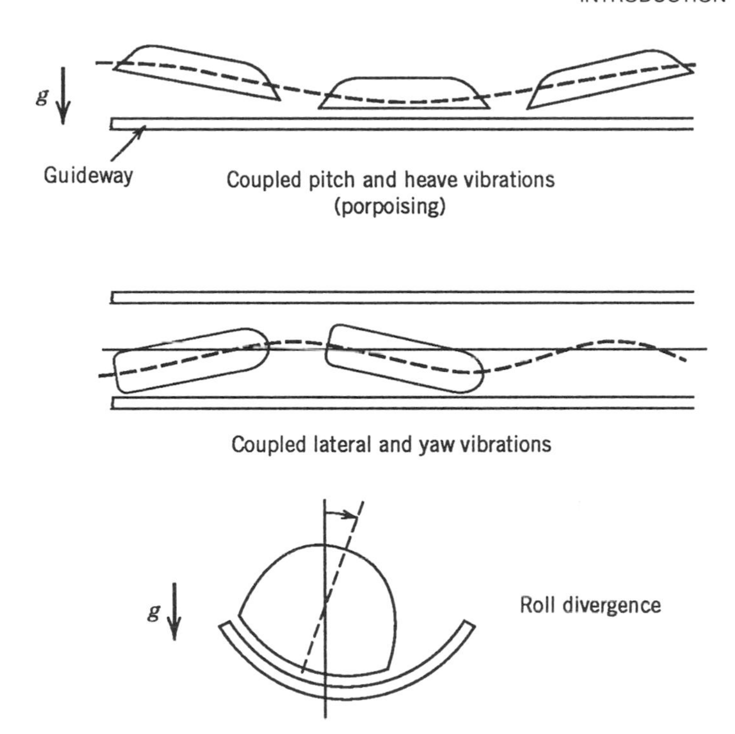

Figure 5-4 Three classes of Mag-Lev vehicle instabilities.

by a Russian group in the Leningrad Railway Engineering Institute also showed the possibility of unstable heave dynamics (Baiko et al., 1980).

While there is no doubt that such magnetomechanical instabilities exist, one of the factors missing in the analysis is aerodynamic damping. Also, linear synchronous propulsion forces might exert some magnetic stiffness. However, the fact that a dynamic stability analysis is rarely presented for the design of full-scale Mag-Lev vehicles is somewhat disturbing. Flight stability of aircraft is always one of the major design considerations [see, e.g., Seckel (1964)].

Ride Quality. What Mag-Lev designers seem first to worry about in the full-scale vehicle dynamics is ride quality. This comes in the form of two types of analyses. The first one assumes a rigid guideway with random departures from a linear track. In the second analysis the effect of a flexible guideway is taken into account. The latter is especially important for the design of an elevated Mag-Lev guideway.

Early tests of EDL dynamics to perturbations in the guideway alignment were reported by the SRI group in California from their superconducting model tests (Coffey et al., 1974). Ride quality and dynamics of a full-scale system were also the subject of an early Ford study (Wilke, 1972). Reports of dynamics tests on the Japanese full-scale Miyazaki vehicles have been given by Yoshioka and Miyamoto (1986).

Recently the U.S. Department of Transportation has sponsored a National Mag-Lev Initiative study to try to regain leadership in Mag-Lev development. In that study, vehicle–guideway interaction was a major component. One study by a group from MIT compared suspension characteristics of both EDL and EML systems on both rough rigid and flexible guideways (Wormley et al., 1992). Another study headed by Parsons, Brincherhoff, Quade & Douglas, Inc. (Herndon, Virginia) looked at a set of specific guideway structural designs (Daniels et al., 1992; see also Chapter 7).

Mag-Lev Damping. While magnetic stiffness and guideway flexibility are important to Mag-Lev dynamics, magnetic and aerodynamic damping can be just as crucial. When a magnetic field source vibrates relative to a stationary conductor, eddy currents usually lead to damped vibrations. However, if the field source is also moving at a steady speed normal to the vibration direction, the magnetic damping can be reduced and may even become negative.

Theoretical evidence for negative damping in an EDL system was first presented by Iwamoto et al. (1974). They looked at the case of a superconducting train magnet moving over a discrete loop track similar to the early Miyazaki test facility in Japan. Further, calculations for a sheet guideway were done by Ooi (1976). His calculations for a finite-width sheet guideway seem to result in positive or negative damping depending on the speed. Another theoretical paper on magnetic damping in EDL systems by Urankar (1976) of Siemens AG (Germany) shows that the damping decreases with speed. In an experimental and theoretical study the author showed that magnetic damping of a magnet, vibrating normal to a moving aluminum sheet conductor, indeed became negative when the aerodynamic damping is accounted for (Moon, 1977; see also Moon, 1984).

These studies seem to show that in EDL systems an increase in forward speed does not increase the damping of vibrations and that some active damping may be necessary in full-scale systems to meet ride quality standards.

5-2 EQUATIONS OF MOTION

An introduction to the dynamics of magneto-mechanical systems may be found in the books by Crandall et al. (1968), Moon (1984) and Woodson and Melcher (1968). In classical dynamic analysis a distinction is made between different types of forces. This classification includes the following:

1. *Conservative*: Forces that are derivable from a potential—that is

$$\mathbf{F} = -\nabla V(q_1, q_2, \ldots, q_n) \tag{5-2.1}$$

   where $\{q_k\}$ are a set of generalized displacements. Forces that are not derivable from a displacement are called *nonconservative*. For magnetic forces, part of the force can be derived from the magnetic energy which acts as a force potential.
2. *Dissipative*: Forces that take energy out of the system—that is

$$\mathbf{F} \cdot \mathbf{v} < 0$$

   where $\mathbf{v}$ is a velocity vector. Examples are eddy-current-induced forces and magnetic drag forces.
3. *Gyroscopic*: Forces that are not derivable from a potential function, but do no work—that is

$$\mathbf{F} \cdot \mathbf{v} = 0$$

   Examples of this force include the moving charge in a magnetic field—for example,

$$\mathbf{F} = Q\mathbf{v} \times \mathbf{B}$$

Also, when the equations of motion are written in body fixed coordinates, Coriolis and centripetal acceleration terms sometimes appear. Coriolis terms viewed as effective forces are gyroscopic.

The importance of whether magnetic forces are conservative or not determines the nature of the dynamic analysis as well as the dynamic phenomena itself. Conservative forces often mean that variational methods, such as Lagrange's equations, can be used. Also, the conservative property implies certain stability possibilities for the magnetic system. For example, a conservative problem cannot become unstable by the development of a limit cycle. The system has a limited number of instability paths; that is, any instability mechanisms are quasistatic in nature. On the other hand, velocity-dependent forces that supply positive work to the system can lead to dynamic instabilities.

Newton – Euler Equations of Motion

In the general theory of rigid bodies, one requires a description of the kinematic equations as well as of the equations of motion expressing the laws of linear and angular momentum. The latter may take the form of either the Newton–Euler formulation or one derived from a variational principle such as Lagrange's equations (see below).

Newton's law for the motion of the center of mass $\mathbf{r}_c(t)$ relates the acceleration $\ddot{\mathbf{r}}_c$ to the magnetic, gravitational, and aerodynamic forces (Figure 5-5*a*, *b*):

$$m\ddot{\mathbf{r}}_c = m\mathbf{g} + \mathbf{F}_m + \mathbf{F}_a \tag{5-2.2}$$

where $\mathbf{g}$ denotes the direction and magnitude of the gravitational constant. The angular motion of the levitated rigid body is determined by the rate of change of angular momentum $\mathbf{L}_c$ about the center of mass and by the moment $\mathbf{M}$ produced by the torques and forces acting about the center of mass:

$$\dot{\mathbf{L}}_c = \mathbf{M} \tag{5-2.3}$$

It is usual to write $\mathbf{L}_c$ in a coordinate system that moves with the center of mass and fixed to the body. The orthogonal unit axes $\{\mathbf{e}_1, \mathbf{e}_2, \mathbf{e}_3)$ are chosen such that the second moments of mass distribution are the principal inertias of the body $\{I_1, I_2, I_3\}$. The angular motion is described by an angular velocity vector:

$$\omega = \omega_1\mathbf{e}_1 + \omega_2\mathbf{e}_2 + \omega_3\mathbf{e}_3 \tag{5.2-4}$$

Using this notation, the angular momentum is given by

$$\mathbf{L}_c = I_1\omega_1\mathbf{e}_1 + I_2\omega_2\mathbf{e}_2 + I_3\omega_3\mathbf{e}_3 \tag{5-2.5}$$

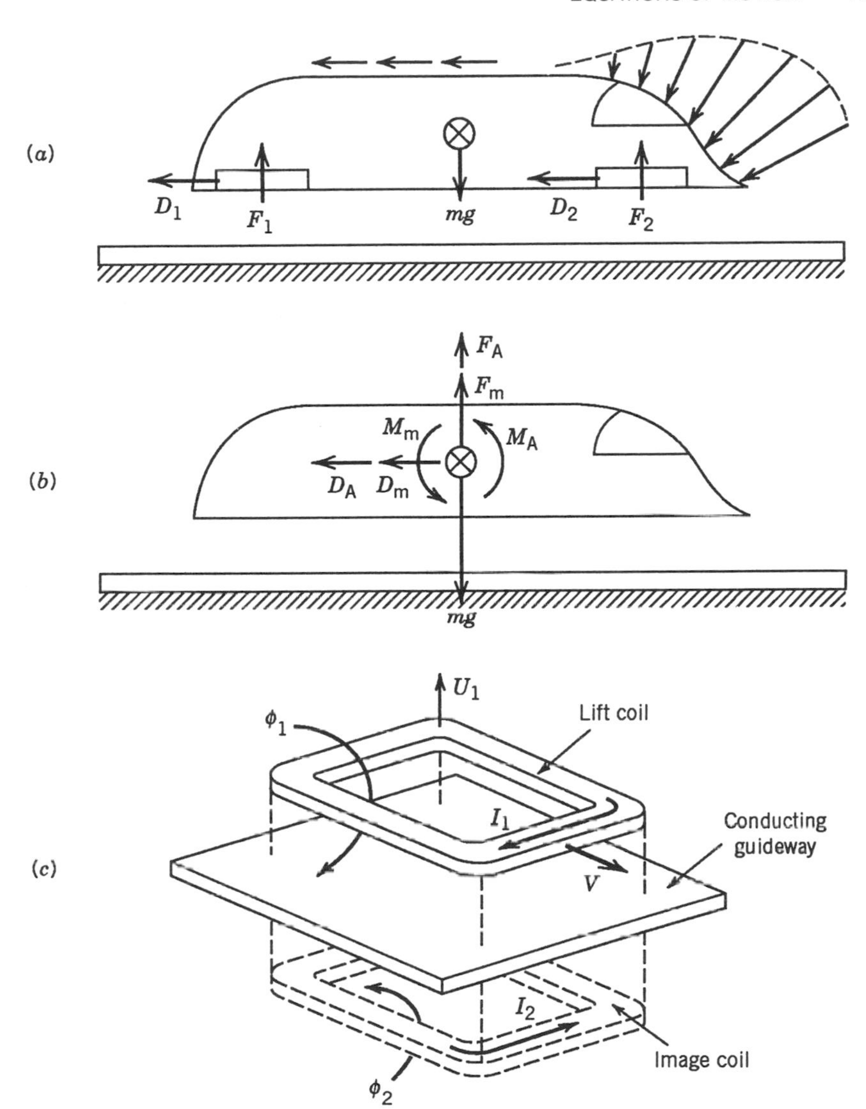

Figure 5-5 (*a*) Sketch of a levitated body showing gravitational, magnetic, and aerodynamic forces. (*b*) Sketch of a levitated body with magnetic and aerodynamic force resultants and moments. (*c*) Vehicle magnet coil and image coil due to induced guideway currents.

The resulting Euler's equations then take the form

$$
\begin{aligned}
I_1\dot{\omega}_1 + (I_2 - I_3)\omega_2\omega_3 &= M_1 \\
I_2\dot{\omega}_2 + (I_3 - I_1)\omega_1\omega_3 &= M_2 \\
I_3\dot{\omega}_3 + (I_1 - I_2)\omega_1\omega_2 &= M_3
\end{aligned}
\tag{5-2.6}
$$

In order to solve specific problems, $\dot{\omega}$, ω, and $\ddot{\mathbf{r}}_c$ must be written in terms of translation and angular position variables. Also, **M** and **F** must be related through the geometry of the body.

Lagrange's Equations for Magnetic Systems

The formulation of equations of motion directly from Newton's law requires a vector representation of the forces in the problem. However, when the forces are related to a potential function, such as the magnetic energy, the equations of motion may be derived from a scalar function called the *Lagrangian* [after the French-Italian mechanician J. L. Lagrange (1736–1813)]. In this method, one identifies a set of independent variables representing the degrees of freedom in the system. For a magnetomechanical problem, one can choose a set of N independent displacements $\{U_n\}$ and M independent magnetic fluxes $\{\phi_n\}$. For example, the fluxes could be associated with M current loops, each carrying currents I_k and supporting a voltage e_k given by

$$e_k = \frac{\partial \phi_k}{\partial t} \tag{5-2.7}$$

where each circuit is assumed to be nonresistive (Figure 5-5*c*).

In the method of Lagrange, a mechanical potential energy function $V(U_k)$ is assumed from which the mechanical forces in the system are derived:

$$F_k = -\frac{\partial V}{\partial U_k} \tag{5-2.8}$$

For example, in classical mechanics, a linear spring force $F = -kU$ has a potential $V = \frac{1}{2}kU^2$, and a gravitational force acting on a mass at a distance z above a reference has a potential function $V = mgz$.

The extension of Lagrange's method to nondissipative magnetic systems involves the assumption of a magnetic potential energy function $W(\phi_n, U_n)$ defined by a variational principle. The variational principle for *static* problems that determines the balance of mechanical and magnetic forces is called *the principle of virtual work*. It states that the change in both mechanical and magnetic energy functions under small changes in the independent variables $\{dU_k, d\phi_k\}$ is equal to the change of work done on the system by external forces. When electric currents provide the energy input, the work done in a small time dt is given by

$$\sum I_k e_k \, dt$$

Using Eq. (5-2.7) we assume that the work done by these currents and voltages produces changes in the mechanical energy function V as well as in the magnetic energy W; that is,

$$\sum I_k \, d\phi_k = dW + dV$$

We next write the variations of $W(\phi_n, U_n)$ and $V(U_n)$ in terms of the variations of the state variables dU_k and $d\phi_n$:

$$\sum I_k \, d\phi_k = \sum \frac{\partial W}{\partial \phi_k} \, d\phi_k + \frac{\partial W}{\partial U_k} \, dU_k + \sum \frac{\partial V}{\partial U_k} \, dU_k$$

Because dU_k and $d\phi_k$ are independent, one can show that

$$I_k = \frac{\partial W}{\partial \phi_k}$$

$$\frac{\partial W(\phi_n, U_n)}{\partial U_k} = -\frac{\partial V}{\partial U_k} \tag{5-2.9}$$

The right-hand side of (5-2.9b) represents the mechanical force F_k, while the magnetic force is given by

$$F_k^m = -\frac{\partial W(\phi_n, U_n)}{\partial U_k} \tag{5-2.10}$$

In some treatments of magnetic forces, the magnetic energy function is written in terms of the currents $\{I_n\}$; that is, the currents—not the fluxes—are chosen as independent variables. This new function $W^*(I_k)$ is called the *co-energy* function and is related to $W(\phi_k)$ by a so-called Legendre transformation:

$$W^*(I_n) = \sum I_k \phi_k - W(\phi_n) \tag{5-2.11}$$

An analysis similar to that of Eq. (5-2.9) leads to

$$\phi_k = \frac{\partial W^*}{\partial I_k}$$

$$\frac{\partial W^*(I_n, U_n)}{\partial U_k} = \frac{\partial V}{\partial U_k} \tag{5-2.12}$$

In this formulation, the magnetic forces are defined as

$$F_k^m = \frac{\partial W^*(I_n, U_n)}{\partial U_k} = -\frac{\partial W(\phi_n, U_n)}{\partial U_k} \tag{5-2.13}$$

Note that in the use of W^*, the currents are fixed while the derivatives act on U_k, whereas in using W, the fluxes are held fixed when deriving the magnetic forces. [See also Crandall et al. (1968) for an extended discussion of these ideas.]

For dynamic problems, the principle of virtual work is extended in what is called *d'Alembert's principle* [see, e.g., Goldstein (1950)]. In addition to the potential energy functions, V and W, one adds another scalar function, the kinetic energy, $T(\dot{U}_n, U_n)$, which is assumed to depend on both the generalized velocities and the displacements. The principle of virtual work extended to dynamic problems then becomes

$$\frac{d}{dt}\frac{\partial T}{\partial \dot{U}_k} - \frac{\partial T}{\partial U_k} + \frac{\partial V}{\partial U_k} + \frac{\partial W}{\partial U_k} = 0 \tag{5-2.14}$$

For a general problem, there will also be equations for the magnetic fluxes and currents in the M circuits. For a flexible body, U_k can also represent a specific elastic deformation mode.

When it is more convenient to use currents, one replaces the term $\partial W/\partial U_k$ with $-\partial W^*/\partial U_k$ in Lagrange's equation [Eq. (5-2.14)].

When neither the current nor the flux is constant, one must add the circuit equations or Maxwell's electromagnetic field equations to the equations of Newton, Euler, or Lagrange, as discussed in Chapter 2.

The Inductance Method

As an example of the use of this method, we assume a linear magnetic device in which the fluxes are linearly related to the currents. This implies that W^* is a quadratic function of the currents $\{I_n\}$:

$$W^* = \frac{1}{2} \sum \sum L_{ij} I_i I_j \tag{5-2.15}$$

where $\{L_{ij}\}$ is known as the *inductance matrix*, which is assumed to depend on the mechanical displacements $\{U_n\}$. Then the magnetic forces are seen to depend on the change of inductance with displacement; that is,

$$F_k^m = \frac{1}{2} \sum \sum \frac{\partial L_{ij}}{\partial U_k} I_i I_j \tag{5-2.16}$$

While the first derivatives of the inductance give the magnetic forces, the second derivatives determine the magnetic stiffnesses and hence the static stability of the system.

Linear Stability Analysis

The dynamic stability of mechanical systems is a well-treated subject [see, e.g., Huseyin (1978)]. The traditional approach has been to linearize the nonlinear terms in Newton's and Euler's laws about either an equilibrium state or about some steady motion, such as the translational motion of a magnetically levitated vehicle along its guideway or the spinning of a levitated rotor about its symmetry axis. When the magnetic forces can be calculated directly (i.e., the dynamics of the electric circuits are implicit, not explicit), the linearized equations often take the form of linear, ordinary differential matrix equations with constant coefficients:

$$\mathbf{M} \cdot \ddot{\mathbf{X}} + (\mathbf{D} + \mathbf{G}) \cdot \dot{\mathbf{X}} + \mathbf{K} \cdot \mathbf{X} = 0 \tag{5-2.17}$$

where: $\mathbf{X} = \{x_1, x_2, \ldots, x_N\}^T$ represents an $N \times 1$ column matrix of generalized position coordinates ($\{\ \}^T$ denotes the transpose); $\mathbf{M}$ is an $N \times N$ symmetric matrix representing masses and moments of inertias; $\mathbf{D}$ is a $N \times N$ symmetric matrix representing dissipation; and $\mathbf{G}$ is an $N \times N$ antisymmetric matrix representing gyroscopic acceleration or force effects usually associated with rotating rigid bodies. Finally, when $\mathbf{K}$ is a symmetric $N \times N$ matrix, it represents both mechanical and magnetic stiffnesses. Such forces can be derived from elastic and magnetic potential energy functions or mutual inductances as discussed in the previous section.

However, some forces, such as magnetic drag, aerodynamic drag, and dry friction, cannot be derived from an energy potential.

When all the forces can be derived from an energy potential, the stiffness matrix **K** is symmetric and only static instabilities can occur such as divergence in aircraft dynamics or buckling in structural mechanics. Dynamic instabilities (i.e., sinusoidal motions that grow exponentially in time) are usually associated with either negative damping coefficients or a nonsymmetric stiffness matrix **K**. Non symmetric stiffnesses can occur in Mag-Lev systems through coupling with the magnetic drag force in the forward direction. (See Section 5-6.)

The model (5-2.17) does not contain explicit nonlinear force terms such as hysteresis in high-T_c superconductors or ferromagnetic materials.

5-3 SINGLE DEGREE OF FREEDOM DYNAMICS

Natural Frequencies of Levitated Bodies

In the analysis of the vibration of levitated bodies, one assumes that there exists an equilibrium state or a steady motion about which the vibration and stability can be determined. This is similar to the classical theory of vibrating mechanical systems and is also similar to aircraft and vehicle dynamics [see, e.g., Seckel (1964)].

In the direct method of calculating natural frequencies of levitated bodies, one calculates the field and forces from Maxwell's equations and determines the magnetic forces **F** as a function of the relative position between the body and its guideway or bearing. The forces are then expanded in a Taylor series in the perturbed displacement variables about the equilibrium position. As an example, consider the vertical motion of a levitated body, often called *heave* (Figure 5-6a). The vertical height of the body above the ground is denoted by

$$h(t) = h_0 + u_z(t) \tag{5-3.1}$$

where h_0 is the equilibrium height. The perturbed force is written as

$$F(h) = F(h_0) + \left.\frac{\partial F}{\partial h}\right|_{h=h_0}(h - h_0) \tag{5-3.2}$$

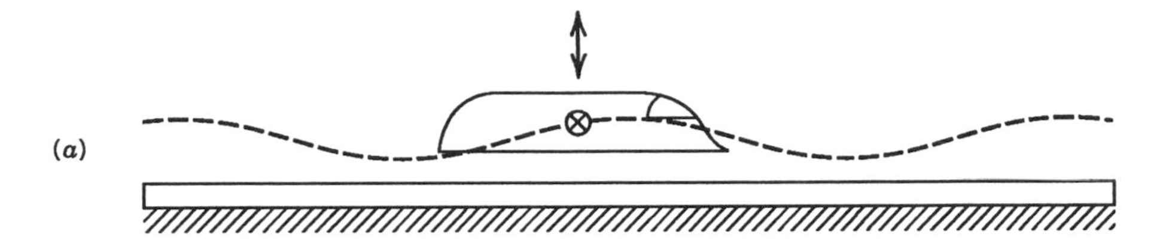

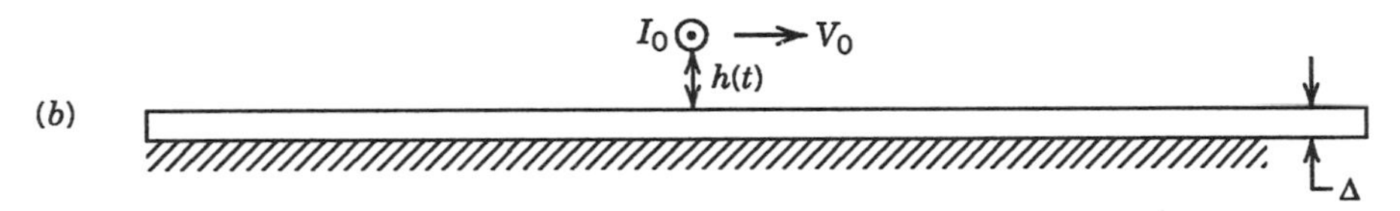

Figure 5-6 (*a*) Levitated body undergoing vertical oscillations. (*b*) Levitated wire above a continuous-sheet guideway.

The equation of motion is then given by

$$m\ddot{h} = F(h) - mg$$

or (5-3.3)

$$\ddot{u}_z - \frac{1}{m}\frac{\partial F}{\partial h}u_z = 0$$

where it is assumed that $F(h_0) = mg$. Also the derivative is evaluated at $h = h_0$. For example, for a body levitated with current-carrying magnets, with ampere-turns I, the magnetic levitation force has the form

$$F = \mu_0 I^2 f\left(\frac{D}{h}\right) \tag{5-3.4}$$

where D is some characteristic dimension of the magnet. At equilibrium, the lift should balance the gravitational force; that is,

$$\mu_0 I^2 f\left(\frac{D}{h_0}\right) = mg \tag{5-3.5}$$

Using the perturbation variable $u_z(t) = h(t) - h_0$, the equation of motion takes the form

$$\ddot{u}_z = \frac{g}{f(D/h_0)} \frac{\partial f}{\partial h} u_z = g \frac{\partial \log f}{\partial h} u_z \tag{5-3.6}$$

For stability, the normalized magnetic stiffness $\partial f / \partial h < 0$. The assumption of undamped linear vibrating solutions implies that

$$u_z(t) = A \cos(\omega t + \varphi_0)$$

From Eq. (5-3.6), the square of the natural frequency ω (in units of radians per second) is given by

$$\omega^2 = -g \frac{\partial \log f}{\partial h} \tag{5-3.7}$$

One can easily see that the natural heave frequency is independent of the mass of the vehicle.

As an example, consider a levitated wire of length L moving with a speed V_0 along a sheet conductor guideway of thickness Δ and conductivity σ (Figure 5-6*b*). It can be shown [see, e.g., Moon (1984)] that the lift and drag forces are given by

$$F_z = \frac{\mu_0 I^2}{4\pi} \frac{V_0^2}{V_0^2 + w^2} \left(\frac{L}{h} \right) \tag{5-3.8}$$

$$F_x = -\frac{w}{V_0} F_z$$

where $w = 2/(\mu_0 \sigma \Delta)$ is a characteristic velocity.

These expressions are good only in the so-called thin-track approximation in which the skin depth is greater than the track thickness Δ. Using the formula (5-3.7), the natural frequency is given by

$$\omega = \left(\frac{g}{h_0} \right)^{1/2}$$

which is the same as that for a pendulum of length h_0.

For levitation forces with more general force–height relationships [e.g., $f = (D/h)^{\beta}$], the frequency is given by

$$\omega^2 = \frac{\beta g}{h_0}$$

In the case of some high-temperature superconducting bearing examples, the force has the form (see, e.g., Chapter 4)

$$f = e^{-\alpha h} \tag{5-3.9}$$

When the force hysteresis is small, one can differentiate this expression to get the magnetic stiffness and, in turn, the natural frequency:

$$\omega^2 = g\alpha \tag{5-3.10}$$

where the constant α has units of inverse length.

Vibration of a Persistent-Current Superconducting Mag-Lev Coil

The influence of the constant flux condition in a persistent-mode superconducting magnet on the natural frequency of vibration is illustrated in the following example (see Moon, 1977). Consider a current-carrying coil moving over a continuous sheet guideway (Figure 5-7). In the high speed limit we use an image magnet below the sheet track to calculate the magnetic forces. The magnetic energy of the coil and its image coil is given by

$$W = \tfrac{1}{2}(L_{11} - L_{12})I^2 \tag{5-3.11}$$

where L_{11} and L_{12} are the self and mutual inductances. For vertical displacement U of the coil, the previous section shows how to calculate the magnetic force:

$$F = -I^2 \frac{\partial L_{12}}{\partial U}$$

Note that as the actual levitation coil vibrates a distance U upward, its image moves a distance U downward. The mutual inductance can be written as a Taylor series in the displacement U; that is,

$$L_{12} = L_0 + L'U + \tfrac{1}{2}L''U^2 \tag{5-3.12}$$

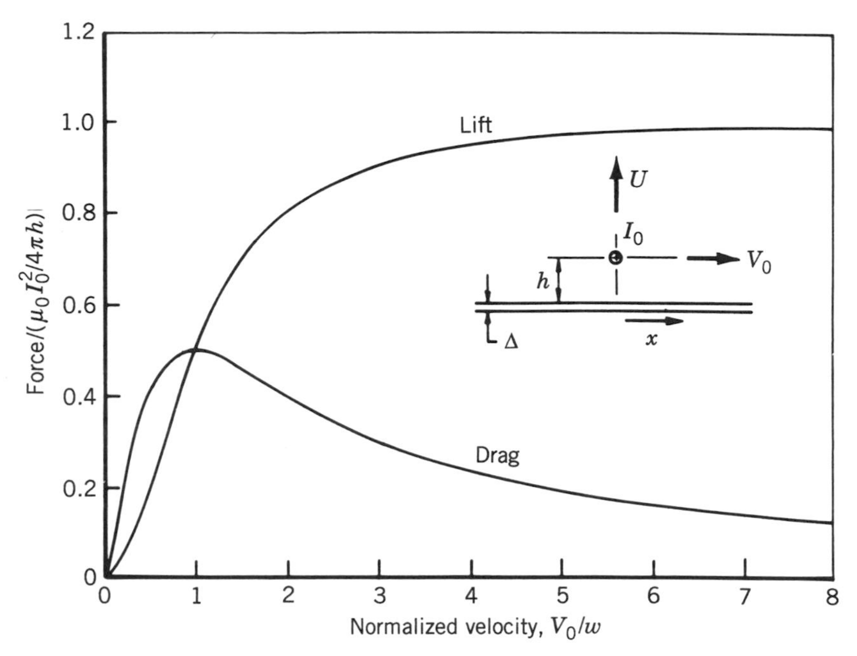

Figure 5-7 Current-carrying coil moving over a continuous-sheet guideway. Lift and Drag as a function of velocity.

The term L' is proportional to the magnetic force that equilibrates the gravity force. Note that the total flux through the magnet is given by

$$\phi = I(L_{11} - L_{12})$$

If the vehicle or coil mass is m, then for a constant-current coil the vertical or heave natural frequency is given by the equation

$$m\ddot{U} + \frac{I^2 L''}{2} U = 0 \qquad (5\text{-}3.13)$$

On the other hand, if the flux ϕ is held constant, the linearized equation of motion becomes [see Eqs. (5-2.13), (5-2.14)];

$$m\ddot{U} + \frac{\phi^2}{2(L_{11} - L_0)^2}\left[L'' + \frac{2(L')^2}{(L_{11} - L_0)}\right]U = 0 \qquad (5\text{-}3.14)$$

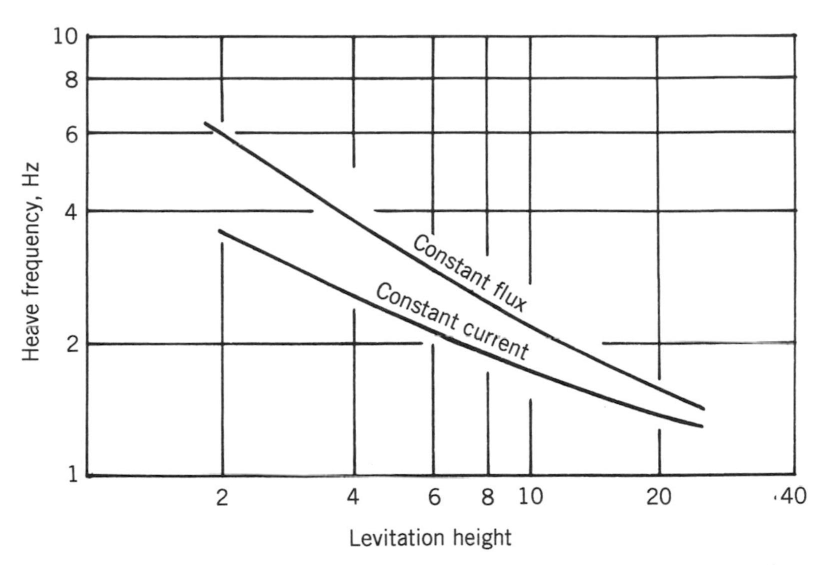

Figure 5-8 Natural heave frequency versus height for a rectangular coil moving over a sheet conductor for constant current and constant flux constraints. [From Moon (1977), with permission.]

Thus for sinusoidal motion $\cos \omega_0 t$, the frequency is given by

$$\omega_0 = \left(\frac{I^2 L''}{2m} \right)^{1/2} \qquad \text{(constant current)}$$

or

$$\omega_1 = \omega_0 \left[1 + \frac{2(L')^2}{L''(L_{11} - L_0)} \right]^{1/2} \qquad \text{(constant flux)}$$

Thus when $L_{11} > L_0$, the constant flux case results in a higher natural frequency. An example for a rectangular coil is shown in Figure 5-8.

Levitated Superconducting Ring

The theorem of Earnshaw concerning the stability of magnetic dipoles in a static magnetic field can be extended to current-carrying coils with steady currents. The stability of rigid current loops levitated in an axisymmetric magnetic field was studied by a group at the Princeton Plasma Physics Laboratory in connection with the design of magnetic

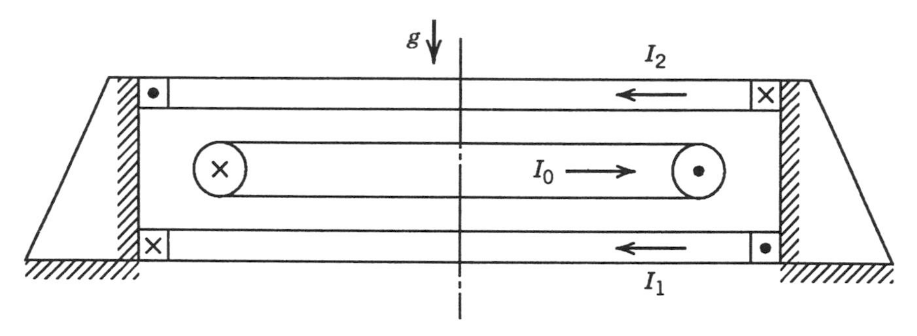

Figure 5-9 Sketch of a levitated superconducting coil. [After Tenney (1969).]

fusion experiments (Tenney, 1969). For a constant-current circular coil, stability conditions were derived for position-feedback-controlled levitation. However, Tenney (1969) also shows that it is possible to levitate a *superconducting* coil in a static magnetic field if the coil is allowed to operate in the persistent-current mode. In this mode the current in the coil will change as it moves in an applied static field so as to keep the magnetic flux constant through the coil. An example is shown in Figure 5-9.

5-4 DYNAMICS OF A SPINNING LEVITATED SUPERCONDUCTOR

A spinning levitated body has applications as a gyroscope or as an energy or momentum storage device. Applications of superconducting bearings to gyro devices go back to the 1950s and 1960s. An example of the dynamic analysis of one such device is discussed below.

The work of Harding (1961) at the NASA Jet Propulsion Laboratory showed that stable levitation of a superconducting sphere was possible. Bourke (1964), in a doctoral dissertation at Stanford University, levitated a spinning axisymmetric superconductor in a static axisymmetric magnetic field (Figure 5-10).

In Bourke's experiment, a 2.4-g aluminum rotor was plated with lead to a thickness of 0.1 mm. It was levitated in a helium cryostat at a temperature of 2 K. Lead exhibits complete flux exclusion at this temperature below 0.06 T. The complete exclusion of flux from the inside of the rotor allowed the use of magnetic pressure to calculate the magnetic force. In fact, the solution of the magnetic field was analogous to methods used in incompressible fluid mechanics. First

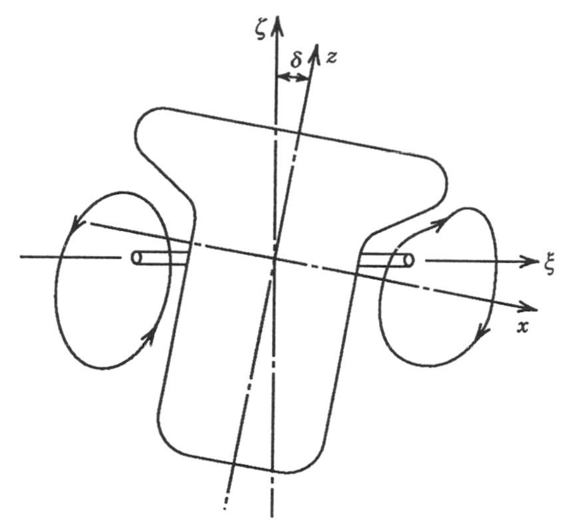

Figure 5-10 Levitation of a Type I superconducting spinning body in the field of a static magnetic field. [After Bourke (1964).]

the magnetic forces on the axisymmetric rotor aligned with the symmetry axis of field were calculated. Then the rotor was assumed to move and rotate a small amount relative to the equilibrium position, and the perturbed magnetic force and moments were found. Bourke used the concept of a center of magnetic pressure, located at a distance z_p below the center of mass, to calculate the torque or moment on the body (see Figure 5-10).

His analysis results in five linear equations of motion for small perturbations from equilibrium. The variables x, y, and z denote the displacement of the center of mass from the static position, and θ and ϕ denote the angular motion of the spin axis from the vertical (Figure 5-10). The equation for the vertical motion $z(t)$ is uncoupled from the other four variables and is in the form of a harmonic oscillator:

$$m\ddot{z} + k_1 z = 0 \tag{5-4.1}$$

where k_1 is the vertical magnetic stiffness.

The other equations can be written in the form of a matrix differential equation:

$$[\mu]\ddot{\omega} + [\delta]\{\omega\} + [k]\{\omega\} = 0 \tag{5-4.2}$$

where $\{\omega\} = \{\phi, \theta, x, y\}^T$ and

$$[\mu] = \begin{bmatrix} I_1 & 0 & 0 & 0 \\ 0 & I_1 & 0 & 0 \\ 0 & 0 & m & 0 \\ 0 & 0 & 0 & m \end{bmatrix}, \qquad [k] = \begin{bmatrix} k_2 & 0 & 0 & -kz_p \\ 0 & k_2 & kz_p & 0 \\ 0 & kz_p & k & 0 \\ -kz_p & 0 & 0 & k \end{bmatrix}$$

$$[\delta] = \Omega_0 I_3 \begin{bmatrix} 0 & 1 & 0 & 0 \\ -1 & 0 & 0 & 0 \\ 0 & 0 & 0 & 0 \\ 0 & 0 & 0 & 0 \end{bmatrix}$$

Here I_1 and I_3 are moments of inertia of the levitated rotor, and Ω_0 is the initial unperturbed spin frequency. The k_{ij} represent the magnetic stiffness terms derived from expressions for the perturbed magnetic forces on the rotor.

These equations have the same form as Eq. (5-2.17). This system can be solved by looking for a solution of the form $\{\omega\} = \{\hat{\omega}\}e^{st}$ and finding the roots of the resulting characteristic equation.

The matrix $[\delta]$ is a gyroscopic effect due to the rotation. When the rotation $\Omega_0 = 0$, one can find an axisymmetric field (or circular

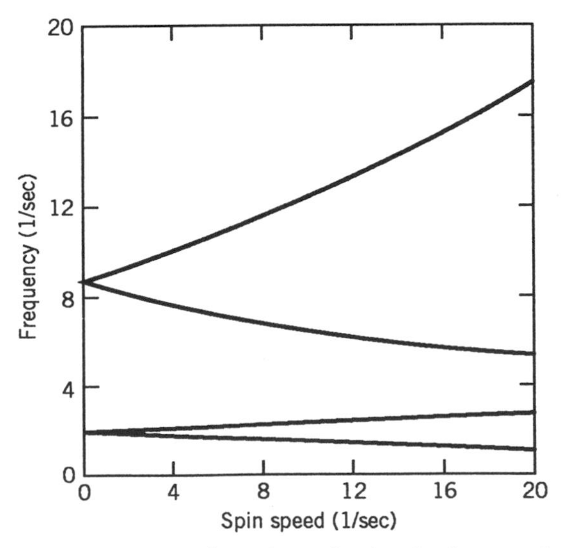

Figure 5-11 Natural frequencies of a spinning body as a function of initial spin velocity. [From Hikihara and Moon (1993), with permission.]

arrangement of current-carrying coils) to achieve stable motion for small disturbances. This means that there are four modes with two distinct frequencies ($s_1 = s_2 = i\omega_1$; $s_3 = s_4 = i\omega_2$). It can be shown that if the system is stable for $[\delta] = 0$, then it will remain stable when it spins. However, the eigenvectors and modal frequencies may change with Ω_0.

Burke (1964) also showed that in general the lateral and rotational motions are coupled in each of the two modes. This may be important in the design of magnetic bearings, because excitation of these coupled modes could lead to the appearance of runout in the levitated spinning rotor. Such runout has been observed in a Japanese levitated rotor (Takaichi et al., 1992). The variation of the natural frequencies with spin speed is shown in Figure 5-11 (Hikihara and Moon, 1993). The gyroscopic effect on the natural frequencies can be seen in this figure. A similar study is reported by Delprette et al. (1992) for a spinning rotor on high-T_c superconducting bearings.

5-5 NEGATIVE DAMPING DUE TO EDDY CURRENTS

The oscillation of a magnetic field source near a stationary conductor normally results in damping forces due to induced circulating currents in the conducting body (eddy currents). However if the field source and conductor have a relative velocity transverse to the vibration, it is possible for the translational motion to add energy into the vibrational degree of freedom through negative magnetic damping as discussed below.

In Chapter 2 we reviewed the generation of lift and drag forces of a moving magnet over a thin-sheet conductor. In particular, we found the levitation forces on a current filament I (at a height h moving over a conductor of thickness Δ, and conductivity σ) (Figures 5-7, 5-12):

$$\begin{aligned} F_L &= \frac{\mu_0 I_0^2}{4\pi h} \frac{v^2}{w^2 + v^2} \\ F_D &= \frac{w}{v} F_L \end{aligned} \tag{5-5.1}$$

where v is the velocity normal to the current filament and w is a characteristic velocity given by $w = 2/\mu_0 \sigma \Delta$. The question that we wish to raise here is, what is the effect of an oscillation in the levitation height on the magnetic forces? For a stationary conductor,

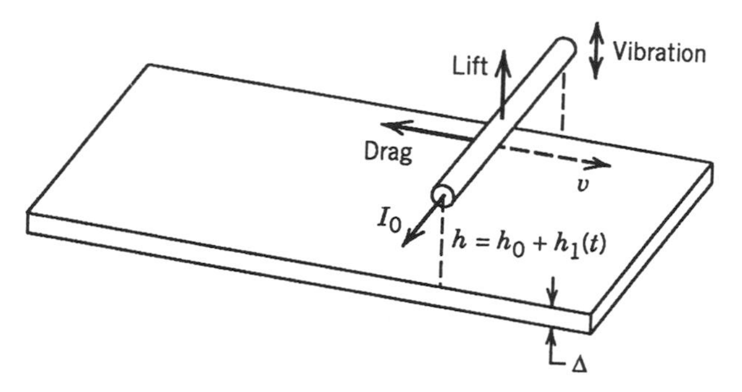

Figure 5-12 Magnetic forces on a levitated filament.

$v = 0$, it is well known that damping forces will be produced due to the creation of eddy currents in the conductor. However, what is not widely recognized is that for a large enough velocity v, this dissipative force can become destabilizing. The subject of eddy current damping in the context of levitation problems was studied by Davis and Wilke (1971), Iwamoto et al. (1974), Moon (1977), and Baiko et al. (1980).

To begin we assume that the levitation height varies sinusoidally; that is,

$$h = h_0 + u_0 e^{i\omega t}, \qquad u_0 \ll h_0 \tag{5-5.2}$$

The perturbed magnetic field at the conductor surface is then assumed to have the same form:

$$\mathbf{B}(x, t) = \mathbf{B}^0(x) + \mathbf{b}^0(x) e^{i\omega t} \tag{5-5.3}$$

It can be shown that the vertical component of $\mathbf{b}^0(x)$ is given by (see Moon, 1984)

$$b_y^0 = -\frac{\mu_0 I u_0 h_0}{\pi} \frac{x}{\left(h_0^2 + x^2\right)^2} \tag{5.5-4}$$

This oscillating field will, in turn, create a time-varying current distribution in the guideway, $J = -\sigma v b_y^0$; or writing $K = J\Delta$, we obtain

$$K(x, t) = K_0(x) + K_1(x) e^{i\omega t} \tag{5-5.5}$$

The perturbed lift force on the wire in the vertical plane is,

$$\mathbf{F} = K(-B_y, B_x)$$

The perturbed lift force component proportional to $e^{i\omega t}$ is found to be

$$F_L^1 = -\frac{\mu_0 I^2}{4\pi h_0}\left[\frac{v^2}{(v^2 + w^2)}\frac{u}{h_0} + \frac{w^2 - v^2}{(w^2 + v^2)^2}\frac{w\,du}{dt}\right] \tag{5-5.6}$$

where we have written $u(t) = u_0 e^{i\omega t}$.

The first term in Eq. (5-5.6) represents a magnetic stiffness term with

$$\kappa = \frac{\mu_0 I^2}{4\pi h_0^2}\frac{v^2}{v^2 + w^2} \tag{5-5.7a}$$

while the second term represents a linear "viscous" damping proportional to the vertical velocity $\dot{u}(t)$.

$$\gamma = \frac{\mu_o I^2 w}{4\pi h_0}\frac{(w^2 - v^2)}{(w^2 + v^2)^2} \tag{5-5.7b}$$

This term is plotted in Figure 5-13. The curve clearly shows that the damping can become negative when $v > w$.

Experiments on negative eddy current damping were first done by Iwamoto et al. (1974), and later by Moon (1977). Data from the latter paper are shown in Figures 5-14 and 5-15.

A magnet was vibrated normal to a moving thin aluminum conductor. In real systems, aerodynamic as well as eddy current damping is present. In our experiments, we first used a nonmagnetic "dummy" magnet to obtain the aerodynamic effects. Subtracting this damping, we indeed observed the negative damping effect as shown in Figure 5-15.

5-6 DYNAMICS OF MAG-LEV VEHICLES

The noncontact nature of magnetic levitation raises a question not normally asked of conventional mechanically supported and guided rail systems: Under what circumstances can the dynamics of the

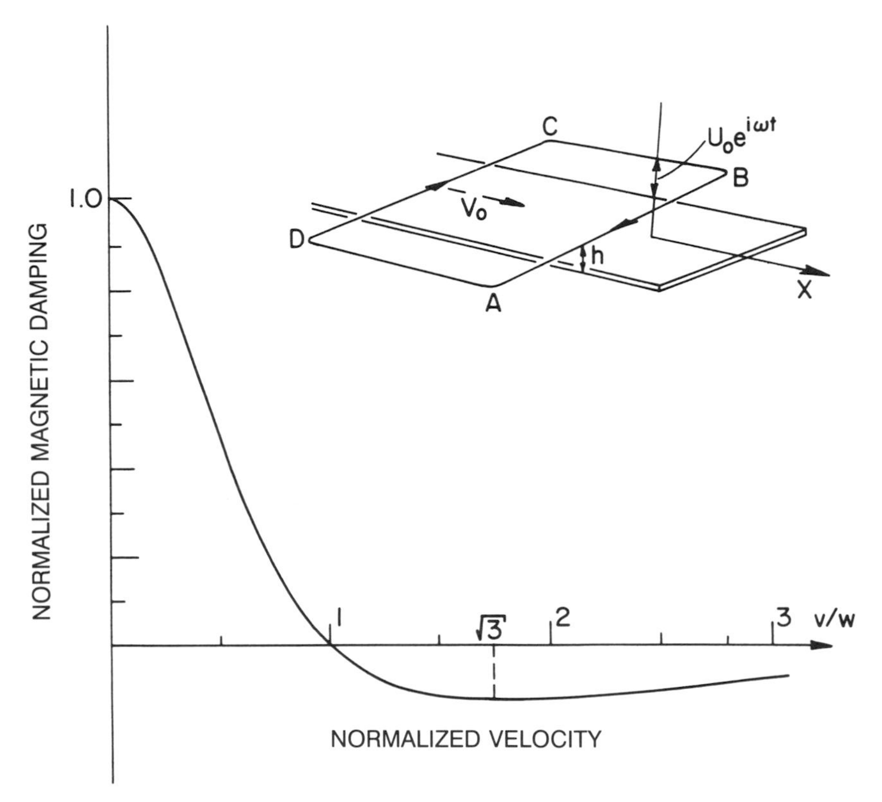

Figure 5-13 Damping of vertical vibrations as a function of forward speed. [From Moon (1977) with permission.]

vehicle close the levitation gap and result in contact of the vehicle with the guideway? To avoid such a catastrophic effect, the Mag-Lev design must be free from any mechanism that could transfer some of the kinetic energy of translation along the guideway into lateral, heave, or rotational dynamics. In this respect, Mag-Lev vehicle dynamics has much in common with aircraft dynamics. And given the speeds of prospective Mag-Lev systems (400–500 km/hr), aerodynamic effects will also play a major role in choosing a safe design.

Besides the issue of stability of the vehicle, there are many other dynamics-related issues that must be addressed in a successful Mag-Lev design. We will not address most of these, but a list of the most important includes the following:

- Dynamic stability
- Ride quality

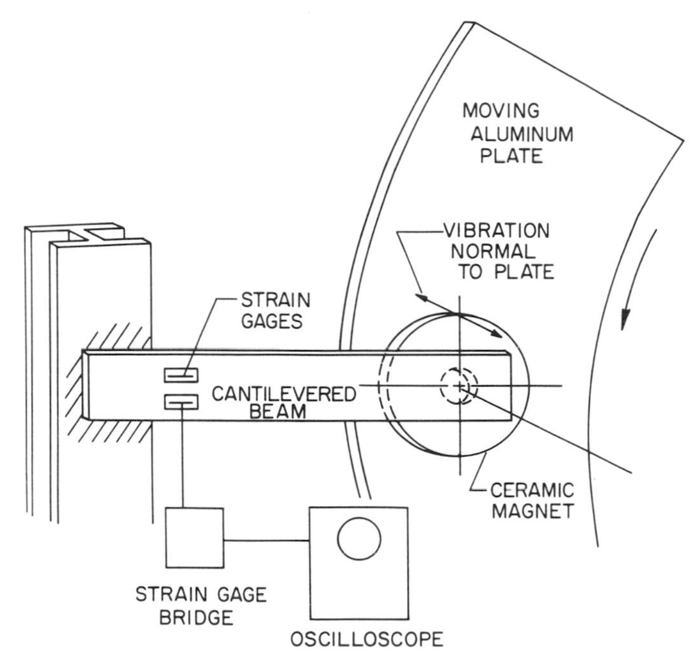

Figure 5-14 Experimental setup for measuring magnetic damping of a vibrating magnet near a moving conductor. [From Moon (1977), with permission.]

- Braking dynamics under emergency
- Response to aerodynamic gusts
- Vehicle–guideway interaction
- Secondary suspension design
- Tilt dynamics and control
- Tunnel entrance and exit dynamics
- Guideway impact recovery
- Loss of magnet dynamics
- Guideway curve entrance and exit dynamics
- Guideway misalignment response (e.g., due to earthquakes)

Some of these problems have been addressed in the studies funded by the U.S. Department of Transportation under the National Mag-Lev Initiative (1992–1993). (See also Chapter 7.) The experience of the Japanese superconducting Mag-Lev project has also been reported in the literature for a few of these problems.

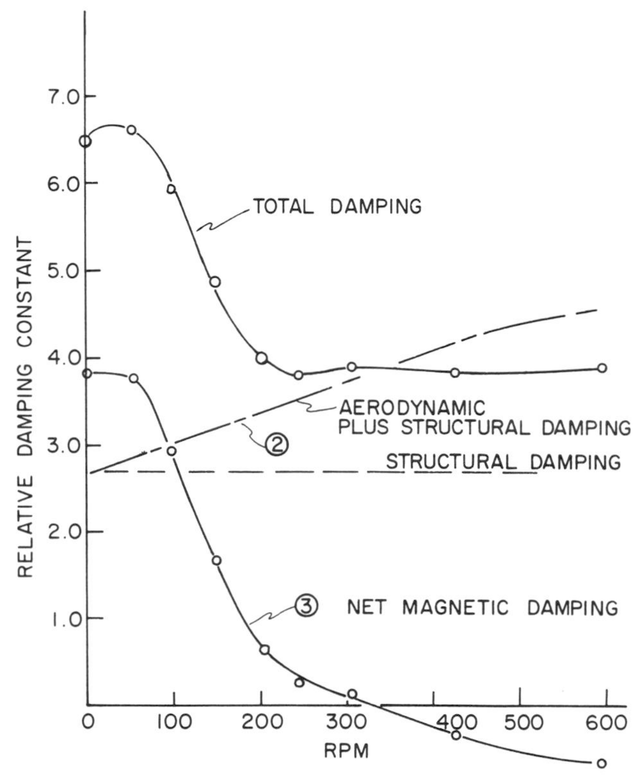

Figure 5-15 Experimental data on magnetic damping versus speed. [From Moon (1977), with permission.]

In the rest of this section, however, we illustrate the analysis of a particular dynamic stability problem that has been reported in small model testing: lateral–yaw instability.

Several of the possible dynamic and static instabilities that can occur in Mag-Lev vehicle systems are illustrated in Figure 5-4. These are described by geometric as well as figurative descriptors:

- Lateral-roll divergence (or list)
- Pitch–heave dynamics (porpoising)
- Lateral–yaw dynamics (snaking)
- Yaw–roll–pitch motion ("Dutch" roll or screw motion)
- Forward motion oscillation (hunting)

It should be noted that the term "instability" refers to the growing departure with time from the quasistatic or steady translation configuration. However, in general, nonlinearities will eventually limit the amplitude of this motion (if it doesn't hit the guideway). Thus, the limiting motion could be a periodic limit cycle, quasiperiodic motion (two or more incommensurate frequencies), or chaotic motion. Or the vehicle could move to a new steady-state motion—for example, tilted to one side in a fixed roll position.

Mag-Lev vehicles are also subjected to dynamic forces of both a deterministic and stochastic nature. Propulsion magnetic fields may introduce periodic magnetic forces that might produce vehicle oscillations. Also, guideway misalignments as well as aerodynamic pressures can produce random excitation forces. Finally, the flexibility of a periodically supported guideway itself may produce resonance effects if the vehicle or train of vehicles traverses it at certain critical speeds. See Popp (1982a) or Wormley et al. (1992) for a detailed analysis of Mag-Lev vehicle-guideway interaction.

Lateral-Yaw Oscillations in Mag-Lev Vehicles

Early studies of magnetically levitated transportation vehicles dealt almost exclusively with equilibrium or static forces of lift and drag. This was especially true for superconducting repulsive levitation. What is ironic is that while superconducting Mag-Lev was portrayed as "flying in a magnetic field," the experience of the dynamics of flight vehicles was not brought to bear on Mag-Lev dynamics. In the history of flight vehicles, dynamic stability has been an issue from the first design of the Wright Brothers and Curtiss planes. In these problems, coupled degrees of freedom were often the source of trouble. The so-called "Dutch-roll" dynamics involved a corkscrew motion of the vehicle involving pitch, roll, and yaw. The techniques of analysis used perturbed lift, drag, and aerodynamic moments and relied on the classical ideas of stability from linear perturbation systems theory. In the following we apply this technique to the lateral or sway dynamics coupled with yaw motions in a magnetically levitated model using repulsive magnetic forces. The analysis is a summary of a paper by Chu and Moon (1983). A more recent study has been performed by a group at Argonne National Laboratories (Cai et al., 1992).

The model is shown in Figure 5-16. It represents an idealization of a vehicle with magnets moving along a conducting guideway. In the laboratory, the model carried small rare earth magnets, whereas in full-scale vehicles, the magnets are superconducting. In full-scale systems, the guideway consists of discrete conducting coils distributed along the track, whereas in rotating-wheel experiments, a continuous-

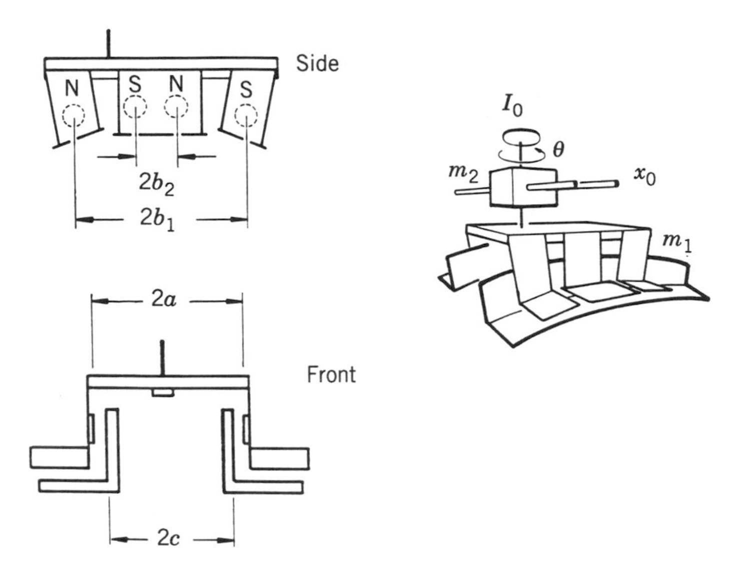

Figure 5-16 Sketch of model for observing Mag-Lev lateral–yaw instabilities. [From Chu and Moon (1983), with permission.]

sheet track as well as discrete track elements have been used (see Chu and Moon, 1983). However, the dynamic perturbations due to the discrete track are neglected because their frequency is usually much higher than the rigid-body dynamics of the vehicle.

Given the six possible degrees of freedom, the general motion of levitated bodies in a guideway can be quite complicated. Pitch–heave or porpoising motions can occur as well as Dutch-roll modes. However, to illustrate the possible motions and instabilities, only two degrees of freedom will be assumed to be active: lateral or sway displacements, denoted by $x(t)$, and yaw rotation about the vertical axis of the vehicle, denoted by $\theta(t)$.

Magnetic forces on the vehicle are assumed to act at the four corners. Also, aerodynamic forces in this problem are neglected, though in full-scale vehicles they could have a major effect. Of concern here is the lateral and drag forces on each corner magnet and the change in those forces with changes in x or θ. It is assumed the gravity force is equilibrated by the vertical lift on the four magnets. Finally, we neglect the magnetic couple that may exist on each of the four magnets because the moment of the magnetic forces about the vehicle center of mass is assumed to be much larger than the moment on one of the magnets.

The source of the variation of the lateral forces L_i and drag forces D_i on each of the four magnets is assumed to come from the change in the gap δ_i between the guideway side wall and each of the magnets; that is,

$$L_i(\delta_i) = L_0 + L'\delta_i$$
$$D_i(\delta_i) = D_0 + D'\delta_i \tag{5-6.1}$$

It is assumed that all four magnets are identical, so that L_0, D_0, L', and D' are identical for all four magnets. Using this assumption, the net force and moment about the center of mass of the vehicle are given by

$$F_x = 4L'(x_0 + \rho\theta)$$
$$M_z = 4(\rho L' - aD')x_0 + 4\left[L'(b^2 + \rho^2) - D'a - aL_0 - \rho D_0\right]\theta \tag{5-6.2}$$

where the geometric terms x_0, ρ, b, and a are defined in Figure 5-17.

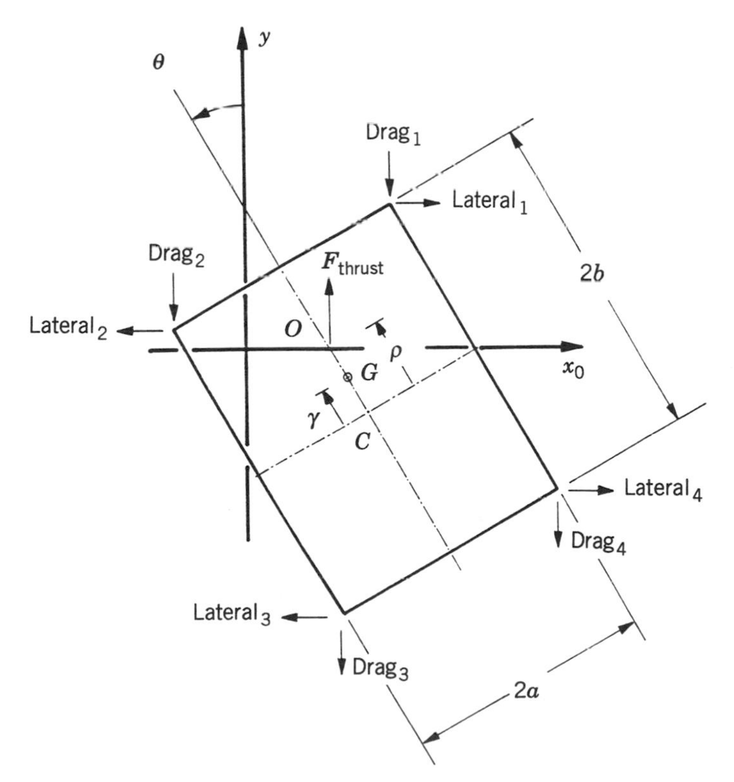

Figure 5-17 Geometry of model for analysis of Mag-Lev lateral–yaw instabilities. [From Chu and Moon (1983), with permission.]

The Newton–Euler equations for the lateral and yaw motions are then given by [see Eqs. (5-2.2), (5-2.6)],

$$[m]\begin{Bmatrix} \ddot{x}_0 \\ \ddot{\theta} \end{Bmatrix} = -[\kappa]\begin{Bmatrix} x_0 \\ \theta \end{Bmatrix} \tag{5-6.3}$$

where the mass and stiffness matrices are

$$[m] = \begin{bmatrix} m_1 + m_2 & m_1\rho \\ m_1\rho & I_G + m_1\rho^2 \end{bmatrix}$$

$$[\kappa] = -\begin{bmatrix} 4L' & 4L'\rho \\ 4(L'\rho - D'a) & 4\left[L'(b^2 + \rho^2) - L_0 a - \rho(D_0 + D'a)\right] \end{bmatrix} \tag{5-6.4}$$

Here m_1 is the vehicle mass and m_2 is the mass of a movable mass in the vehicle.

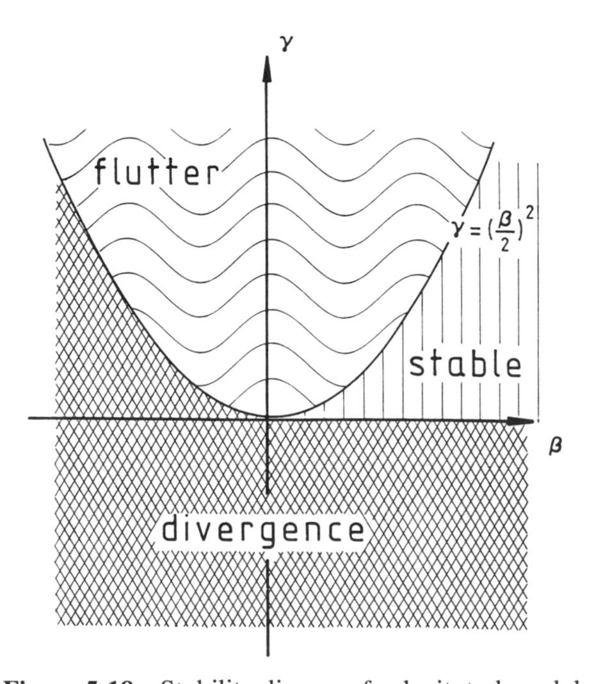

Figure 5-18 Stability diagram for levitated model.

The concept of stability here is identical to that in linear systems theory. The system is assumed to have solutions of the form

$$\begin{Bmatrix} x_0 \\ \theta \end{Bmatrix} = e^{st} \begin{Bmatrix} X \\ \Theta \end{Bmatrix} \tag{5-6.5}$$

where s can be real, imaginary, or a complex number. When Real(s) > 0, the system is said to be *unstable*. Real(s) $= 0$ implies neutral stability, whereas Real(s) < 0 is considered stable. This notion of stability holds only for the local departures from equilibrium. It is possible, however, for a system to be stable for small perturbations, but become unstable for large disturbances from equilibrium.

Substitution of the solution (5-6.5) into the equation of motion leads to a characteristic equation of the form (Figure 5-18)

$$s^4 + \beta s^2 + \gamma = 0 \tag{5-6.6}$$

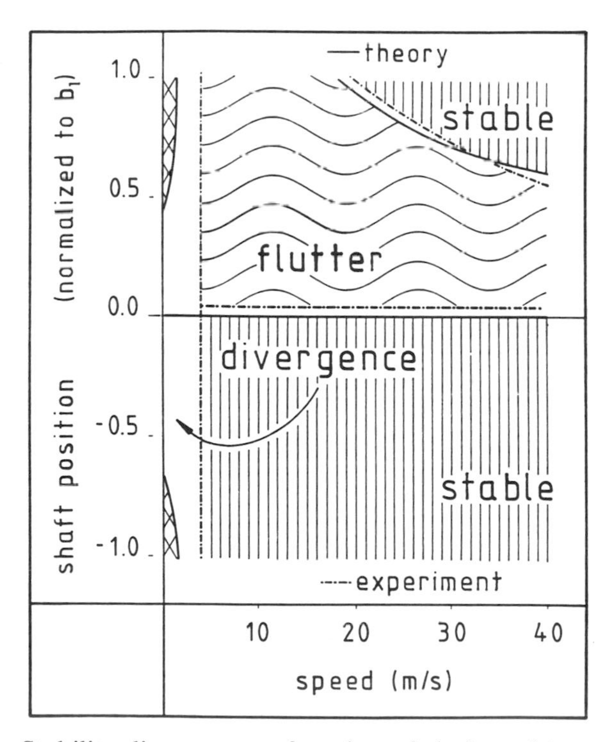

Figure 5-19 Stability diagram as a function of shaft position and speed for theoretical and experimental models. [From Chu and Moon (1983), with permission.]

where the constants β and γ depend on the components of the m_{ij} and κ_{ij} matrices.

A few general remarks can be said about the dynamics, however, before looking at numerical solutions.

Divergence and Flutter Instabilities There are two different types of instabilities corresponding to different types of eigenvalues, s, of the solution. A solution corresponding to s real and positive is one whose amplitude grows monotonically in an exponential way. In aircraft and aeroelastic stability problems, this is known as a *divergence* solution, and is also similar to buckling in elastic structures. The system moves away from the equilibrium position in a monotonic way. Of course, nonlinearities in real systems come into play to restrict the amplitude (e.g., hitting the guideway).

When the eigenvalue is complex (i.e., $s = \alpha + i\omega$, where α is positive and $\omega \neq 0$), the system moves away from the equilibrium state in an oscillating manner with exponentially increasing amplitude until nonlinearities limit the motion. This type of instability is called *flutter* in aeroelastic dynamics or is known as a *dynamic instability*.

The values of the mathematical parameters β and γ in Eq. (5-6.6) that lead to stable, divergence, and flutter solutions are shown in

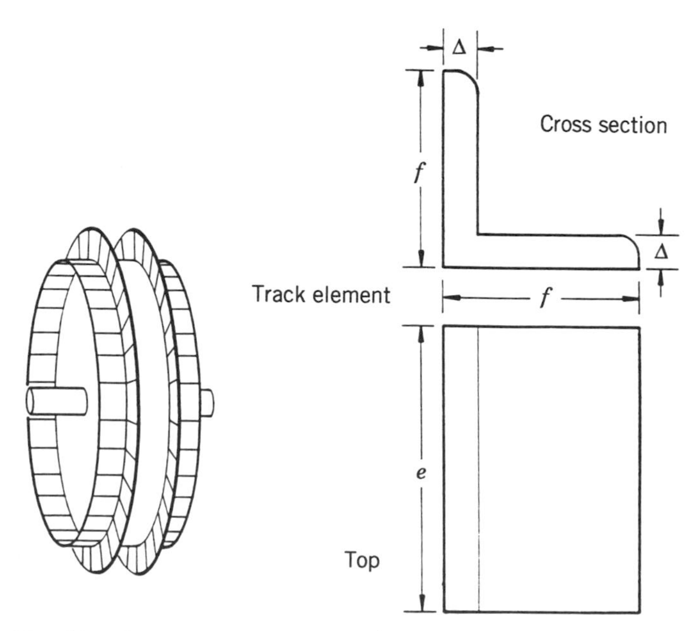

Figure 5-20 Sketch of rotating wheel guideway with "L"-shaped aluminum track elements. [From Chu and Moon (1983), with permission.]

Figure 5-18. Of course, β and γ depend on the system parameters. If we choose two such parameters, such as the speed, V_0, and the position of the center of mass, r_{cm}, then one can remap the stability regions in the β–γ plane onto the V_0–r_{cm} plane as in Figure 5-19. This shows that for positive center-of-mass position, low speeds can result in an instability of either the flutter or divergence type.

To compare these calculations with experiments, a 0.73-kg model was built using rare earth magnets and was levitated on a circular wheel track with "L"-shaped aluminum track elements as shown in

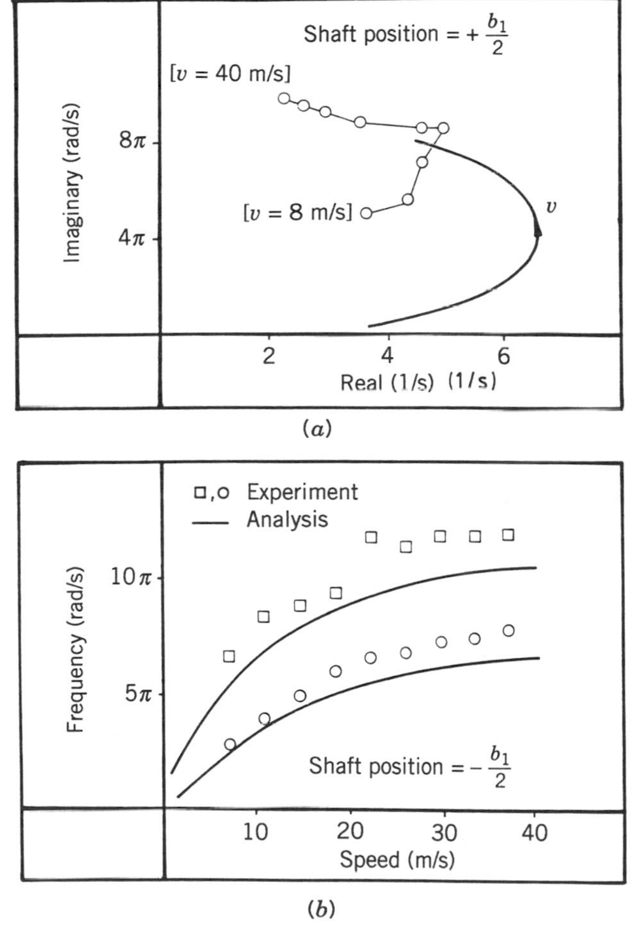

Figure 5-21 Eigenvalues of the two-degree-of-freedom vibration as a function of speed. (*a*) Comparison of theory and experiment in the unstable regime; (*b*) Stable regime, comparison of theory and experiment for two modes. [From Chu and Moon (1983), with permission.]

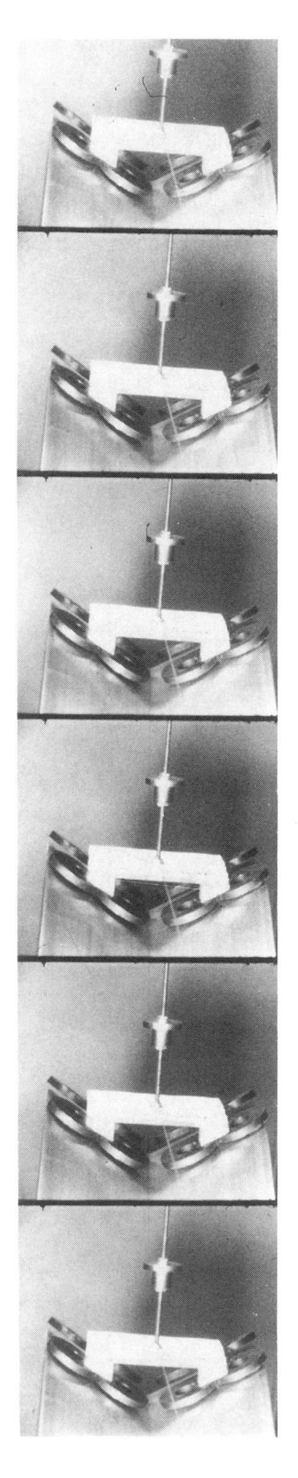

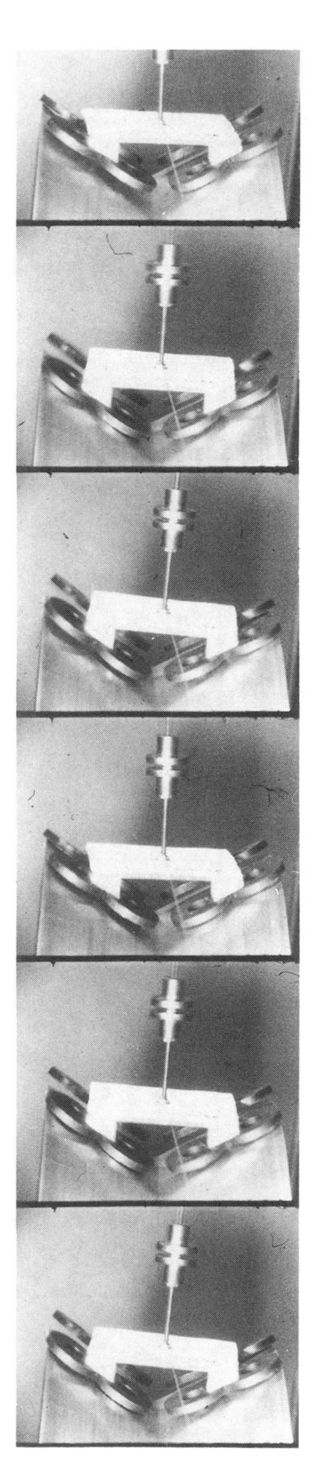

(a) Stable motion

(b) Unstable Yaw–lateral motion

Figure 5-22 Sequence of photos from a film showing yaw–lateral–roll vibrations. (a) Stable model levitation above a rotating "V"-shaped guideway. (b) Unstable vibrations. [From Moon (1977), with permission.]

Figure 5-20. The lift, drag, and lateral or edge forces for the magnets moving past the conducting track elements were measured. From these data the values of L_0, L', D_0, and D', as functions of speed and magnet guideway gap δ, could be determined.

Another way to show the effect of parameters on the stability is a root locus diagram shown in Figure 5-21. Here the path of one of the eigenvalues as a function of vehicle speed is shown in the complex plane. As the speed is increased, the eigenvalue moves toward the real axes from an unstable state to a stable one.

Another study using a "V"-shaped aluminum guideway on a rotating wheel was performed in the 1970s which showed a yaw–lateral–roll instability for certain geometric arrangements of magnets on a levitated model. A sequence of motions from films of these instabilities is shown in Figure 5-22. The left sequence shows steady levitated motion, while the right sequence shows a yaw oscillation.

5-7 ACTIVE CONTROLLED LEVITATION DYNAMICS

In Chapter 1, we discussed two types of magnetic levitation. Active controlled magnetic levitation devices provide a stable magnetic force equal and opposite to the gravitational force on a body. Such devices are used to suspend models in wind tunnels, to provide noncontacting bearings for gyroscopes and rotating machinery, and to suspend passenger-carrying vehicles. The electromagnetic levitation method (EML) uses magnetization forces that are induced by electric currents around a ferromagnetic circuit. For example, in Figure 1-16*a*, electromagnets in the vehicle are attracted to a ferromagnetic rail. EML can provide suspension at zero forward speeds in contrast to electrodynamic or repulsive levitation involving superconducting magnets. However, the forces created by electromagnets are inherently unstable if the currents are constant. Therefore, feedback forces are required to stabilize EML devices.

Superconducting levitation systems can provide stable magnetic forces, but often suffer from low magnetic stiffness or low damping. Therefore, one can imagine hybrid levitation schemes employing the best of ferromagnetic levitation, active levitation, and superconducting levitation. Such a hybrid superconducting EML Mag-Lev vehicle system has been recently proposed by a group headed by the Grumman Aerospace Corporation in 1992 (Figure 7-18). [See e.g., U.S. Dept. of Transportation, (1993a).]

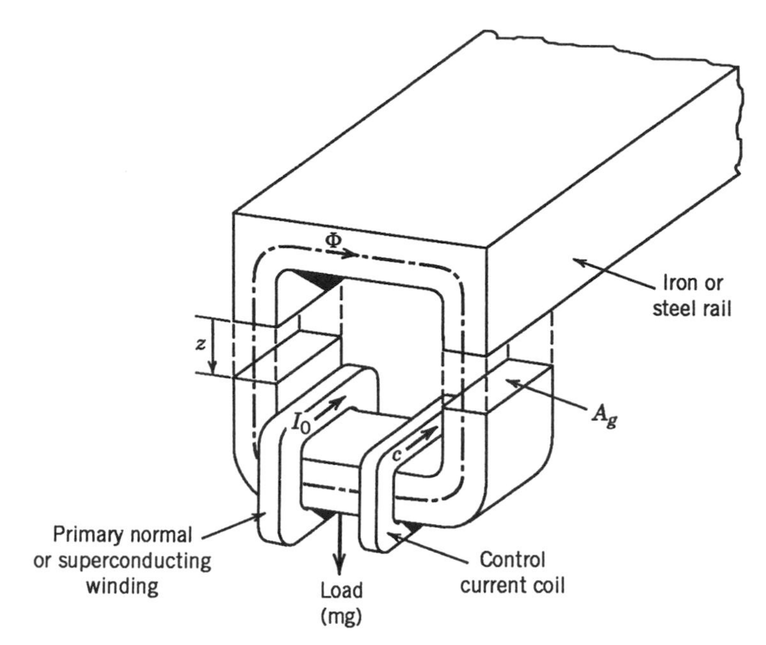

Figure 5-23 Geometric arrangement of electromagnet and guideway for EML-feedback-stabilized suspension.

The discussion below is only meant to illustrate how feedback control can produce stiffness and damping following a paper by Meisenholder and Wang (1972). In real technical systems, the problem involves several degrees of freedom with multiple sensors, actuators, and more sophisticated control schemes. The discussion below is based on linear analog control theory. More modern treatments are based on nonlinear, digital control methods which are beyond the scope of this book. The reader is referred to the book edited by Schweitzer (1988) or the paper by Popp (1982b).

In this section we present an analysis of a simple linear feedback system for magnetic suspensions. We write dynamical equations for both the suspended mass (using Newton's law) and the magnet circuit (using Maxwell's equation). The one-degree-of-freedom system is shown in Figure 5-23. Applying Newton's law and the magnetic circuit equations for the gap z and the magnetic flux Φ, the resulting equations have the form:

$$\begin{aligned} m\ddot{z} &= mg + F \\ N\dot{\Phi} &= -RI + V_0 + V_c \end{aligned} \tag{5-7.1}$$

where V_0 and V_c are the dc and dynamic control voltages, respectively.

From Eq. (2-4.6) the flux Φ is related to the sum of the magnetic reluctances:

$$\Phi = \frac{NI}{\mathfrak{R}} \tag{5-7.2}$$

where the Ampere-turns, NI, can be produced by a superconducting coil, if desired, and a control current coil, and

$$\mathfrak{R} = \mathfrak{R}_{\mathrm{fe}} + \frac{2z}{\mu_0 A_g}$$

Here $\mathfrak{R}$ is the sum of the iron path reluctance, $\mathfrak{R}_{\mathrm{fe}}$, and the gap reluctance. (A_g is the gap area.) The magnetic force can be derived either from the Maxwell stress tensor or from the magnetic energy method, (2-2.21) or (2-3.5),

$$F = \frac{-\Phi^2}{\mu_0 A_g} \tag{5-7.3}$$

In (5-7.1) the constant voltage V_0 provides the primary current I_0. The control voltage V_c generates the control current which will stabilize or add damping and stiffness to the suspended mass. In linear control theory, V_c is linearly proportional to the state variables. Following Meisenholder and Wang (1972), we choose the following control law:

$$V_c = G_1(z - z_0) + G_2\dot{z} + G_3\ddot{z} \tag{5-7.4}$$

Here the control voltage is proportional to the measured acceleration, velocity and position departure of the levitated object from the steady height z_0. (This is equivalent to choosing the state variables as position, velocity and flux.)

The goal of a magnetic levitation design is to choose the gains G_i such that the system is dynamically stable for small perturbations from equilibrium. Toward this end we define the following perturbation variables:

$$\begin{aligned} z &= z_0 + h(t) \\ I &= I_0 + c(t) \\ \mathfrak{R} &= \mathfrak{R}_0 + \mathfrak{R}_1 h(t) \end{aligned} \tag{5-7.5}$$

where $c(t)$ represents the control current. Also the following equilibrium relations hold:

$$V_0 = RI_0$$
$$mg = \frac{(NI_0)^2}{\mu_0 A_g \Re_0^2} \tag{5-7.6}$$

Next, we expand the nonlinear magnetic flux and force expressions (5-7.2) and (5-7.3) in a Taylor series in the perturbation variables; that is,

$$\Phi = \frac{NI_0}{\Re_0}\left(1 + \frac{c}{I_0} - \frac{\Re_1}{\Re_0}h\right)$$
$$F = -mg\left(1 + \frac{2c}{I_0} - \frac{2\Re_1}{\Re_0}h\right) \tag{5-7.7}$$

The coupled linearized equations then take the form

$$\ddot{h} - 2g\frac{\Re_1}{\Re_0}h = -\frac{2gc}{I_0}$$
$$\frac{N^2}{\Re_0}\dot{c} = -Rc + \frac{N^2 I_0}{\Re_0^2}\Re_1\dot{h} + G_1 h + G_2\dot{h} + G\ddot{h} \tag{5-7.8}$$

These equations can be rewritten in simplified notation:

$$\ddot{h} - \alpha^2 h = -\beta c$$
$$\dot{c} + \gamma c = \Gamma_1 h + (\Gamma_2 + \delta)\dot{h} + \Gamma_3\ddot{h} \tag{5-7.9}$$

where Γ_1, Γ_2, and Γ_3 are position, velocity, and acceleration feedback gains, respectively. This system of linear differential equations can be solved either by Laplace transforms or, more simply, in terms of the functions

$$\begin{bmatrix} h(t) \\ c(t) \end{bmatrix} = e^{\lambda t}\begin{bmatrix} \bar{h} \\ \bar{c} \end{bmatrix} \tag{5-7.10}$$

Substituting Eq. (5-7.10) into Eq. (5-7.9), we obtain an expression for the Laplace transform of the control current of $\bar{c}$,

$$\bar{c} = \frac{[\Gamma_1 + (\Gamma_2 + \delta)\lambda + \Gamma_3\lambda^2]\bar{h}}{\lambda + \gamma} \tag{5-7.11}$$

The gains Γ_1, Γ_2, and Γ_3 (or G_1, G_2, and G_3) are chosen so that the Laplace transform of the control force $-\beta\bar{c}$ appears as a restoring spring and damper force:

$$-\beta\bar{c} = -a\bar{h} - b\lambda\bar{h} \tag{5-7.12}$$

where α is similar to a spring constant and b is a damping constant. Using Eq. (5-7.12) in Eq. (5-7.11), we obtain equations for the gains in terms of a and b; that is,

$$\begin{aligned} \Gamma_1 &= \frac{a\gamma}{\beta} \\ \Gamma_2 &= -\delta + \frac{a + \gamma b}{\beta} \\ \Gamma_3 &= b/\beta \end{aligned} \tag{5-7.13}$$

Because α^2 is the negative stiffness of the β uncontrolled system [see (5-7.9).] a necessary condition for stability is that $a > \alpha^2$; that is, the control stiffness should exceed the negative magnetic stiffness. In terms of our original physical variables we obtain

$$G_1 > \frac{V_0\Re_1}{\Re_0}$$

If only position and velocity feedback are used (i.e., $G_3 = \Gamma_3 = 0$), then the linear dynamics cannot be made to correspond to a simple damped spring mass system and further analysis of the linear stability must be made. Such analysis would involve a root locus of all three roots λ as functions of the gains G_1 and G_2. Further discussion on the dynamics of electromagnetic levitation as applied to passenger-carrying vehicles may be found in Meisenholder and Wang (1972), Cai et al. (1992), Weh (1989) or Popp (1982b).

TABLE 5-1 Comparison of Linear and Nonlinear Phenomena

Linear Dynamics	Nonlinear Dynamics
Resonance	Subharmonics
Instability	Limit cycle
Periodic motion	Chaos
Robust with respect to initial conditions	Sensitive to initial conditions
Predictable	Unpredictable
Unique solution	Multiple solutions

5-8 NONLINEAR DYNAMICS AND CHAOS IN LEVITATED BODIES

As discussed in the introduction to this chapter, the term "linear" refers to the dependence of the magnetic forces to the first power of the state variables. In general, however, most magnetic phenomena are nonlinear in position or angular state variables, in velocity or angular velocity variables, or in the magnetic field or electric circuit variables. In spite of the reality of nonlinear forces, most analyses of dynamics of levitated bodies use linear models in order to simplify the mathematics. However, these models fail to capture important physical phenomena. Nonlinear phenomena include the following: amplitude-dependence of natural frequencies; jump and hysteretic behavior in forced vibration problems; limit cycle periodic motions; subharmonic generation; and the most recently observed phenomena of chaotic dynamics and unpredictable motions. The differences between linear and nonlinear models are summarized in Table 5-1. Introductory books in nonlinear dynamics include Hagedorn (1988) and the

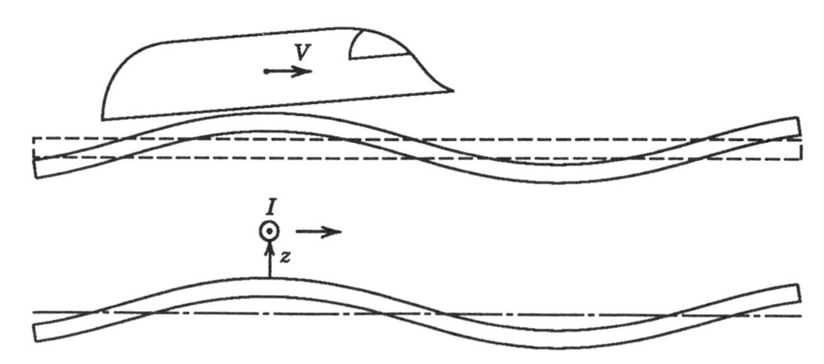

Figure 5-24 Levitated vehicle and magnet wire moving over a guideway with a periodic vertical departure from flatness.

classic book by Stoker (1950). An introduction to chaotic phenomena in nonlinear dynamics can be found in the recent book by the author (Moon, 1992a).

In spite of the obvious nonlinear properties of magnetically levitated systems, very little analytical or experimental work is reported on nonlinear dynamics of levitated systems. This is very strange considering that full-scale people-carrying Mag-Lev systems are close to deployment at the time of publication of this book (late 1993). One would think that safety considerations would demand a more realistic study of the dynamics of such systems.

A few studies of chaotic dynamics of levitated bodies have been reported, however. These include several by the author (Moon, 1988; Moon and Li, 1986, Hikihara and Moon, 1994), and a few from Japan (Gafka and Tani, 1992; Kuroda et al., 1992; Tamura et al., 1992). We shall discuss two simple cases in this section:

1. The vertical motion of a levitated coil moving over a conducting guideway
2. The lateral vibration of a magnet over a YBCO superconducting bearing

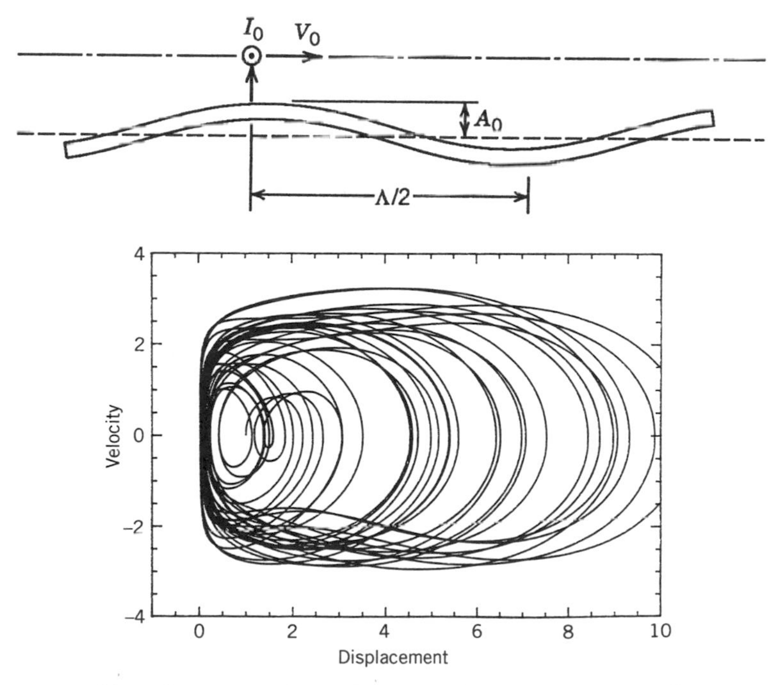

Figure 5-25 Chaotic heave dynamics of a levitated magnet. [From Moon and Li (1986), with permission.]

Vertical Heave Dynamics of a Mag-Lev Vehicle

Consider a coil element of length, β, carrying constant current I moving over a continuous-sheet guideway as shown in Figure 5-25. In the high speed limit, the force on the coil will be given by the field due to an image coil below the guideway of opposite current direction.

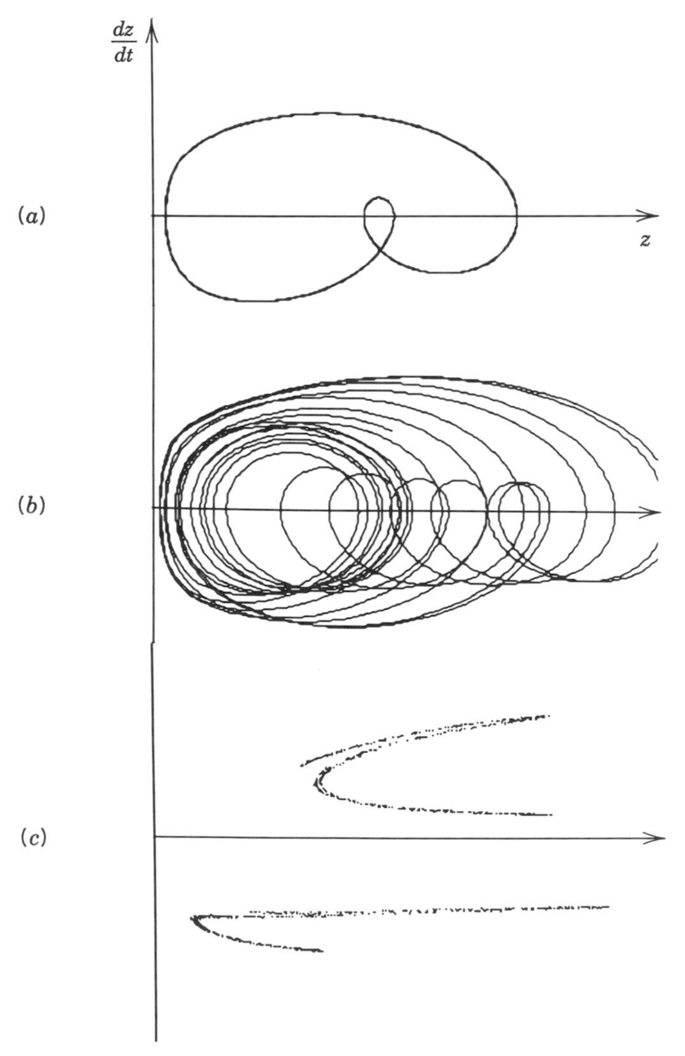

Figure 5-26 (*a*) Periodic motion, (*b*) chaotic motion, (*c*) Poincaré map of chaotic heave dynamics in Figure 5-25*b*. [From Moon and Li (1986), with permission.]

When the coil is close to the sheet conductor, the force is inversely proportional to the height of the coil above the conducting sheet, z; that is,

$$F = \frac{\mu_0 I^2 \beta}{2\pi z} \tag{5-8.1}$$

In this model we have assumed that the force acts principally on the coil elements transverse to the horizontal velocity. We also assume that the guideway has a vertical wavelike deformation pattern of amplitude $A_0 \cos kx$ and wavelength Λ, where $\Lambda \gg z$. If the coil moves with a horizontal velocity (i.e., $x = V_0 t$) and a wave number $k = 2\pi/\Lambda$, then the wavelike track will produce a sinusoidal forcing term on the levitated coil proportional to $\cos \omega t$, where $\omega = 2\pi V_0/\Lambda$.

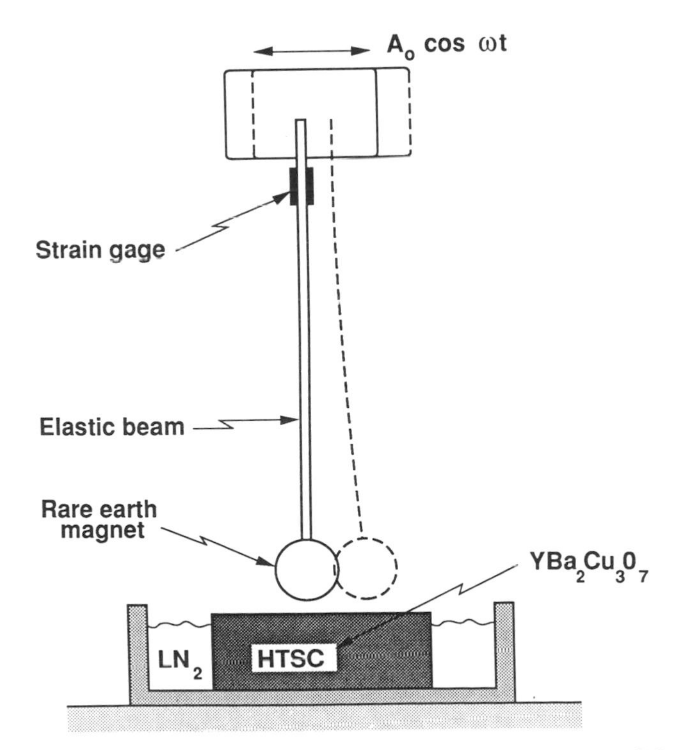

Figure 5-27 Sketch of a rare earth magnet vibrating over a YBCO superconductor. HTSC, high-temperature superconductor. [From Moon (1988), with permission.]

Under these very generous assumptions we can derive an equation for the vertical motion (called *heave*) of the form

$$m\ddot{z} + \delta\dot{z} - \frac{\mu_0 I^2 \beta}{2\pi z} = -mg + m\frac{V_0^2 4\pi^2 A_0}{\Lambda^2}\cos\omega t \qquad (5\text{-}8.2)$$

where $\omega = 2\pi V_0/\Lambda$ and an arbitrary damping term has been added. This system can be written in the form of a third-order autonomous system of first-order differential equations:

$$\begin{aligned} \dot{z} &= v \\ \dot{v} &= -cv + \frac{b}{z} - a + f\cos\phi \\ \dot{\phi} &= \omega \end{aligned} \qquad (5\text{-}8.3)$$

This system of equations can easily be numerically integrated in time using a Runge–Kutta or other suitable algorithm. The trajectory

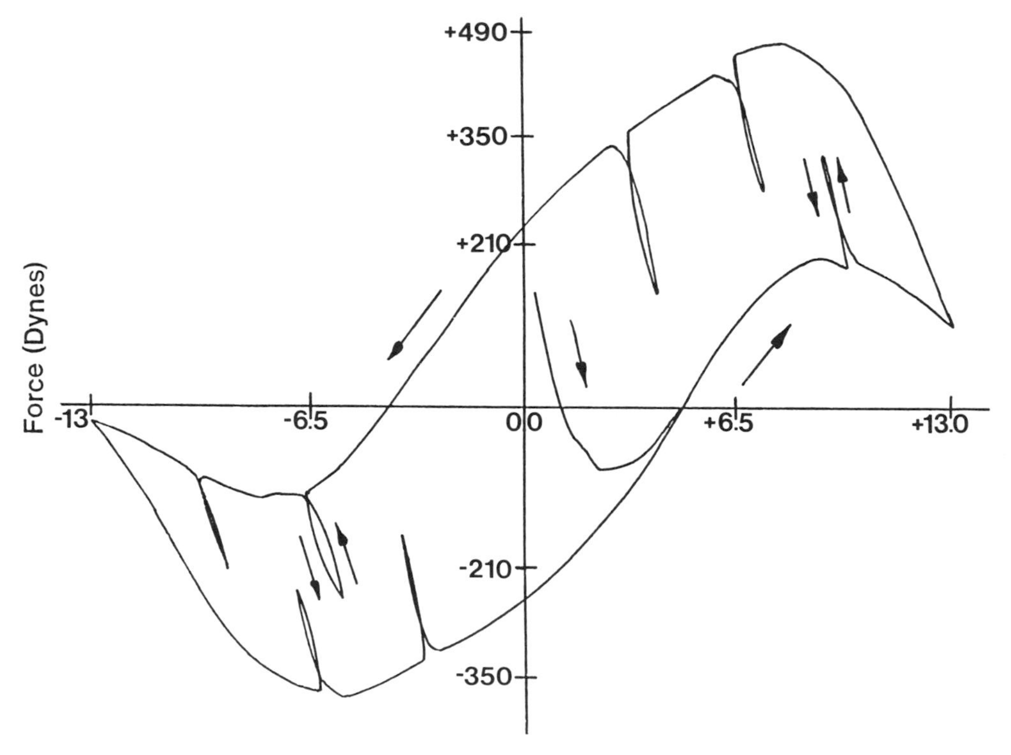

Figure 5-28 Lateral force versus displacement showing nonlinear and hysteretic behavior. [From Moon (1988), with permission.]

is easily projected onto the phase plane of z versus v. As the amplitude of the guideway waviness is increased, one can see a change in the geometry of the motion from elliptical to a distorted ellipse to chaotic motion as shown in Figure 5-25. The chaotic motion is better viewed by looking at a strobescopic view of the dynamics by plotting (z_n, v_n) at discrete values of the phase $\phi = \omega t$ or $t_n = 2\pi n/\omega$. This picture is called a *Poincaré map* [see, e.g., Moon (1992a)]. The Poincaré map of the chaotic motion of the levitated vehicle is shown in Figure 5-26. In contrast to the unordered continuous time plot in Figure 5-25, the Poincaré map shows a fractal-like structure. This type of chaotic motion with fractal structurc is called a *strange attractor*. It indicates that the dynamics are very sensitive to the initial conditions.

Chaotic-like dynamics in a levitated model moving over a rotating-wheel guideway have been observed by Moon (1982) and by Chu and Moon (1983).

Chaotic Lateral Vibration of a YBCO Magnetic Bearing

In Chapter 4 we saw that the magnetic force between a permanent magnet and a high-temperature superconductor such as YBCO is

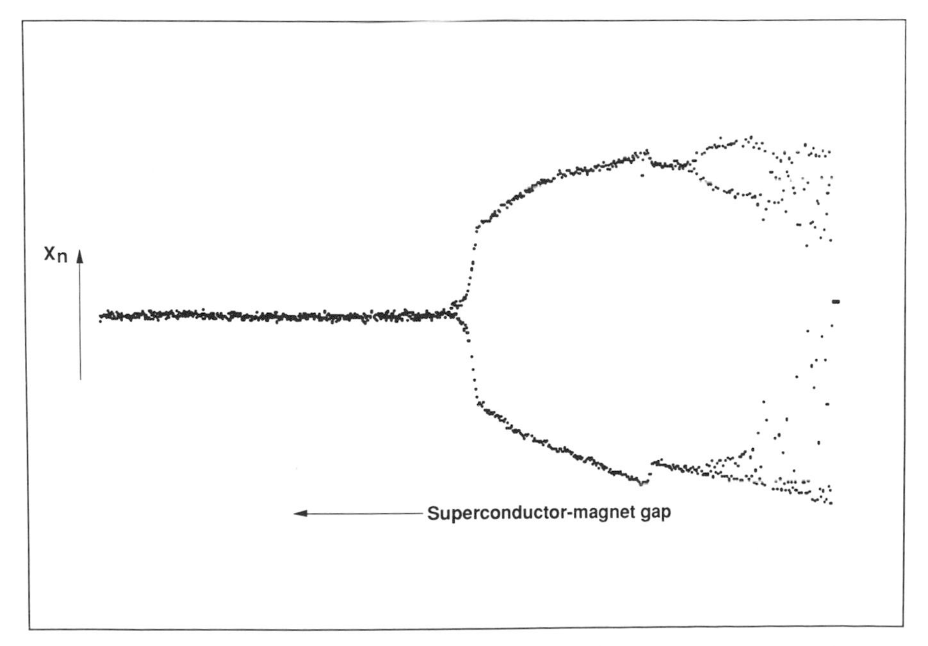

Figure 5-29 Period doubling in the Poincaré map as a function of magnet–superconductor gap. [From Moon (1988), with permission.]

hysteretic near the critical temperature (e.g., Figure 4-15). Hysteretic forces are both nonlinear and dissipative and can produce complex nonlinear dynamics. An example is shown in Figure 5-27 from a paper by the author (Moon, 1988). A permanent magnet is restrained to move laterally over a YBCO superconductor. The force–displacement behavior is shown in Figure 5-28 and exhibits strong hysteresis. As the gap between the magnet and the superconductor is decreased, the dynamics of the magnet become increasingly complex in a pattern called *period doubling* shown in Figure 5-29. Subharmonic frequencies appear in the spectrum of the form $m\omega/n$, where $n = 2, 4, \ldots, 2^k$. This bifurcation behavior is shown in the Poincaré map (see previous section) as a function of the magnet–YBCO gap. At a critical gap, the motion becomes chaotic. Another tool for observing chaotic motions is to plot a *return map* on one of the state variables, say X_{n+1}, versus X_n, where X_n is the displacement of the magnet at discrete times synchronous with the driving amplitude—that is, $t_n = 2\pi n/\omega$. This return map is shown in Figure 5-30 and shows a simple parabolic

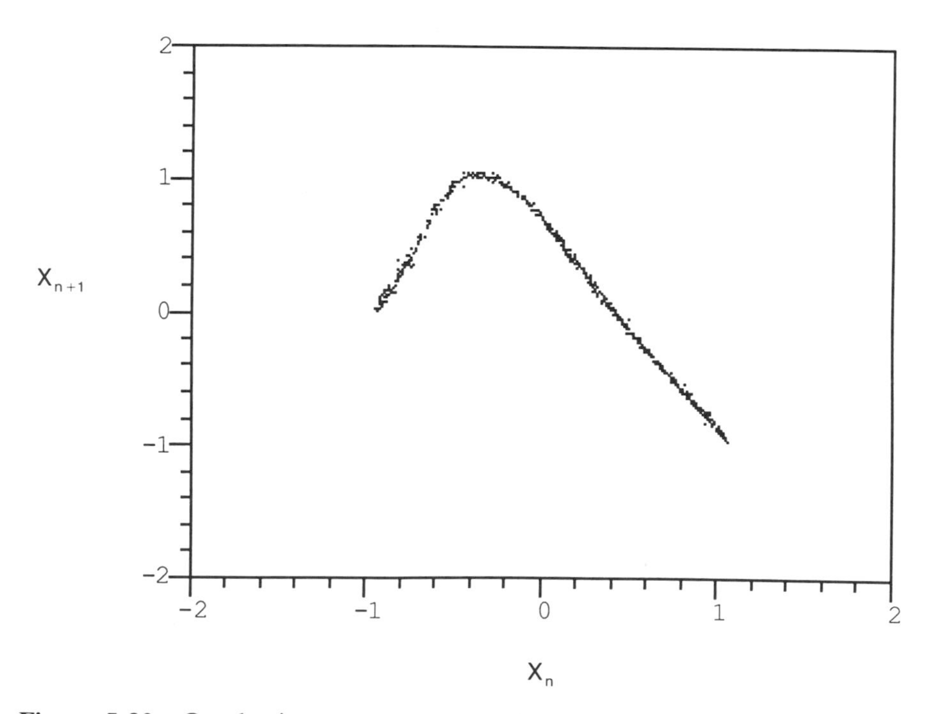

Figure 5-30 Quadratic-type return map of the chaotic motion of a vibrating magnet near the surface of a YBCO superconductor. [From Moon (1988), with permission.]

shape. This map is similar to a very famous equation of chaos known as the *logistic map* [see, e.g., Moon (1992a)]:

$$X_{n+1} = aX_n(1 - X_n) \tag{5-8.4}$$

For $a > 3.57$ the dynamics may become chaotic, and this equation generates a probability density function.

This simple experiment again indicates that although magnetically levitated bodies are governed by deterministic forces, the nonlinear nature of the forces can generate complex and sometimes unpredictable dynamics which are sensitive to initial conditions and changes in other system parameters. Thus, care in the design of such systems should include exploration of the possible nonlinear behavior of levitation devices.

CHAPTER 6

APPLICATIONS OF SUPERCONDUCTING BEARINGS

> It is impossible to imagine the height to which may be carried in a thousand years, the power of man over matter. We may perhaps learn to deprive large masses of their gravity, and give them absolute levity, for the sake of easy transport.
>
> —Benjamin Franklin

6-1 INTRODUCTION

Bearings make relative motion possible. Bearings are critical components of machines of the twentieth century, be they disk drives, gyros, pumps, generators, or jet engines. To most people, bearings are an invisible technology in their contact with modern machines. The performance of bearings, however, often limits the performance of machines. Solid- and fluid-element bearing systems have classic performance limitations such as wear, noise, vibration generation, lubrication needs, and so on. Because of these problems, the technology of noncontact, active magnetic bearings has continued to mature and make progress in many application areas [see, e.g., Schweitzer (1988)]. Already active magnetic bearings now lift 50-ton Mag-Lev vehicles and suspend 1-ton rotors in gas pipeline pumps. In spite of this progress, the basic problems of active bearings remain passive instability, complex sensors and actuators, the need for a power supply and cost. However, the development of new superconducting materials offers a challenge to the role of active magnetic bearings through the development of passive, stable, magnetic bearings. From small cryorefrigeration devices in spacecraft on 15-year missions to large-energy storage flywheels for power utility systems, both large and small applications are possible. However, passive superconducting bearings are at least a decade behind in development experience compared with active magnetic systems.

Early Application: Superconducting Gyro

Proposals for building superconducting levitated gyro elements go back to the 1950s. Simon (1953) proposed using diamagnetic forces to levitate a hollow lead sphere, and he wrote a paper on the forces on superconductors in magnetic fields. Culver and Davis (1957, 1960) of the Rand Corporation (Santa Monica, California) proposed the use of a superconducting gyro for inertial navigation. Both General Electric and Minneapolis Honeywell, around 1960, proposed designs for a cryogenic gyro (Buchhold, 1960). Harding and Tuffias (1960) of the Jet Propulsion Laboratory built and tested a levitated niobium sphere. The 2.5-cm-diameter sphere was spun in a vacuum (10^{-6} mm Hg), and the decay constant was estimated to be 600 days. A later experimental study at Stanford University was conducted by Bourke (1964), who levitated a cylindrically shaped lead-plated body at 4.2 K using superconducting coil (Figure 6-1). The cylinder weighed 24 g and was spun to 700 rpm. This study was prompted by a proposal to test the theory of general relativity by designing a low-decay superconducting gyro. This study also presents a very nice dynamic analysis of the levitated rigid-body dynamics under calculated magnetic stiffnesses. A good agreement between theory and experiment was obtained. Modern studies of high-temperature superconducting levitation have not matched the thoroughness of this 30-year-old study.

By the mid- to late 1960s, good low-temperature superconducting wire (niobium–titanium) was at hand and the technical focus of attention on magnetic levitation shifted to high-speed vehicles [see, e.g., Powell and Danby (1966)]. The subject of levitation of transportation vehicles will be reviewed in Chapter 7. The rest of this chapter will review rotary and linear bearing levitation applications.

6-2 ROTARY MOTION BEARINGS

Almost immediately after the announcement in January 1987 of the discovery of the first liquid nitrogen superconductor, yttrium–barium–copper oxide (YBCO), laboratories all over the world

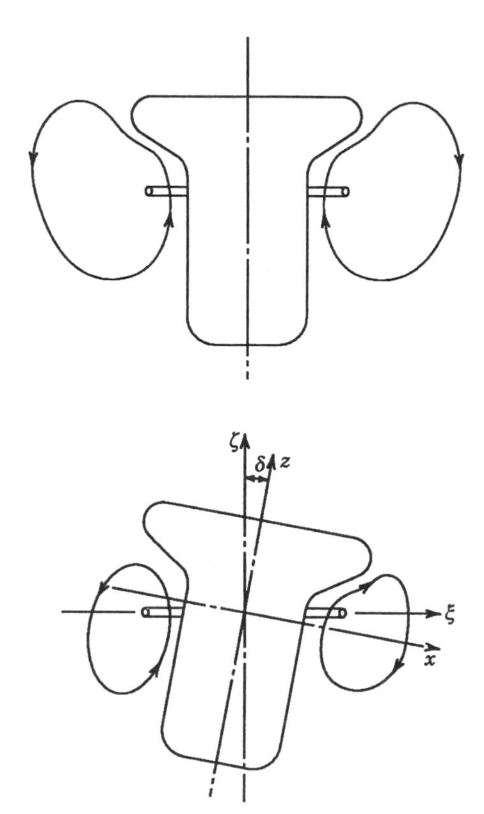

Figure 6-1 Sketch of spinning levitated superconducting body in a static magnetic field. [After Bourke (1964).]

were levitating small permanent magnets. In the early summer of that year at Cornell University, R. Raj, J. D. Wang, and the author built a 5-g levitated rotor with machined YBCO pellets as journal bearings and spun the rotor up to 10,000 rpm (Figure 6-2). Within a year the speed was pushed up to 60,000 rpm, and later up to 120,000 rpm (Moon and Chang, 1990). While we were reporting on these measurements at a meeting on superconductivity, a famous theoretician speculated that application of superconductors to rotary bearings would be impractical because of flux drag. It is now recognized that this conclusion is wrong. Namely, rotation of a perfectly symmetric magnetic field source about its axis of symmetry near a Type II superconducting body will not result in any drag torque unless there are deviations from the field symmetry. To further prove this point, engineers at Allied Signal Corporation have recently reported rotation of a small 14-g rotor with two permanent magnets on YBCO bearings at a speed of 520,000 rpm (Rigney and Trivedi, 1992). Other studies of the application superconductors to motors and rotating machinery include Crapo and Lloyd

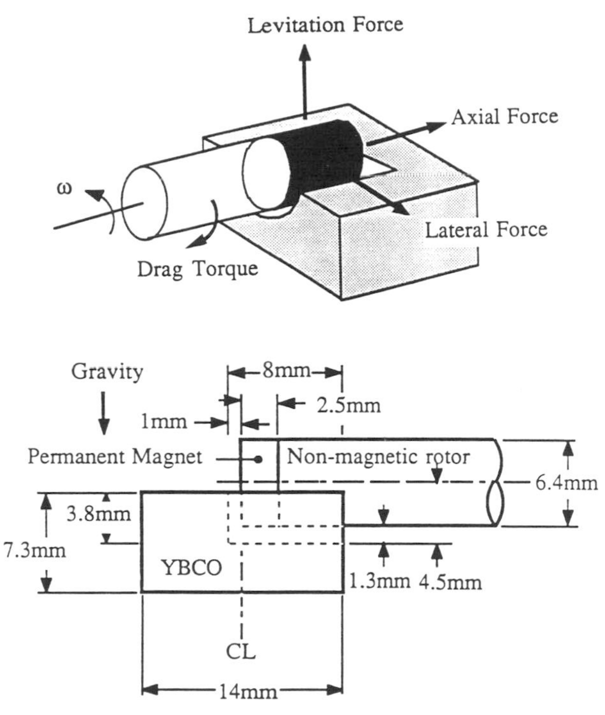

Figure 6-2 Sketch of a YBCO journal bearing. [After Moon and Chang (1990).]

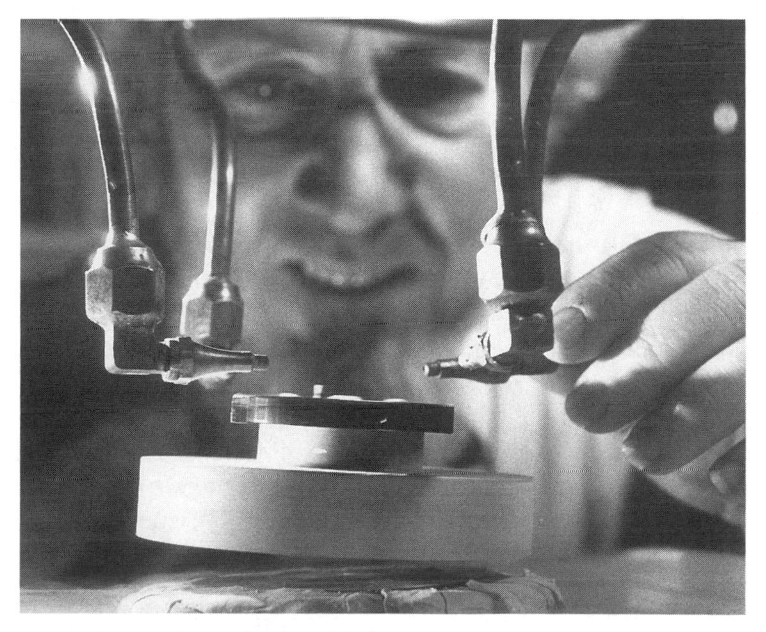

Figure 6-3 Photo of a spinning 0.8 kg rotor supported by YBCO magnetic bearing (Cornell University).

(1990), Delprette et al. (1992), Fukuyama et al. (1991) and Rao and Bupara (1992).

The photo in Figure 6-3 shows a 0.8 kgm rotor that was spun up to 30,000 rpm with a nitrogen gas turbine at Cornell University (Moon et al., 1993). Such size rotors could have application to energy storage flywheels for satellites and possibly electric automobiles.

Cryomachine Applications

The first technical area to express interest in passive superconducting bearings was the aerospace community. Because providing a cryogenic environment for these bearings is a necessity, it was natural to look for applications that already meet this requirement. Aerospace machines that have low-temperature operation include:

- Cryopumps for liquid hydrogen and for liquid helium
- Cryocoolers
- Cryoflow meters

One application that involved the use of a low-temperature superconducting bearing was reported by Rivetti et al. (1987) for a turbine flowmeter operating in liquid helium. The 8-g rotor had conical solid niobium rotor ends to provide combined radial and thrust support. An 800-turn superconducting coil at each end provided the field source (Figure 6-4).

The experiment at Allied Signal mentioned above was for a prototype cryocooler design. It is estimated that the superconducting bearings will reduce power loss by 70% in the reversed Brayton cycle cryocooler.

Another miniature cryocooler design study in 1992 was reported by Creare, Inc. (Hanover, New Hampshire) using magnetic bearings to support a 6.4-mm-long compressor shaft running at 480,000 rpm (Figure 6-5). (Iannello, 1990) These small cryocoolers are targeted for cooling infrared sensors for spacecraft applications. Figure 6-5 shows a 1-10 W turbo expander which would operate at 10 K.

In another design study, Dill et al. (1990) of MIT, Inc. (Albany, New York) discuss possible applications to bearings for rocket engine turbopumps. Another possible device is small high-speed turbopumps for the National Aerospace Plane (NASP) which might require liquid hydrogen as a fuel for the scramjet portion of the engine cycle. Dill et al. (1990) conclude, however, that direct-current, coil wound super-

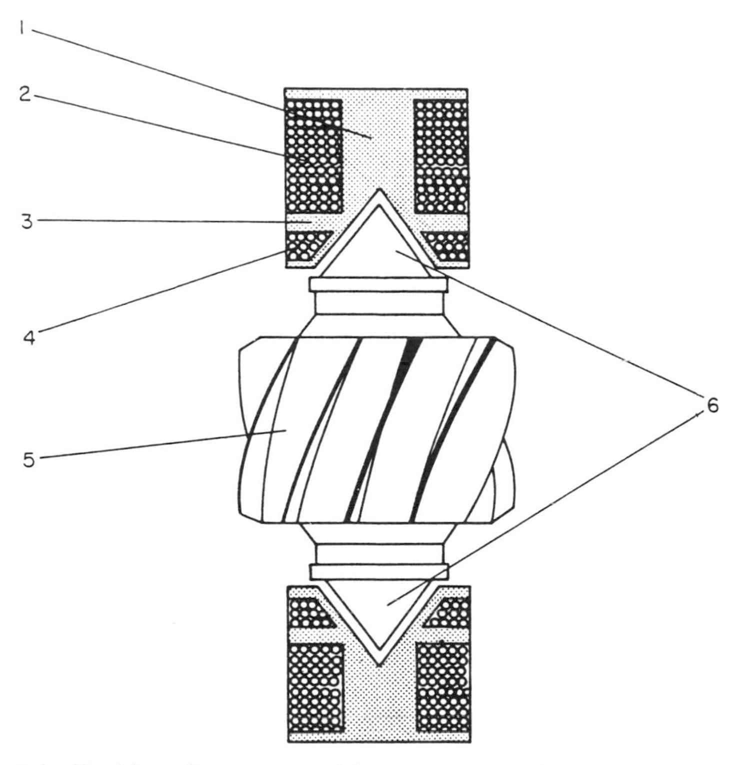

Figure 6-4 Turbine flowmeter with superconducting bearings. 1, Stainless steel fixed support; 2, main coil; 3, clearance between coils; 4, radial coil; 5, aluminum rotor; 6, niobium caps. [From Rivetti et al. (1987), © Butterworth–Heinemann Ltd. with permission.]

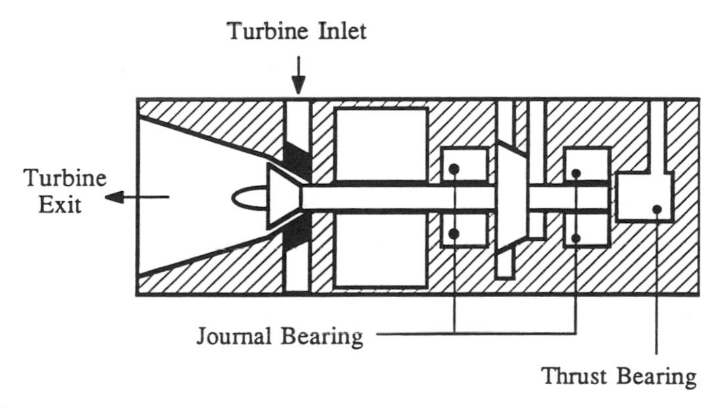

Figure 6-5 Design for a miniature cryocooler using high-T_c superconducting bearings. [Courtesy of Creare, Inc., Hanover, New Hampshire.]

conducting bearings might be required in order to increase the field and the magnetic pressure above 150 N/cm^2 which might be required of the bearings for volume limitation.

Other aerospace applications might include:

- Cryoalternators
- Laser blower motors
- Vibration isolation systems
- Point gimbals on antennae

Energy and Momentum Storage

High-speed rotors provide a natural storage system for energy and angular momentum. One of the limits to the use of rotating bodies as energy storage has been losses in conventional bearings. Potential applications include:

- Angular momentum reaction wheels for spacecraft attitude control
- Flywheels for energy storage in electric vehicles
- Large flywheels for power utility energy storage
- Energy storage for pulsed power applications in laser devices
- Backup power storage flywheels for critical computer systems

At the time of this writing (late 1993), the levitated loads of rotary devices using bulk high-temperature superconductors was increasing rapidly from 1-kg to 20-kg rotors. It is expected that a demonstration of a 100-kg levitated rotor or larger will be made by 1994, probably in Japan.

A 0.8-kg rotor rotating at 30,000 rpm was demonstrated at Cornell University using a ring magnet and five YBCO pellets (Figures 6-3, 6-6). The rotor was spun using an air turbine. Takahata et al. (1992) of Koyo Seiko Bearings in Japan have demonstrated a 1-kg rotor at 5000 rpm using eight pellets of YBCO processed by the MPMG method (see Chapter 3) shown in Figure 6-7. They reported a static axial stiffness of 40 N/mm and a lateral or radial stiffness 8 N/mm. They report an 80-min decay time and an initial speed of 100 rpm. They used a 90-mm-outer-diameter and 68-mm-inner-diameter ring magnet manufactured by Sumitomo Special Metals with a field of 0.4 T.

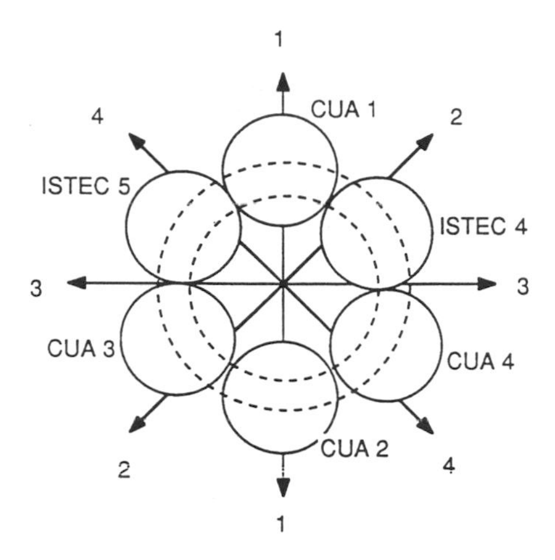

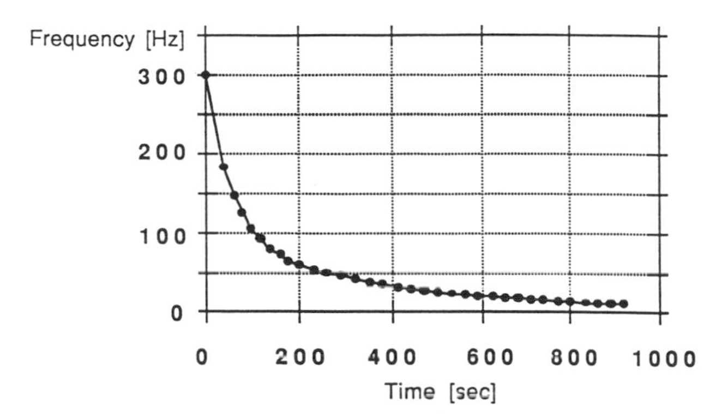

Figure 6-6 Thrust bearing configuration with a ring magnet on the rotor and a discrete element YBCO superconducting stator bearing. [From Moon et al. (1993), with permission.]

In a joint project between ISTEC Tokyo and NSK Bearings in Japan, a 2.4-kg vertically mounted rotor was levitated and spun up to 30,000 rpm (Figure 6-8). This system uses both YBCO rings and rare earth permanent magnet rings (Takaichi et al., 1992). The unique feature about this design is that it utilizes the attractive suspension force between a YBCO ring which is field-cooled in the field of the permanent magnet. The authors noted, however, that they encountered lateral vibration in the rotor near 30,000 rpm. The rotor was driven, however, by a high-speed motor which had a ferromagnetic stator which may contribute to the whirling.

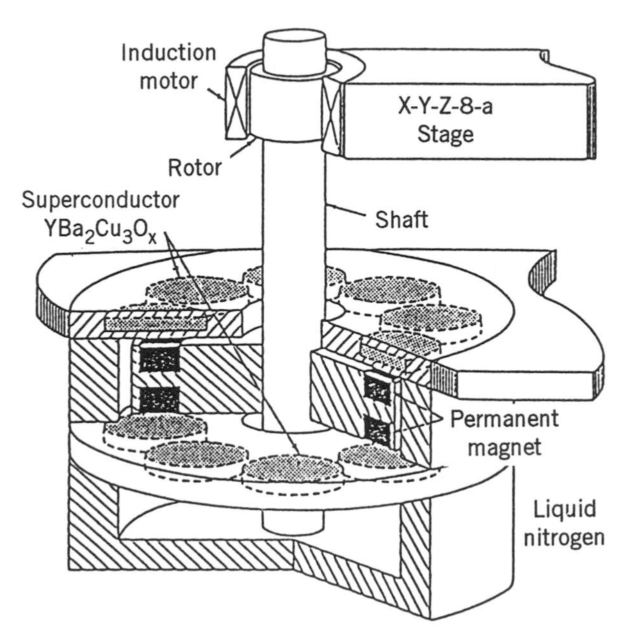

Figure 6-7 Discrete element high-T_c superconducting bearing. [From Takahata et al. (1991), with permission; Koyo Seiko Bearings, Japan.]

In the United States the MTI Corporation has reported suspending a 7-kg rotor which was spun up to 9000 rpm. This vertical rotor is levitated on a single YBCO thrust bearing and supported laterally with permanent magnet bearings. This illustrates a hybrid levitation scheme.

Another hybrid levitation scheme is reported by a group at the University of Houston (McMichael et al., 1992) (Figure 6-9). In this device they used permanent magnets to provide axial levitation, and they stabilized the rotor with a bulk high-temperature YBCO bearing. They reported an axial stiffness of 4.26 N/mm which is due, in part, to the permanent magnets.

The study of the dynamics of high-speed rotating machinery on conventional high-stiffness mechanical bearings is a mature area as is the dynamics of gyroscope rotors with zero stiffness. The dynamics of rotating bodies on relatively low stiffness, passive magnetic bearings has not received much study, especially under the nonlinear, hysteretic magnetic forces encountered in the new bulk superconductors. It is not thought that high-speed rotary dynamics on these new bearings will be a technology constraint, except for the centrifugal acceleration induced stresses. However, further study to rule out

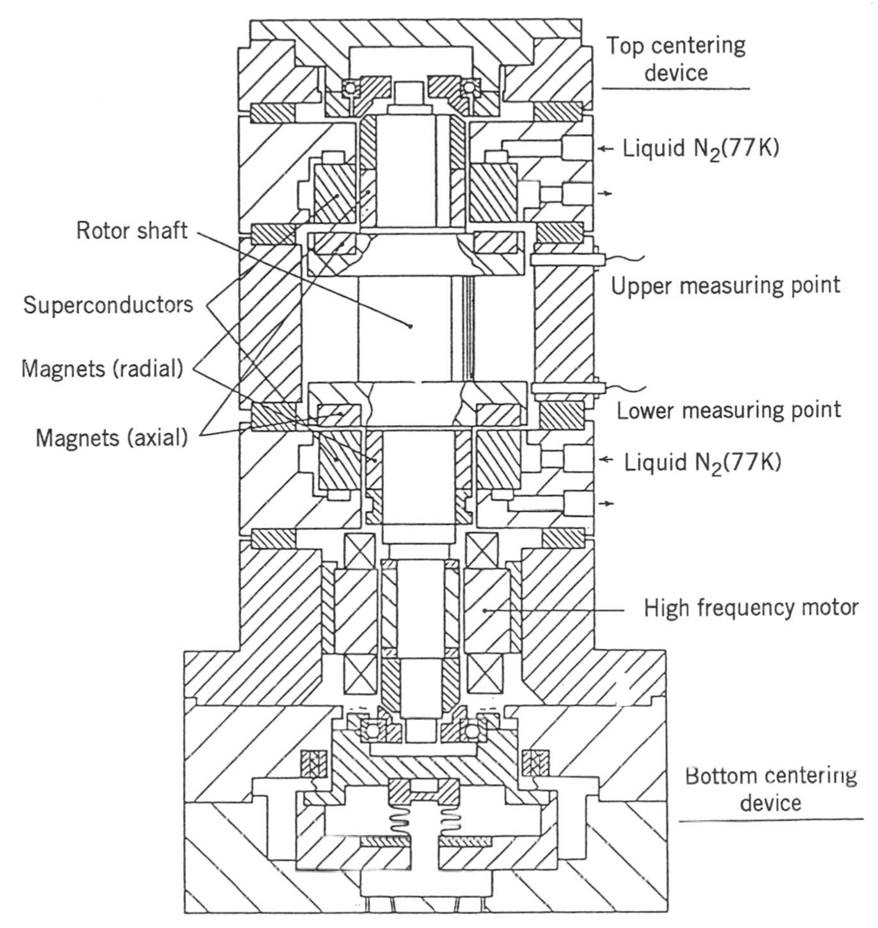

Figure 6-8 Cross section of a design for a thrust and radial bearing using YBCO and a 2.4-kg rotor at 30,000 rpm. [From Takaichi et al. (1992), with permission; ISTEC, NSK Bearings, Japan.]

potential problems has yet to be done at this time (see also Chapter 5).

Sat Con Technology of Cambridge, MA has reccntly proposed using a superconducting coil along with cryo-cooled aluminum control coils for a control mument gyro for large payload space applications (Downer et al., 1990).

Finally superconducting bearings have been proposed for a lunar telescope application by Chen et al. (1992) of the NASA Goddard Space Center.

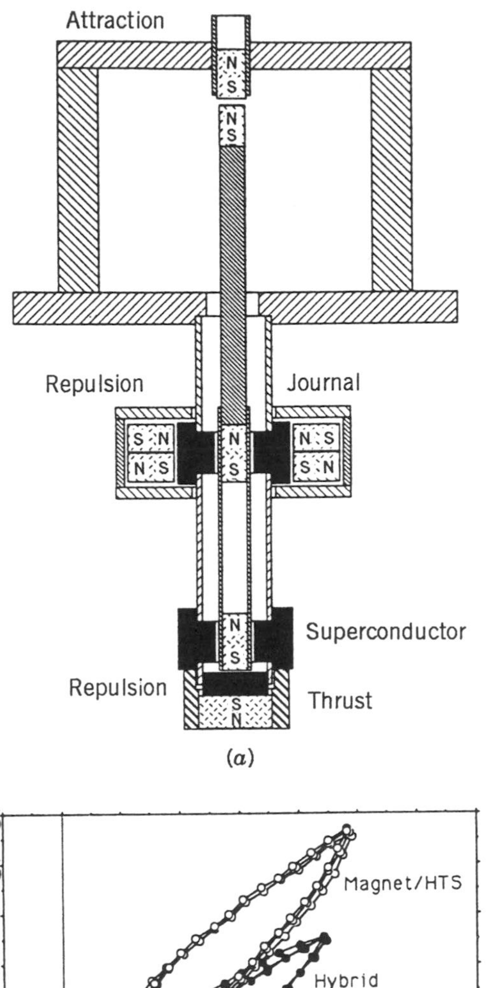

(*a*)

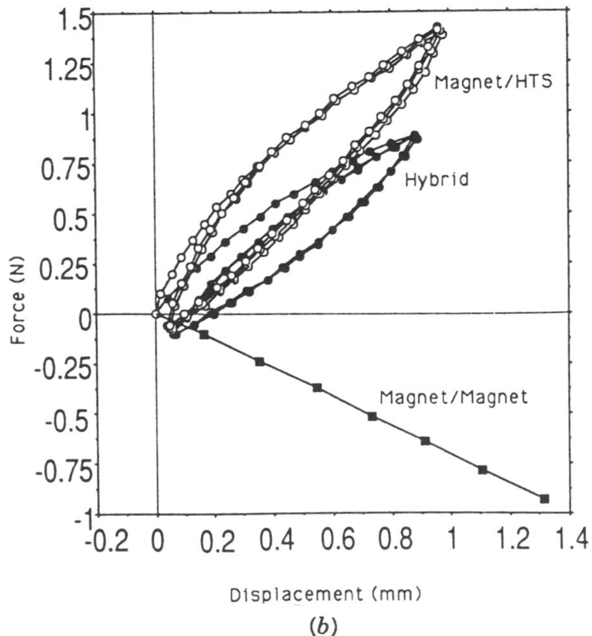

(*b*)

Figure 6-9 (*a*) Hybrid magnetic bearing using high-T_c superconductor and permanent magnets. (*b*) Force–displacement properties of the hybrid bearing showing the decrease in magnetic stiffness in the hybrid scheme. HTS, high-temperature superconductor. [From McMichael et al. (1992), with permission.]

Figure 6-10 Linear magnetic bearing using high-T_c superconductor. [From Azukizawa et al. (1992), with permission; Toshiba Ltd., Japan.]

6-3 LINEAR MOTION BEARINGS

Of course, Mag-Lev vehicles discussed in Chapter 7 are the ultimate levitated linear bearing. Here we mention smaller-scale devices. Of great interest is horizontal motion devices for moving silicon wafers for chip manufacturing. One prototype shown in Figure 6-10 is reported by Toshiba, Ltd. in Japan (Azukizawa, et al., 1992). They designed and built a levitated linear carrier (1.3 kg) with YBCO superconductors on the carrier platform, permanent magnets in the guideway, and an electric linear motor primary underneath to move the platform.

Another example of applications to the electronics manufacturing is the design and testing of a silicon wafer carrier by a group at SEMATECH in the United States (Wolfshtein et al., 1989). In this design, permanent magnets are on the carrier and the YBCO pellets are placed on a robotic horizontal positioning arm (Figure 6-11). Thus, the device serves as a vibration isolation device.

Another example of a linear motion actuator is a device designed for a micromachine actuator (Kim et al., 1989). In this example, parallel strips of thin-film superconductor (1 μm thick) are placed on the stator, and strips of permanent magnets are placed on the moving part. The levitated plate is moved by driving some of the thin-film

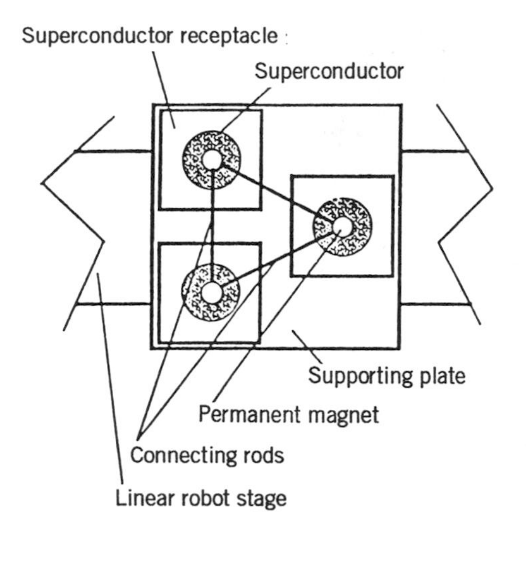

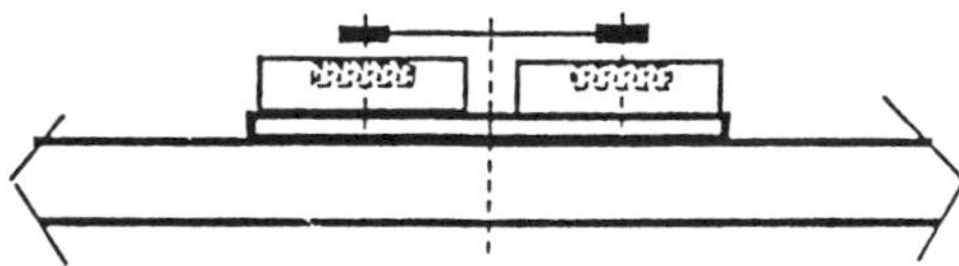

Figure 6-11 Design for a levitated silicon wafer carrier. [From Wolfshtein et al. (1989), with permission; SEMATECH.]

strips normal using transport currents, thus creating an imbalanced lateral force.

A large scale application of high temperature superconductors to railguns or linear electromagnetic launchers has been proposed by Weldon (1992) of the University of Texas. In the 1970s, proposals were made by O'Neill and Kolm (1978) to build moon-based stations with a superconducting linear mass-driver launching system.

6-4 VIBRATION DAMPING AND ISOLATION

Experiments by the author and colleagues have shown damping and drag force hysteresis of magnetic levitation forces in high-T_c materials (Moon et al., 1989; Moon, 1990b), and other laboratories have reported similar measurements. This raises the possibility of using magnetic flux drag and damping as a vibration absorber or vibration isolation device as shown in Figure 6-12. This might have use in satellite applications where traditional room-temperature damping

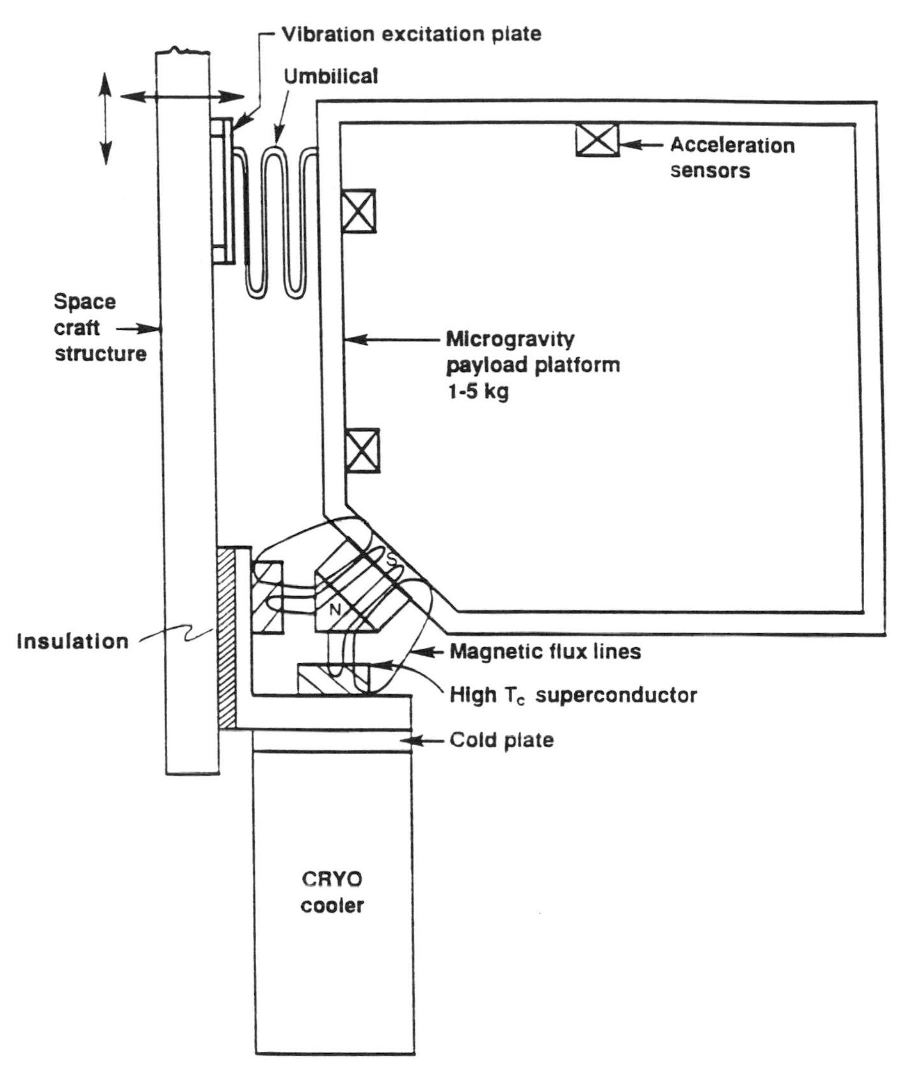

Figure 6-12 Design for a superconducting magnetic damper for microgravity space application.

material cannot be used, for example, to damp out vibrations caused by a cryopump or cryocooler. Another possibility is in the isolation of microgravity experiments from spacecraft acceleration due to, say, crew motion or thruster pulses or other machine excitation on the spacecraft.

One recent study of using YBCO as a vibration absorber has been reported by Lamb et al. (1992) at the University of Houston. They reported significantly better magnetic damping with flux drag in YBCO than by using, say, eddy current damping in copper. In the Section 6-3 we reported applications to the manufacture of silicon chip processing in which a levitated carrier serves as a vibration absorber (Wolfshtein et al., 1989). Another possibility may be in vibration isolation in vacuum processing of certain materials.

Vibration damping using flux drag in superconductors is an area which should see further development in the next few years.

CHAPTER 7

MAGNETIC LEVITATION TRANSPORTATION

Trains would run in a vacuum and be held in suspension by the repulsion of opposing magnets on the cars and tube respectively.

—Robert Goddard, from "Travelling in 1950," *Scientific American* (1909).

7-1 INTRODUCTION

In the late 1960s there began a search for a dramatic improvement in the speed of land transportation. Three different technical paths were taken: improved steel wheel on rail, air-cushion suspension, and magnetic levitation (presently referred to as Mag-Lev). This latter path almost immediately branched into the feedback-controlled electromagnetic levitation (now called EML) and superconductor-based electrodynamic levitation (now called EDL). One of the technical developments that inspired the latter was the availability of low-temperature (< 10 K) superconducting wire. Another technology that enhanced ferromagnetic-rail-based levitation or EML was the transistor and solid-state power electronics and later the integrated circuit. These technologies gave designers the capability to create active EML systems with compact sensor, actuator, control, and power circuits.

By the early 1970s, however, air-cushion technology development was halted even though promising programs were underway in the United States, Great Britain, and France. Noise, power consumption, and a weight penalty of on-board air-moving machines were the apparent factors that contributed to its demise. Although U.S. funding of Mag-Lev virtually ceased in the mid- to late 1970s, Germany and Japan continued research and development on both the EDL and EML Mag-Lev systems. Steel-rail technologies also received substantial R&D in France, Great Britain, Germany, Japan, and Sweden, perhaps spurred on by potential competition from a future Mag-Lev technology. There was a real incentive to develop high-speed wheel-on-rail systems (400–500 km/hr) because existing rights of way and other supporting capital investments could be used. In the case of Mag-Lev, however, a completely new investment would have to be made in rights of way, guideways, power conditioning, stations, and so on.

As mentioned in Chapter 1, as of the early 1990s, after 20 years of development, Mag-Lev seems poised to challenge its wheel–rail competitor. However, in 1970 the promise of 500 km/hr speeds for Mag-Lev compared to existing rail at 200 km/hr (e.g., the Japanese Shinkasen) represented an obvious improvement, whereas in 1993 the speed benefit ratio is now only 500/300.

High-Speed Wheel – Rail Systems

Both the automobile and airplane, which revolutionized travel in the twentieth century, have begun to experience the limits of technology, environment, and terminus and line capacity in many parts of the industrialized world. To meet the growth in transportation demands, many governments have invested funds into reinventing the transportation technology of the nineteenth century—the steel-wheeled railroad system. The steel-wheeled train of the early twentieth century had several technical limits to higher speeds which, 30 years ago, were thought to be limited to around 250 km/hr (155 miles/hr). Principal amongst these limits were limited traction, unstable wheel-bogie dynamics, wear and vibration-induced fatigue failure (high maintenance), excessive noise, and ride comfort in the passenger compartment. In the United States, efforts to improve passenger rail technology by the rail companies were undercut by the introduction of the Boeing 707 and the growth of air travel which significantly decreased inter-city rail usage as well as the new interstate highway system. However, by the mid-1960s, other industrialized countries began to make dramatic improvements in steel-wheel rail transportation. A brief review is given here to serve as a point of comparison with potential Mag-Lev systems in the future. The only quasi-high-speed rail service in the United States is the Amtrak Metroliner service between New York and Washington which, in a few sections, can reach speeds of 201 km/hr (125 miles/hr).

Japan: Shinkansen (Bullet Train) Japan introduced its famous Bullet Train in 1964 between Tokyo and Osaka. These trains have dedicated guideways, continuous welded rail to minimize vibration, and daily maintenance at night when the system does not run. Hitting speeds of 200 km/hr (124 miles/hr), they significantly cut travel time. Subsequently, additional lines were constructed. To date, the Shinkansen has carried 2.5 billion passengers without injury. The system is presently being upgraded with new vehicle designs to increase the speed to 250–350 km/hr. One problem that continues to plague high-speed rail is excessive trackside noise which, in a densely populated country, can become a political as well as a technical problem.

France: Train à Grand Vitesse (TGV) French engineers have truly revolutionized train design with their fleet of high-speed trains. Since the introduction of these trains in 1981, ridership has increased steadily, thereby eliminating the need for added air capacity between the cities it serves. A new line called the Atlantique, which runs east and west, has vehicles that can run at 300 km/hr (186 miles/hr). The system uses continuous welded rail and reinforced concrete crossties. It has two power cars, one at each end, and ten trailers. It can climb a 3.5% grade. A TGV franchise has recently been awarded by the state of Texas for service between Dallas, Houston, and the state capitol, Austin, and the search for financing is now underway.

Germany: Intercontinental Express (ICE) Not to be outdone by the French, German engineers have recently introduced a high-speed train with speeds of up to 250–300 km/hr between Hamburg, Frankfurt, and Munich. In the late 1980s, ICE was a top candidate for a line in Florida from Tampa to Miami, but lack of financing canceled the project.

Tilt Trains Rail service in populated areas or through valley topography involves many curves which limit train speeds. Curves create lateral accelerations on the passenger which can be ameliorated by either banking the track or tilting the car body relative to the track or both. New rail designs which create a tilt effect during circular motion include the Swedish X-2000 train, the Spanish Talgo Pendular, and the Italian Pendorino. The X-2000 has been tested in the U.S. Amtrak route between New York and Boston and is expected to appreciably cut travel times. These trains have speeds of approximately 200 km/hr (125 miles/hr) and are not considered to be high-speed trains.

Both French TGV and the German ICE are pushing the 300-km/hr operating speed, and in a special test run a TGV vehicle exceeded 500 km/hr. Thus, at this time it remains to be seen whether Mag-Lev will eventually challenge wheeled systems in the marketplace in the twenty-first century or whether the technology will be seen by historians as the technical foil that forced wheel–rail engineers to improve wheel–rail train speed beyond the 150-km/hr limits of the early twentieth century. This writer still sees significant advantages to the Mag-Lev transportation system, some of which will be highlighted.

In this chapter we will review the major levitation technologies as well as their particular realization in either proposed designs or existing prototypes. We will also discuss the generic technical problems that face Mag-Lev transportation, especially those based on superconducting technology.

7-2 PRINCIPAL LEVITATION SCHEMES

EML — Ferromagnetic Rail

The basic geometry of this system is shown in Figure 7-1. The electromagnetic levitation is based on trapping magnetic flux lines to produce magnetic tension stresses between the ferromagnetic guideway and the vehicle magnets. As discussed in Chapters 1, 2, and 5,

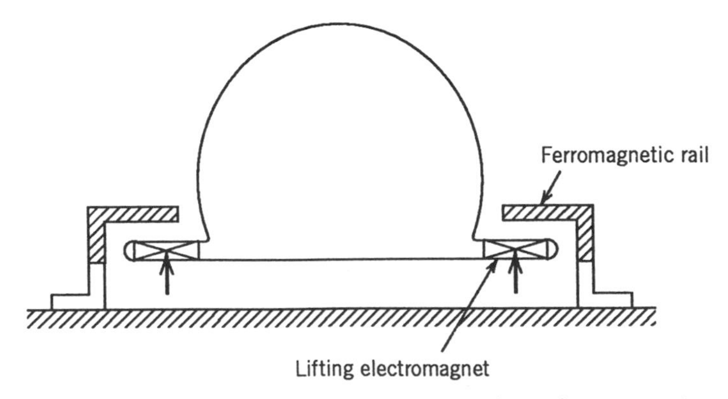

Figure 7-1 Sketch of electromagnetic levitation (EML) concept for a moving vehicle.

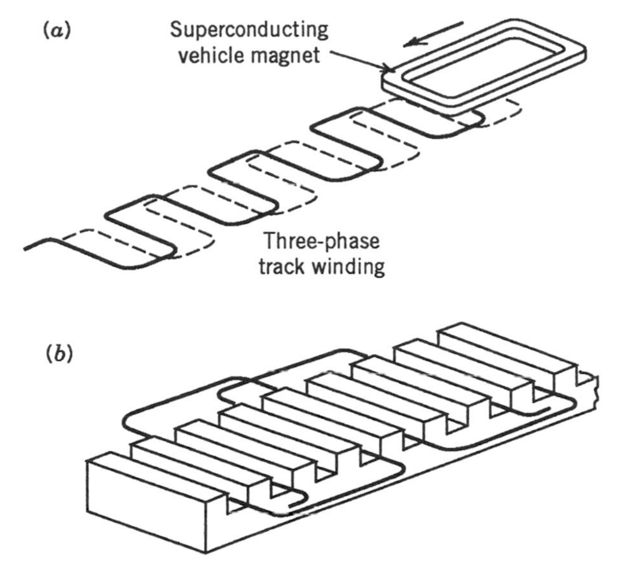

Figure 7-2 (*a*) Linear synchronous motor for propulsion of Mag-Lev vehicles. (*b*) Linear motor concept for German Transrapid EML design (TR07).

these systems are usually unstable without feedback. Existing prototypes use normal electromagnet coils and small gaps ($\sim$ 1 cm). However, new designs have been proposed for which the magnet coil is superconducting and the gap is larger ($\sim$ 5–10 cm). Flux-trapping magnet systems have an advantage over flux-repelling systems because the stray flux is low at some distance from the ferromagnetic guideways. In repulsive flux systems, large stray flux in the passenger section of the vehicles presents serious design problems in attempts to lower or shield these stray fields from the passengers. Propulsion in EML systems uses a linear motor scheme as illustrated in Figure 7-2.

EDL — Continuous-Sheet Track

This is a simple repulsive, eddy current levitation system as shown in Figure 7-3. The guideway can be shaped to provide both vertical and lateral guidance forces. The magnets in this system are usually wound from superconducting wire operated at less than 10 K (See Figure 1-19). Both lift and drag forces depend on speed as in Figure 7-4, and the magnetic lift to drag ratio (L/D) increases with speed. At high speeds, the flux is excluded from the conductor track, thereby creating magnetic pressure forces. These systems can usually be designed to be inherently stable at all speeds.

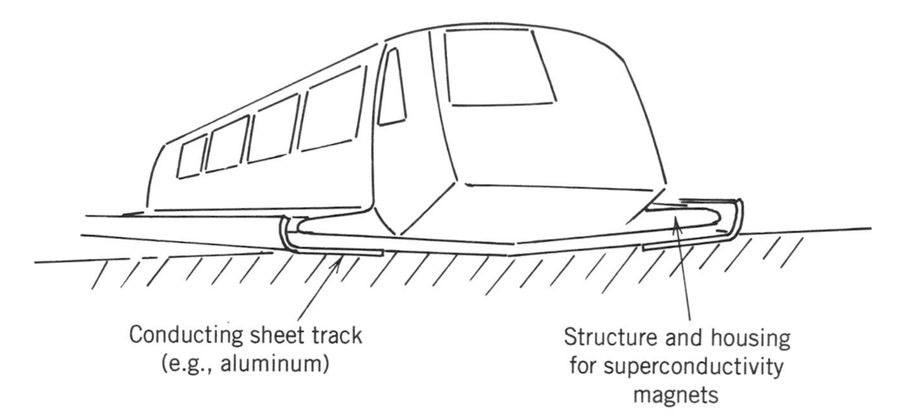

Figure 7-3 Sketch of electrodynamic levitation (EDL) concept for a moving vehicle using superconducting magnets as a field source.

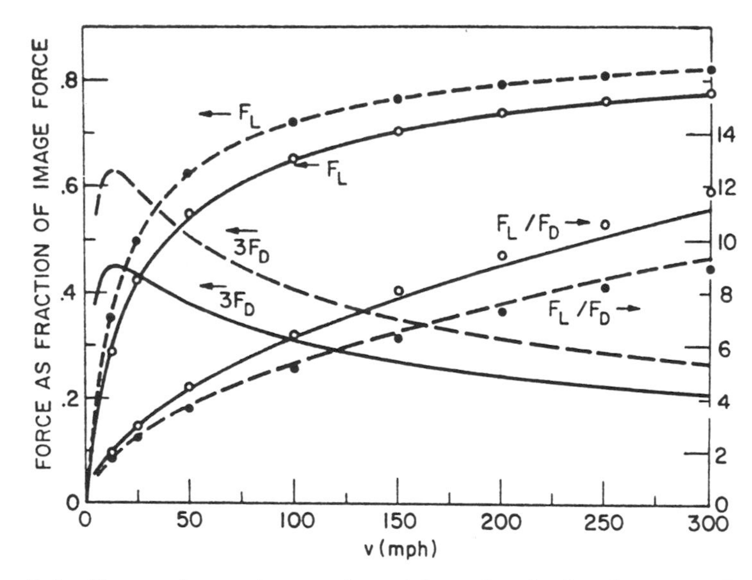

Figure 7-4 Comparison of experimental and calculated magnetic lift to drag ratio for a rectangular coil moving over a thick aluminum guideway. ○: Experimental data for a 5 cm × 10 cm rectangular coil, long side in the direction of motion. ●: Short side in the direction of motion. (Ford Motor Corporation, Borcherts et al., 1973 © 1973 IEEE.)

One of the advantages of this continuous track concept is that the forces on the magnet wire element are continuous. Discrete track systems (see below) will generate oscillating forces. The total force on one magnet or on the vehicle may often filter out the dynamic force components. However, individual wire or magnet elements may experience large dynamic force components which may present fatigue failure problems.

Techniques for calculating forces in continuous-sheet track include image magnet methods, Fourier analysis, and numerical codes [see, e.g., Reitz and Davis (1972)].

EDL — Ladder Track

This concept is similar to the continuous track concept, but uses a periodically perforated sheet or continuous wire element track as shown in Figure 7-5. This concept is believed to result in higher lift to drag ratios as well as using less conductor. The disadvantages of this system include (a) higher manufacturing costs of the track and (b) fatigue that would damage dynamic force components on the levitation coils.

EDL — Discrete Coil Track

This system has been employed in several JNR prototype vehicle designs (Kyotani, 1975, 1986, 1988), as well as in small model tests. The discrete track elements can be solid conductors or wire or coil wound, shorted coils as in the JNR design (Figures 1-19, 7-6*b*). Forces are generated when the moving flux in the superconducting train coils moves past each of the coils. The current generated in each guideway coil acts to repel the vehicle coil flux, thus creating a repulsive magnetic force. As in the ladder track, the vehicle magnet coil elements can experience dynamic force components in discrete track systems. This scheme was first proposed by Powell and Danby (1967).

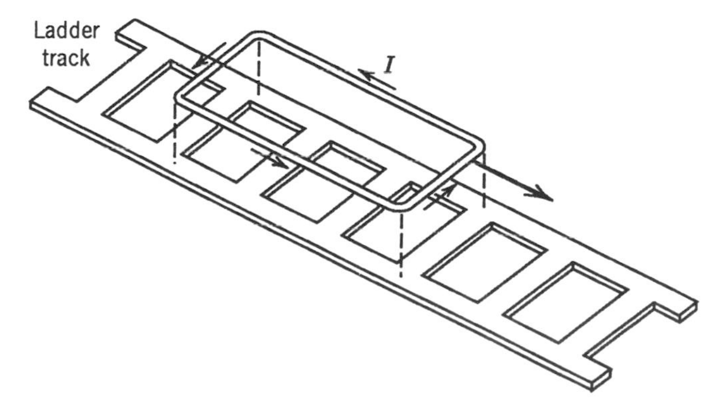

Figure 7-5 Ladder truck concept for EDL Mag-Lev.

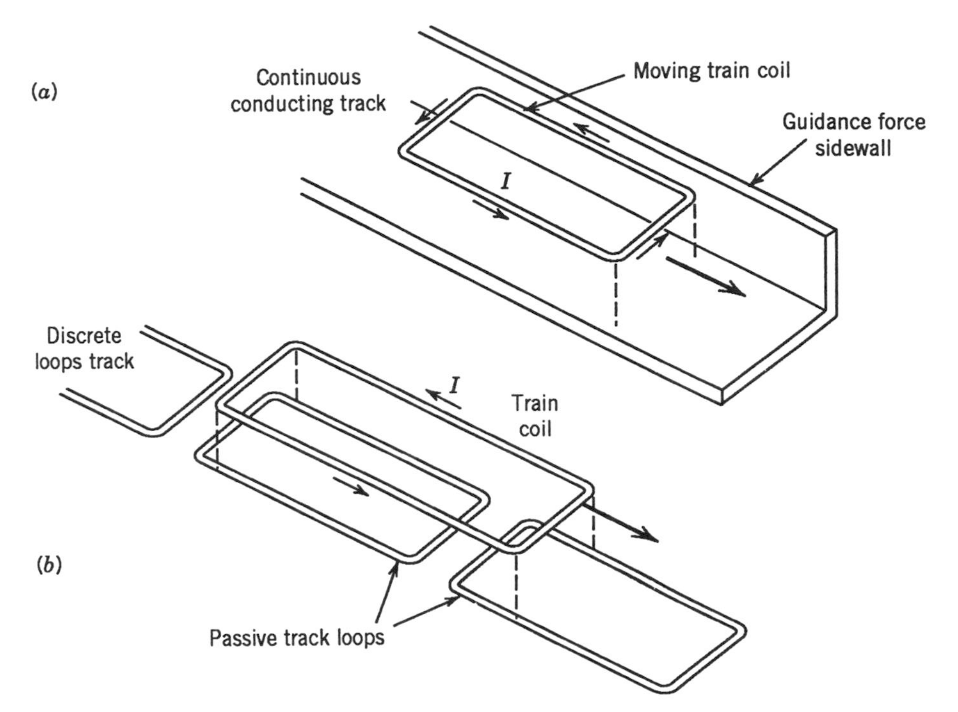

Figure 7-6 (*a*) Continuous conductor track with guidance sidewall. (*b*) Discrete track coil concept for EDL Mag-Lev.

EDL — Null-Flux Guideway

Another EDL idea presented by Powell and Danby (1968) was the null-flux track shown in Figures 7-7 and 7-8. The vehicle magnet moves past pairs of guideway coils which are connected so that the

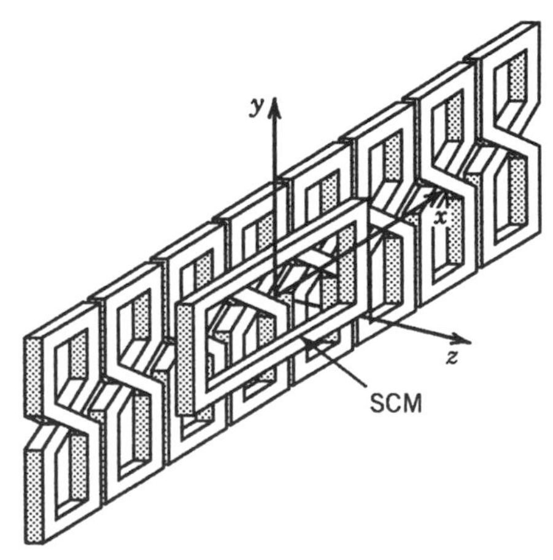

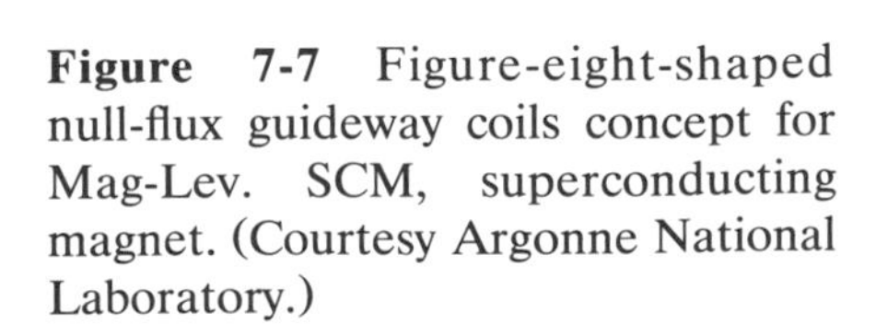

Figure 7-7 Figure-eight-shaped null-flux guideway coils concept for Mag-Lev. SCM, superconducting magnet. (Courtesy Argonne National Laboratory.)

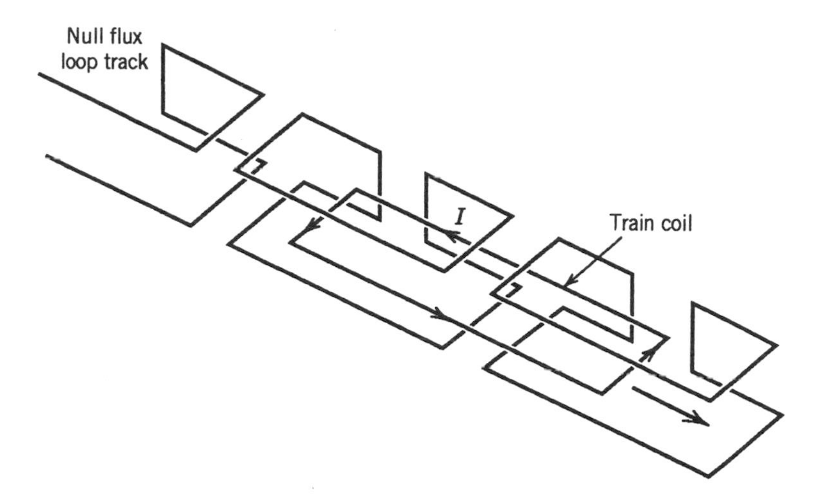

Figure 7-8 Null-flux coil guideway concept for EDL Mag-Lev (Powell and Danby, 1968).

induced magnetic flux produced at the vehicle magnet cancels or is zero. Thus, in order to develop lift, the vehicle coil must ride unsymmetric with respect to the pairs of guideway coils.

This system has advantages of lower guideway currents and lower magnetic drag losses. Thus, one can achieve very high lift to drag ratios. The system also generates a higher magnetic stiffness as compared to the pendulum-type suspension of the continuous or discrete track concept (see Chapter 6). The disadvantages are the more complex coil geometry and added costs to manufacture the guideway coils. (See also Danby et al., 1972.)

7-3 MAG-LEV DESIGN CONCEPTS

Earlier Mag-Lev Research and Prototypes

A good review of pre-1980 research and development in magnetically levitated transportation can be found in the monograph by Rhodes and Mulhall (1981). Not referenced in that book, however, is an article by Robert Goddard, the American rocket scientist. In a 1909 issue of *Scientific American* he describes a gravity gradient tunnel transport system between New York and Boston with vehicles suspended by magnetic forces. His ideas, however, lacked specific detail.

In 1912 a French engineer, Emile Bachelet, proposed an eddy-current repulsive levitation scheme and actually built a small model. In one version, electromagnets in the guideway created eddy currents (or Foucault currents as they are known in France) in an aluminum plate under the vehicle. In another concept, electromagnets were carried on the vehicle as outboard winglets and generated eddy currents in two aluminum plates along the guideway. Bachelet was unable to secure financial backing for implementation of his ideas, however, and estimates of the power needed for his system were very high. Today, however, his electromagnets are replaced with superconducting magnets which were not available in his time. Bachelet later emigrated to the United States and became a citizen.

Work on other levitated devices continued into the 1950s and 1960s, including a magnetically levitated superconducting gyro (see Harding and Tuffias 1960). A review of magnetic and electric suspension may be found in the monograph by Geary (1964), and other nontransportation levitated devices are described in Chapter 6 of this book. In his book, Geary describes a patent by a British engineer, G. R. Polgreen, who proposed a levitated transportation system using barium ferrite permanent magnets in the guideway and under the vehicle. The unstable lateral motions were stabilized by guidance wheels. A small model was built and tested, and was propelled by a linear electric motor (see also Polgreen, 1966).

The modern age of Mag-Lev transportation, however, came into being with the development of high-current-density superconducting wire in the early 1960s (Nb–Ti), and with the rapid availability of transistors, integrated circuits, and solid-state power electronics. One of the first proposals for superconducting levitated vehicles was made by two physicists, James Powell and Gordon Danby, of Brookhaven National Laboratories (Long Island, New York) in 1966. At around the same time, a group at Sandia Laboratories proposed using a levitated rocket sled in an evacuated tube using superconducting magnets to achieve speeds of 5000 m/sec (see Guderjahn et al., 1969). This, in essence, was a variation of the Goddard proposal in 1909. Within a few years there was an explosion of activity around the world in Mag-Lev research and development.

Before describing full-scale Mag-Lev prototype development, it is appropriate to review some of the early basic research using small model experiments.

Ford Motor Corporation Research engineers at Ford published some of the basic calculations of magnetic forces for repulsive levitation [see, e.g., Reitz and Davis (1972) and Borcherts (1975)]. Experiments were carried out using a small superconducting coil placed over a 0.60-m-diameter rotating solid aluminum wheel. The wheel circumference served as a moving guideway and was able to reach speeds of 134 m/sec (300 miles/hr). Good agreement was obtained between theoretical and experimental magnetic forces. In 1971 Ford was awarded a large contract to build a high-speed superconducting magnet test sled. However, the U.S. government canceled the contract before work was finished (see *Business Week*, January 13, 1975; also Reitz and Borcherts, 1975). Before ending its Mag-Lev research, Ford presented an outline for a full-scale Mag-Lev design vehicle (see Wilke, 1972).

Stanford Research Institute (SRI) In the early 1970s a group at SRI (Menlo Park, California) built and tested what may have been the first fully levitated superconducting vehicle (see Coffey, 1974). The 4.25-m × 1-m vehicle was supported by four small superconducting coils (0.32 in. long × 0.27 in. wide) over an aluminum sheet guideway. The 480-kg vehicle was towed to speeds of 8–10 m/sec along a 100-m guideway. A particular feature of these tests was a detailed study of the vehicle dynamics under magnetic forces (see Coffey et al., 1974).

Massachusetts Institute of Technology (MIT) In the early 1970s a group at MIT proposed an integrated levitation and propulsion concept called the *magneplane* (see Kolm and Thornton, 1972). This group, under Henry Kolm and Richard Thornton, obtained National Science Foundation funds to build a linear test track in Cambridge, Massachusetts. Their concept included a circular-shaped aluminum sheet guideway which permitted the vehicle to roll naturally as it went into turns. The track was also designed as an active linear synchronous motor and used the superconducting magnets on the 0.6-m-long vehicle to generate the propulsion thrust forces. Coordinated levitation and propulsion tests were accomplished with this unique facility. However, films of the lateral dynamics of the vehicle showed severe yaw–lateral instabilities (see Moon, 1987). Design for a full-scale system were proposed at the time. About 20 years later, a new magneplane concept has been proposed as part of the U.S. Mag-Lev Initiative (see below) and Dr. Kolm has started a company called Magneplane International to develop a full-scale system.

Canadian Mag-Lev Group Parallel with the U.S. effort in the 1970s, researchers at the University of Toronto, McGill University, and Queen's University collaborated in a Mag-Lev development program sponsored by the Canadian Institute of Guided Ground Transport [see, e.g., Atherton et al. (1980), Atherton and Eastham (1974), Eastham and Atherton (1975), Slemon (1975), and Hayes (1979)]. At Queen's University a large horizontal rotating steel guideway, 7.6 m in diameter, was built to test various superconducting levitation and propulsion concepts. Also, the group presented an overall design concept for a 100-passenger, 480-km/hr vehicle sponsored by the National Research Council. The guideway was a flat split-sheet conductor supported on a reinforced concrete box-beam structure (see Figure 7-9). The flat guideway was chosen to allow snow to be easily blown off. The Canadian Group has remained active as consultants to several present day Mag-Lev project studies.

United Kingdom — University of Warwick Like the United States and France, high-speed ground transportation in the United Kingdom in the late 1960s was focused on air-cushion or hovercraft-type technology. Development of a full-scale vehicle was carried out by Tracked Hovercraft, Ltd., but this effort ended in 1973 [see Laithewaite, (1977a, b)]. However, research on Mag-Lev was carried out at the University of Warwick under Professor R. G. Rhodes (see Rhodes and Mulhall, 1981). This facility included both a 3-m-diameter rotating wheel and a 550-m-long linear test guideway. The wheel achieved speeds of 45 m/sec, whereas towed model speeds on the linear track reached 50 m/sec. The track consisted of two parallel flat aluminum strips (Figure 7-10). The small model vehicle contained two 0.4-m × 0.4-m superconducting coils. Work ended on this project in the early 1980s.

Princeton and Cornell Universities During the 1970s, the author conducted experimental research on the dynamics of magnetically levitated vehicles using a 1.2-m-diameter rotating wheel facility. A "V"-shaped aluminum guideway was wrapped around the wheel (Figure 7-11) in one set of tests, and a discrete aluminum "L"-shaped guideway was built in another set of tests. A flat-sheet guideway was also used. Rim speeds of up to 40 m/sec were achieved. The levitated models were 1–2 kg in weight and were supported using large ferrite and rare earth permanent magnets attached to the models. These experiments revealed many types of static and dynamic instabilities [see Moon (1974, 1977, 1984) and Chu and Moon (1983)] which depended on both the speed and geometric configuration of the model. Experiments were also conducted on the direct observation of

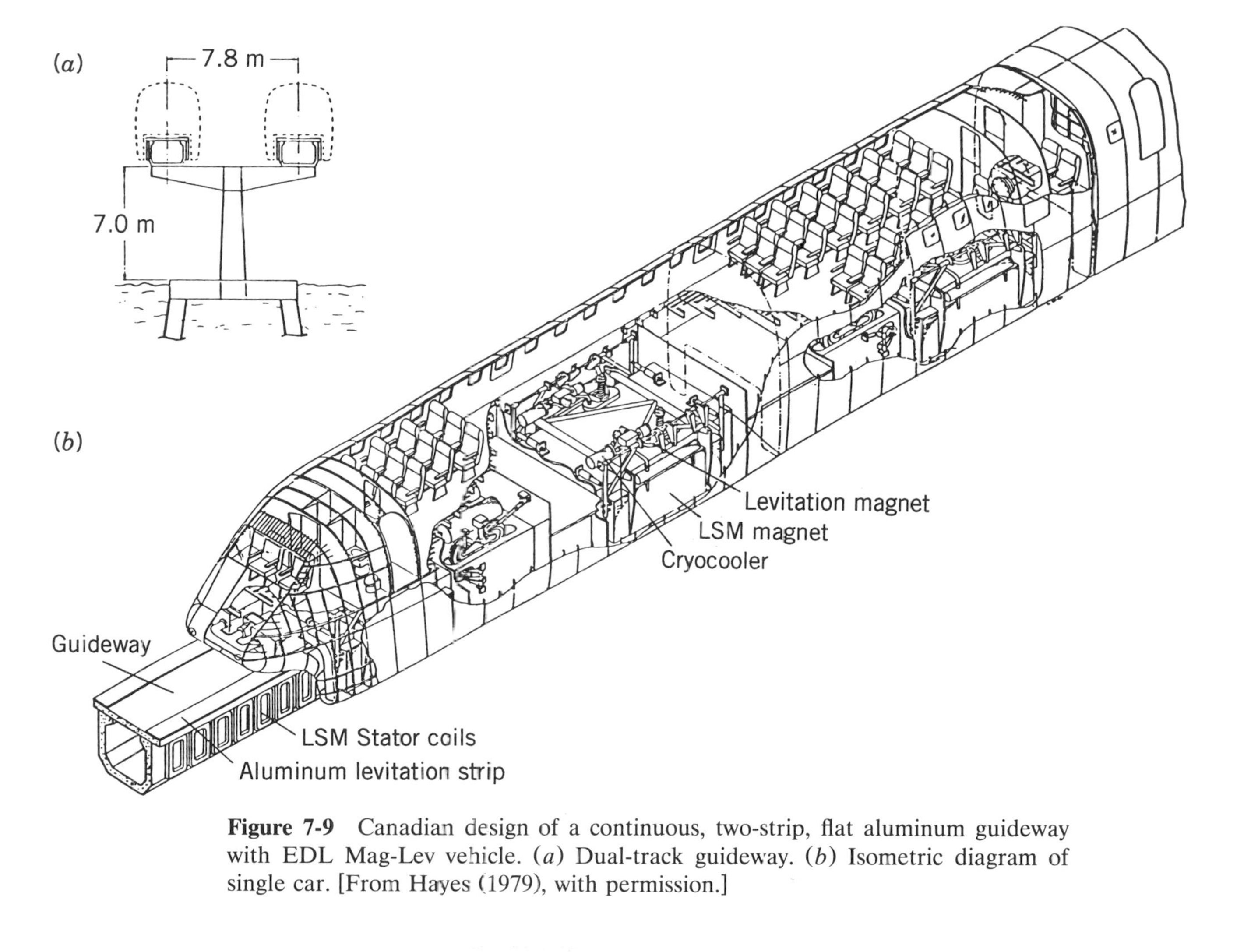

Figure 7-9 Canadian design of a continuous, two-strip, flat aluminum guideway with EDL Mag-Lev vehicle. (*a*) Dual-track guideway. (*b*) Isometric diagram of single car. [From Hayes (1979), with permission.]

Figure 7-10 Photograph of a small test model of an EDL vehicle on a split-sheet guideway at the University of Warwick. [From Rhodes, (1982), with permission.]

Figure 7-11 Photograph of a levitated EDL model on a "V"-shaped aluminum guideway at Cornell University, 1977.

eddy currents using both liquid-crystal physics and infrared scanning technology (Moon, 1974, 1984). This wheel is now being used at Argonne National Laboratory for continuing Mag-Lev dynamics studies (see Cai et al., 1992).

Japan: EDL Program Japanese interest in magnetically levitated transportation goes back to the late 1960s shortly after the Shinkansen or Bullet Train was introduced in 1964. Both EDL and EML systems were investigated. In the early 1970s the research laboratory of the Japanese National Railway (JNR) under Y. Kyotani introduced a large-scale EDL prototype, the ML100, which used four superconducting magnets (Figure 7-12*a*). By the late 1970s, JNR had developed a full-scale vehicle, the ML500, which eventually achieved a world record of 517 km/hr at a 7-km test-track facility at Miyazaki in Kyushu Prefecture in the southern part of Japan. The guideway was an inverted "T" shape with discrete coils laid horizontally along the track on each side of the inverted "T." In 1980 JNR built a passenger-carrying EDL vehicle MLU-001, which carried four superconducting magnets on each side in a "U"-shaped guideway at Miyazaki. This vehicle achieved a speed of 305 km/hr in 1982. The length of the track limited the speed of this heavier vehicle. Eventually a three-car set was built which ran at slower speeds but which made many hundreds of kilometers of test runs. These tests formed the basis of their current design for a commercial, revenue vehicle, the MLU-002.

In 1994, construction is nearing completion on a 43-km test track, 128 km west of Tokyo in Yamanashi Prefecture. This test facility is envisioned to become part of a complete Mag-Lev line between Tokyo and Osaka in the first decade of the next century. This test facility will be 80% tunnels and will test more realistic conditions than did the facility at Miyazaki, including a grade of 4%.

The MLU-002 prototype revenue vehicle has been designed to be levitated using a null-flux "U"-shaped guideway. This will use vertical,

Figure 7-12 (*a*) Photograph of Japanese National Railway EDL Mag-Lev vehicle ML100 (early 1970s). (*b*) Japanese design for a full-scale superconducting magnet EDL vehicle. Dimensions in mm. [From Kyotani (1988), with permission.]

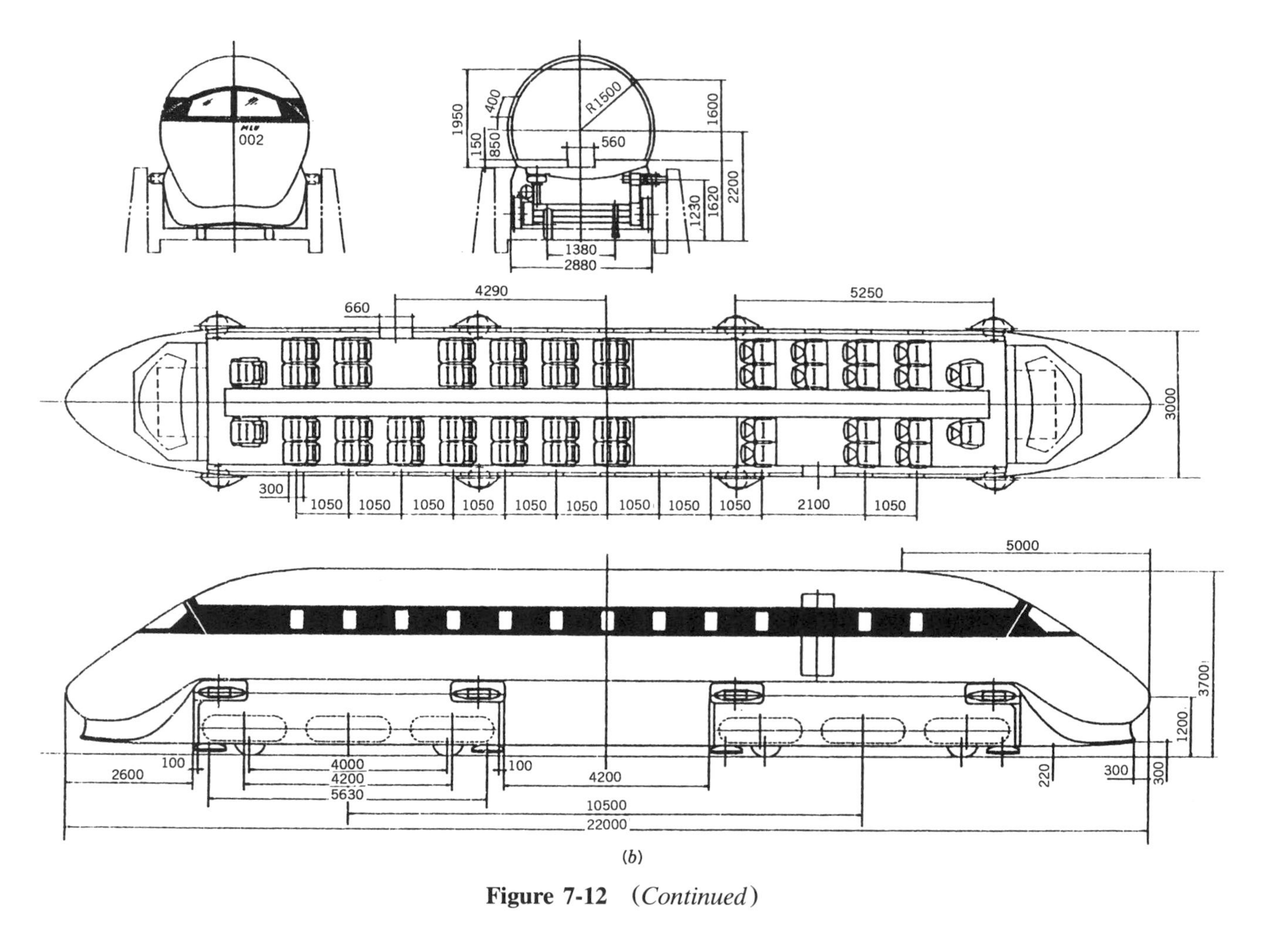

(b)

Figure 7-12 *(Continued)*

Figure 7-13 Full-scale Japanese HSST EML vehicle on an elevated guideway.

plane superconducting coils on the vehicle and will utilize vertical, plane reaction and propulsion coils in the guideway (Figure 7-12*b*). The vehicle car length will be 22 m and will carry 44 passengers and weigh 17 tonne. The superconducting magnets will be grouped in bogies at each end of the vehicle in order to decrease the magnetic field in the passenger cabin to less than 10 gauss (10^{-3} T).

Japan: EML Program (HSST) This prototype program was started by Japan Air Lines (JAL) after studies in the late 1960s revealed a need for a high-speed link between the new Tokyo airport at Narita approximately 64 km from Tokyo. In 1973, JAL started to design a test vehicle called HSST-01. This vehicle was tested on a 1.3-km track and achieved a speed of 307.8 km/hr (191 miles/hr) in 1978. This nonsuperconducting attractive suspension levitation system uses electromagnets and a ferromagnetic rail with a 1-cm gap. HSST-03, a passenger-carrying version, was exhibited at expositions in Canada and Japan and carried over 1 million people along a short track (Figure 7-13) (Nagaike and Takatsuka, 1989).

The development of this EML system is now being carried out by the HSST Corporation of Japan, which is trying to market three vehicle concepts for passenger-carrying urban (100 km/hr), suburban (200 km/hr), and interurban (300 km/hr) service. Thus it would be a competitor to the existing TGV and ICE steel-wheel systems in Europe.

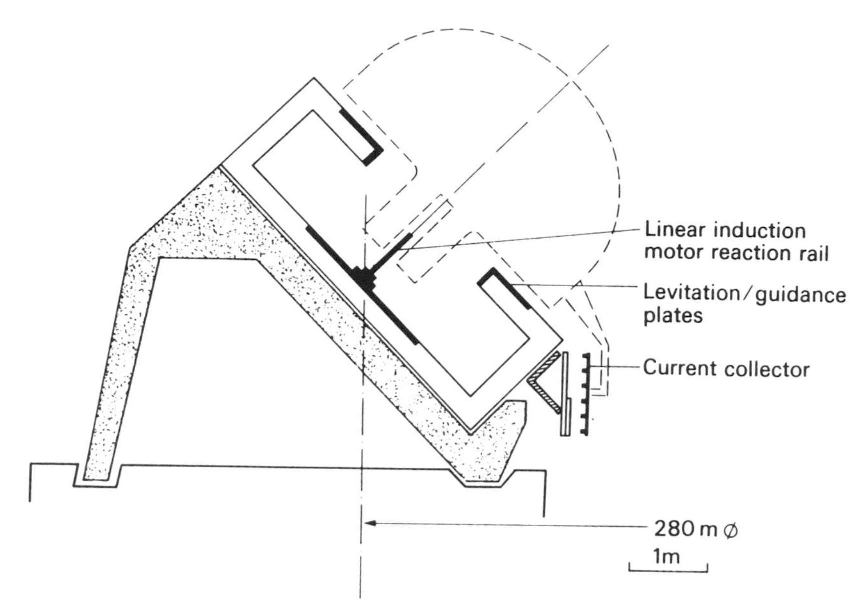

Figure 7-14 Cross section of German circular guideway and EDL vehicle at Erlangen. (From Rhodes and Mulhall, 1981 by permission of Oxford University Press.)

The HSST Corporation has been trying to conclude a contract to build a short-line (7 km) urban EML system in Las Vegas, Nevada. It is also trying to build a system in Nagoya, Japan.

Germany: EDL Program Beginning in 1969, the Federal Ministry of Research and Development in the Federal Republic of Germany (FRG) (then West Germany) began a comprehensive study of high-speed ground transportation options. While air-cushion technology was originally on the list, R&D was stopped in 1973 because of excessive energy consumption, noise, and maintenance problems. Permanent magnet suspension was also explored initially, but was also dropped in 1974 in favor of the EDL and EML levitation systems (Rogg, 1986).

The superconductor-based EDL system received extensive development in FRG during the period of 1970–1978. The showpiece of this program was a 280-m-diameter circular test-track facility in Erlangen in the state of Bavaria (Figure 7-14). The German EDL group built a 17-tonne vehicle in 1976 which achieved speeds of 230 km/hr. This group included the companies of Brown-Boveri, AEG-Telefunken, Linde, and Siemens Research Laboratories in Erlangen. The 17-tonne test vehicle was 12 m long and was levitated at a gap of 10 cm with a linear induction motor which generated a thrust of 22 kN. The facility at Erlangen also had a 5.8-m-diameter rotating wheel which could produce rim speeds of 40 m/sec. Experiments on linear synchronous motor concepts were carried out with this wheel. In 1978, however, the FRG decided to focus all of its high-speed development efforts on the EML attractive system. The technology for this system appeared to be further along in development than the EDL system, and the 400-km/hr speeds of the EML system seemed more suited to European topography and population distribution.

Germany: EML Program Electromagnetic levitation vehicle development began in the FRG in 1970 and has continued to the present. A consortium of companies under the title Magnetschnellbahn includes Krauss Maffei, Messerschmidt-B.B., Thyssen Henschel, Diamler-Benz, AEG, and Siemens. The current EML system is called the Transrapid-07 or TR07 and is being aggressively marketed around the world (Glatzel et al., 1980; Rogg, 1986).

In 1971 the German group began to test two vehicle concepts. The one known as MBB-Prinzip-Fahrzeug weighed 5.8 ton and achieved a speed of 90 km/hr on a 660-m track. The second vehicle concept had an integrated levitation, guidance, and propulsion system and was called the Transrapid 02 (TR02). The TR02 weighed 12 tonnes and achieved a speed of 163 km/hr on a 930-m track. In the late 1970s a more advanced prototype, the TR04 (weighing 18.5 tonnes and carrying 12 passengers), was able to reach a speed of 250 km/hr on a 2.9-km track. In 1976 a special vehicle called the KOMET, equipped with hot water rockets, was able to reach a speed of 401.3 km/hr. In 1979 a 70-passenger Transrapid vehicle was exhibited at a transportation fair in Hamburg on a 970-m track. More than 40,000 passengers rode this vehicle at a speed of 90 km/hr.

In the early 1980s the Transrapid group in FRG built an extensive test facility in Emsland near the Dutch border. This facility included a 31-km elevated guideway with a 12-km straight section and closed loops on the ends for continuous operation (Figure 7-15*a*). In a decade of operation, the German EML vehicles have logged over 100,000 km of tests. The TR06 vehicle was a two-car set (54 m long) with a 196-passenger capacity and an upper speed of 400 km/hr (Figure 7-15*b*). The guideway uses a synchronous iron core long stator propulsion which can generate 85 kN of thrust and an acceleration of the 100-tonne vehicle of 0.8 m/s^2 (Miller and Ruoss, 1989).

The current design called TR07 has been optimized to decrease both aerodynamic drag and noise. At the time of this writing (late

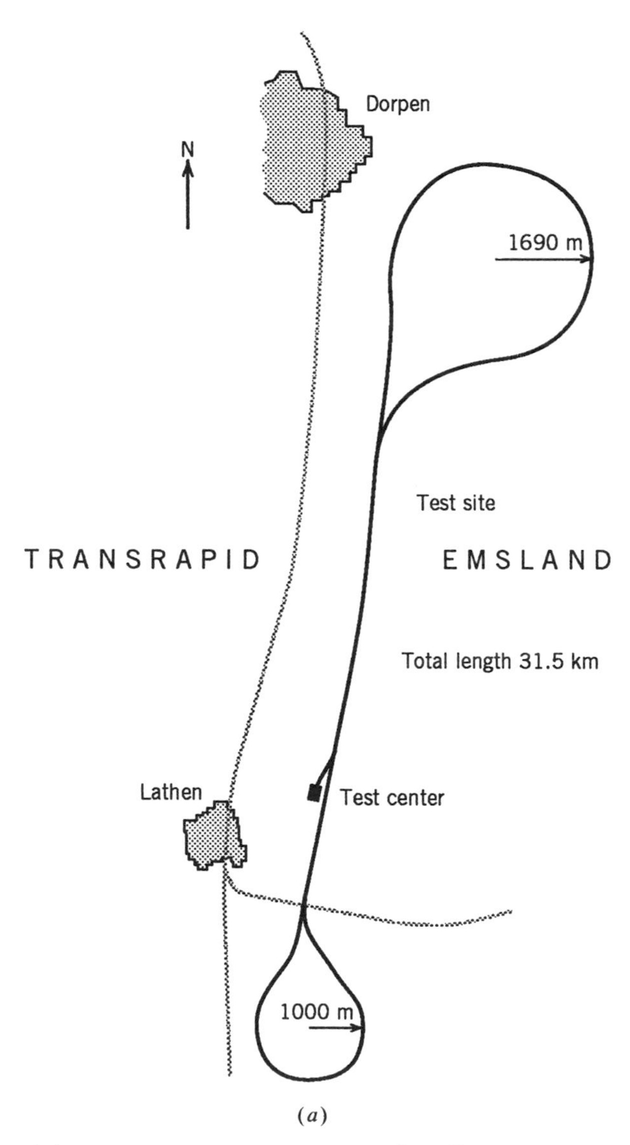

(*a*)

Figure 7-15 (*a*) Plan view of test center for German EML Transrapid vehicle development at Emsland. (*b*) Photograph of German Transrapid 0-6 EML vehicle on an elevated guideway.

(*b*)

Figure 7-15 (*Continued*)

1993), the marketing company Transrapid International has a contract to build a short revenue line in Orlando, Florida (21 km).

In the late 1980s the FRG has decided to build a Transrapid line between Dusseldorf and Bonn, but after unification these plans have been changed to build a line from Hamburg to Berlin. However, as of 1993, these plans have been put on hold because of a lack of capital and a recession, as well as competition from the German high-speed steel-rail system ICE.

Germany: M-Bahn Project The M-Bahn concept uses a combination of permanent magnets attached to the vehicle and mechanical springs to achieve stable levitation below active ferromagnetic guideway rails on each side of the vehicle. The system is built by AEG, a subsidiary of Daimler-Benz. In the United States, AEG has teamed up with Westinghouse Corporation. M-Bahn is designed for low-speed (40–80 km/hr) urban or airport applications and uses a synchronous linear motor for propulsion (Dreimann, 1989). A 600-m elevated demonstration line was installed in Berlin in 1984. As of 1991, 2.5 million people have ridden on these vehicles before the system was closed. The 9-tonne vehicles could carry 130 passengers. The demise of this system came after the opening of the Berlin wall, which changed the city's plans with regard to transportation needs. A test facility is located in Braunschweig. AEG has plans to build an M-Bahn passenger-carrying system at Frankfurt's Rhine–Main Airport. One of the features of this system is a low noise level of 65 db, which is slightly above human conversation.

Birmingham, England: EML Project The Research Division of British Rail began a study and built a low-speed EML demonstration vehicle in the mid-1970s. In the early 1980s a group of British companies built a 620-m, elevated, passenger-carrying EML line in Birmingham Airport connecting the terminal with the rail station (Pollard and Riches, 1985; Dalgleish and Riches, 1986). The 8-tonne vehicle was designed to carry 3.2 tonnes of standing passengers and has a speed of 42 km/hr. A linear electric motor generates a maximum thrust of 4 kN with a 15 to 20-mm air gap. The vehicle carries eight electromagnets, two at each corner of the vehicle. Each magnet pair is staggered so that active control of lateral motion can be effected.

New Design Concepts

In 1991 the U.S. government, through the Department of Transportation, created the National Mag-Lev Initiative. This was in response to two related developments: (1) the interest of a half dozen states and regions in new high-speed ground transportation and (2) the aggressive marketing of European and Japanese Mag-Lev and conventional steel-wheel technology with no potential American competition in sight. Thus, the real transportation needs of automobile- and air-congested regions and the potential loss of jobs prompted a renewal of American interest in high-speed technology and, in particular, Mag-Lev. One of the early champions in the U.S. Senate was D. P. Moynihan of New York State, who was inspired by the persuasive arguments of U.S. Mag-Lev pioneers such as H. Kolm of MIT, G. Danby and J. Powell of Brookhaven National Laboratory, and others.

These contracts were awarded in two categories: 27 small technology component studies and four overall system concept definition studies.

National Mag-Lev Initiative

Twenty years of high-speed vehicle development in Europe and Japan has resulted in aggressive marketing of new transportation systems in North America and has challenged U.S. technologists to come up with new Mag-Lev designs. The NMI studies of 1991–1992 showed that by using new materials and new microelectronic and power electronic devices a U.S. Mag-Lev system could be designed and built that met or exceeded the capabilities of existing designs (U.S. Department of

TABLE 7-1 General Performance Parameters

Parameter	TGV-Atlantique	TR07	Bechtel	Foster–Miller	Grumman	Magneplane
Concept	Steel Wheel-on-Rail	EMS, Separate Lift and Guidance	EDS, Ladder Levitation	EDS, Sidewall Null-Flux	EMS, Common Lift and Guidance	EDS, Sheet Levitation
Vehicles/consist	12	2	1	8	2	1
Seats/consist	485	156	120	600	100	140
Speed (mph) (kph)	186 310	311 519	300 500	300 500	300 500	300 500
Switch design speed (mph)	143	125	72	112	145	224
Air gap (inches) (cm)	N/A	0.4 1.0	2.0 5.1	3.0 7.6	1.6 4.1	5.9 15

Source: U.S. National Mag-Lev Initiative Final Report, September 1993.

TABLE 7-2 Cost and Performance of Different Systems

Parameter	TGV[a]	TR07[c]	U.S. Mag-Lev
Elevated guideway ($ millions/mi)[b]	22.3	19.6	17.6
At grade guideway ($ millions/mi)[b]	3.3	17.4	13.0
Range of initial capital costs ($ millions/mi)[d]	17.2–33.4	30.4–49.3	26.6–45.4
Cruise speed (mph)	200	311	300
Maximum grade (%)	5.0	10	10
Acceleration time(s) with full thrust (sec)			
0–186 mph	408	107	36
0–300 mph	—	240	58
Total bank angle (°)	7	12	30

Source: U.S. National Mag-Lev Initiative Report, September 1993.
[a] Modified *Train a Grand Vitesse* (TGV) proposed for the Texas HSR System.
[b] Includes only distance-related technology costs.
[c] German Mag-Lev System.
[d] A construction financing cost is included in these estimates using the 7% discount rate.

Transportation, 1993). These features included:

- Vehicle tilt capabilities to negotiate curves at higher speeds
- Lighter-weight vehicles to improve energy use

- Aerodynamic drag and noise reduction features including control surfaces
- Possible use of high-temperature superconductors
- Higher-grade climb capability
- Lower magnetic field in passenger cabin
- Unit costs comparable to those of TGV or TR07 systems

The four NMI system design studies were all superconductor-based and can be grouped into three categories:

- An EDL system with a circular-shaped guideway with self-banking vehicles (the Magneplane)
- An EML system with superconducting windings in the electromagnets which produces larger gaps than do conventional EML designs (Grumman group)
- Two EDL systems with vehicle tilt capabilities and null-flux guideway coils. One system uses a locally commutered synchronous motor (Foster Miller and Bechtel teams)

The Magneplane

In 1972 and 1973, H. H. Kolm and R. D. Thornton of MIT published articles and were awarded a patent on a new Mag-Lev concept that they called "flying in a magnetic field or magneplane" (Kolm and Thornton, 1972, 1973). The inventors proposed to use a guideway with a circular arc shape to allow the vehicle to self-bank as the vehicle entered a curve. The superconducting magnets in the bottom of the vehicle which generate the levitation forces would also serve as a field source for a linear synchronous motor propulsion system. A 1/25th scale model was built and tested at MIT, with promising results. However, by the late 1970s, funds were cut along with other Mag-Lev projects in the United States and the project lay dormant.

Dr. Henry Kolm, however, was a tireless campaigner for magnetic transportation in the United States, and in the late 1980s he and other Mag-Lev pioneers convinced Senator Daniel Patrick Moynihan to sponsor legislation to renew U.S. participation in the developing Mag-Lev technology. This legislation led to the U.S. National Mag-Lev Initiative in which four system design concepts were supported by the U.S. Department of Transportation. By this time, Dr. Kolm had retired from MIT and formed Magneplane International. This company was awarded one of these design contracts.

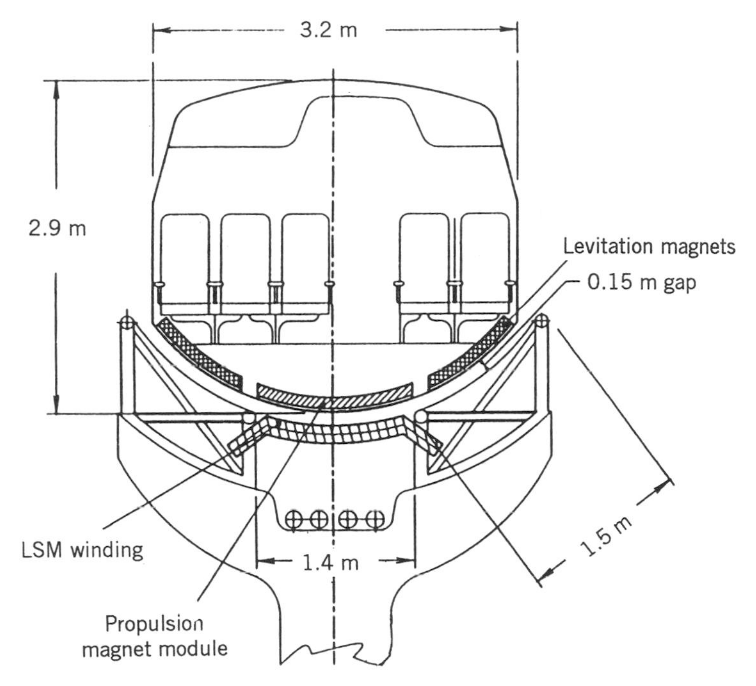

Figure 7-16 Magneplane International System concept EDL design for U.S. National Mag-Lev Initiative, 1992 (U.S. Dept. of Transportation, 1993a).

The 1992 version of the Magneplane design is similar in concept to its predecessor of 20 years earlier. The basic design can be seen in Figures 7-16 and 7-17. The Magneplane vehicle is designed to run at speeds in excess of 134 m/sec (500 km/hr), and the system is envisioned to carry 25,000 passengers per hour. Four superconducting levitation magnets are grouped at each end of the vehicle. Induced eddy currents in the 2-cm-thick aluminum-sheet conductors on the high sides of the circular guideway provide the lift forces. The guideway propulsion coils are laid in the center of the circular guideway, and a traveling wave of flux interacts with the vehicle superconducting magnets at each end of the vehicle to produce the propulsion force. All the superconducting magnets use Nb_3Sn cable-in-conduit conductors carrying supercritical liquid He at 6–8 K. The linear synchronous motor (LSM) is activated in several-kilometer blocks along the guideway. The designers also envision active damping to quench any unwanted lateral or yaw type instabilities (see Chapter 5). There is no secondary suspension. The system would use single vehicles carrying 40–150 passengers rather than using a multi-car consist. Each vehicle is capable of self-banking as the guideway executes a turn. Banking

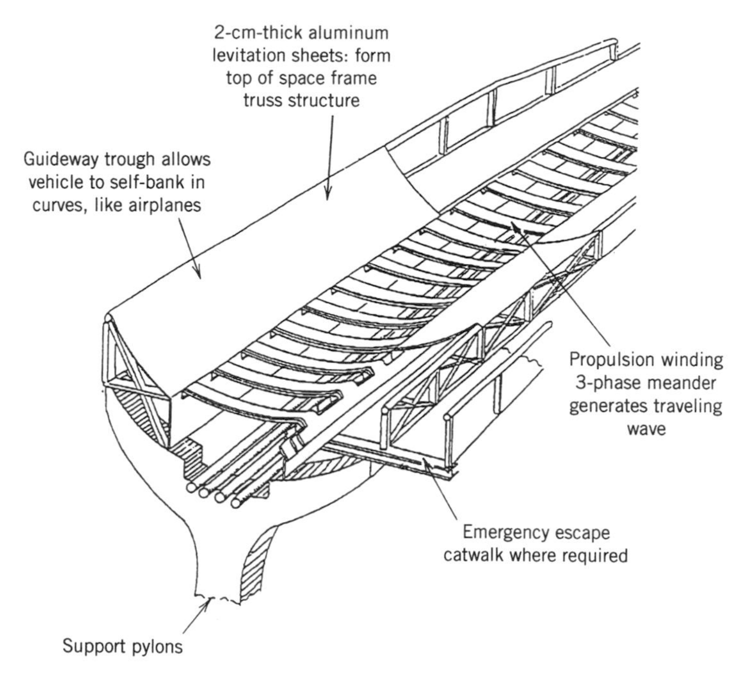

Figure 7-17 Detailed sketch of EDL guideway of the Magneplane International System Concept Design, 1992 (U.S. Dept. of Transportation, 1993a).

angles of 30° or more would be possible, therefore enabling the vehicle to maintain high speeds while negotiating curves.

Superconducting Electromagnetic Levitation System

One of the principal drawbacks of the EML system such as the TR07 or HSST is the small gap (1 cm) between the vehicle and the guideway rail. This small gap is a consequence of the limitation of using normal conductors in the electromagnets to generate the flux. However, if one were to use superconducting windings in the electromagnet, then higher ampere-turns would permit a larger gap, say 3–4 cm. This was one of the ideas presented in one of the four Mag-Lev designs developed for the U.S. National Mag-Lev Initiative by an industrial team headed by the Grumman Corporation (Long Island, New York).

This team has developed a clever hybrid of magnets and EML levitation. The system, shown in the sketch in Figure 7-18, uses both superconducting and normal windings in the vehicle electromagnets. The superconducting winding generates a constant field source which provides sufficient lift to levitate the vehicle. The normal conductor

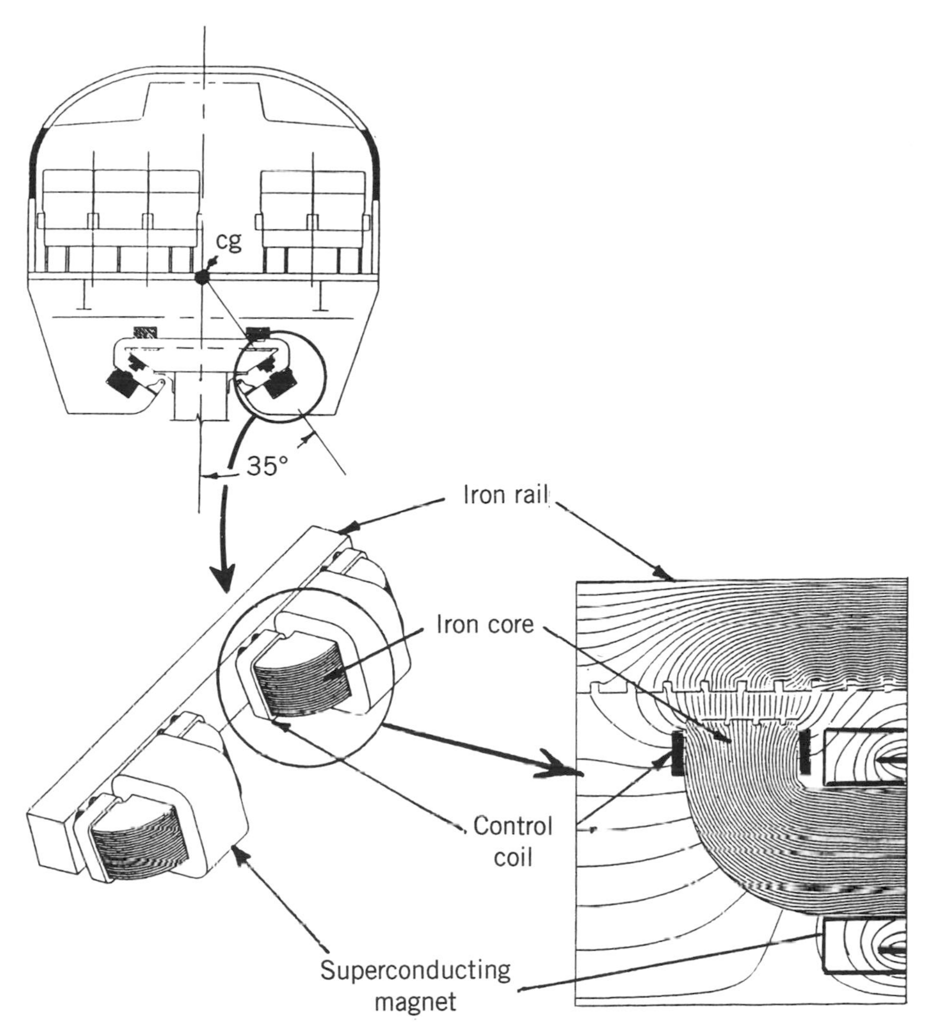

Figure 7-18 Grumman Team System concept EML design for U.S. National Mag-Lev Initiative, 1992, using superconducting magnets in the levitation electromagnets (U.S. Dept. of Transportation, 1993a).

coils in the elcctromagnet are then used to stabilize the negative stiffness of the direct current (dc) field in the magnet.

The design uses 24 hybrid electromagnets on each side of the vehicle. The set of magnets on each side of the vehicle are alternately offset 2.0 cm to the left and right to provide for lateral and roll control forces. Propulsion is created by three-phase linear synchronous motor windings in the guideway. The guideway has an inverted "V" shape (Figure 7-18), and the vehicle is wrapped around the guideway similar to the TR07. The Grumman team chose a two-vehicle consist with 100-passenger capacity. The vehicles have a secondary suspension and a 9° tilt capability.

New U.S. EDL Designs

Foster-Miller NMI Team A sketch of this design is shown in Figure 7-19 and is close in spirit to the current Japanese vehicle MLU002 which also uses a null-flux coil guideway. However, the Foster–Miller vehicle design has a 12° tilting capability for taking curves at higher speeds. As in the new Japanese design, superconducting magnets are grouped at each end of the vehicle to lower the magnetic field in the passenger cabin. The "U"-shaped guideway has vertically mounted null-flux lift coils and active propulsion coils secured to the inner sides of the guideway. The design gap between the vehicle and guideway is 0.1 m. The null-flux coils are connected in a figure-eight pattern (Foster–Miller, 1992), (see also Figure 7-7).

A unique feature of the linear synchronous motor is local commutation. In this design each propulsion coil is switched by a local semiconductor device. Thus only dc power is provided along the guideway. The linear magnetic propulsion wave is generated in the switching wave of the semiconductor devices. A possible candidate for these local LSM switches is an insulated-gate bipolar transistor (IGBT). Present-day devices can handle 800 A at 1400V. Other potential

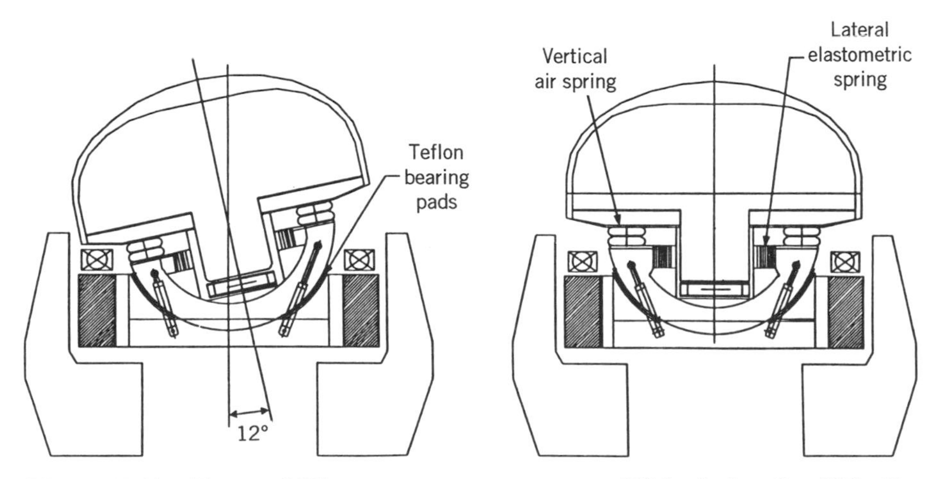

Figure 7-19 Foster–Miller team system concept EDL design for U.S. National Mag-Lev Initiative, 1992, showing a split "U"-shaped guideway with vehicle tilt mechanism (U.S. Dept. of Transportation, 1993a).

switching elements are gate-turn-off thyristors (GTOs) or microcontrol thyristors (MCTs). Because thousands of these switches would be needed, a key requirement for the Foster–Miller concept to be cost-feasible is to get the unit costs of these LSM switches down by an order of magnitude from present-day unit costs.

Bechtel Team Concept This concept is shown in Figure 7-20. It has some of the features of the earlier Canadian design (Figure 7-9), which also had a box-beam guideway structure. This design also has 15° vehicle tilt capability. The designers have chosen a maximum speed of 150 m/sec (540 km/hr or 336 miles/hr) and operational acceleration capability of 2 m/sec^2 (0.2 g). The vehicle has a length of 36.1 m and a passenger capacity of 120. The designers claim a

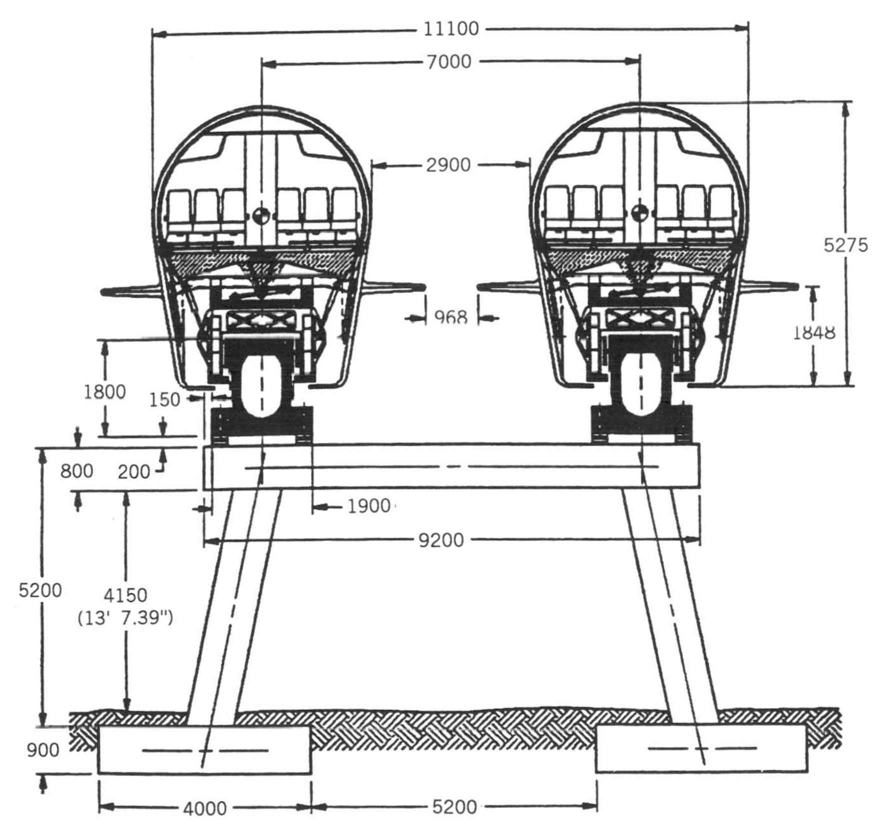

Figure 7-20 Bechtel Team system concept EDL design for U.S. National Mag-Lev Initiative, 1992, showing elevated dual guideway with box-beam and tilt mechanism. All dimensions are in millimeters (U.S. Dept. of Transportation, 1993a).

proprietary "flux-canceling" electrodynamic suspension using a ladder conductor structure along the guideway. The designers claim extremely low magnetic drag and very low liftoff speeds ranging between 5 and 10 m/s (22 miles/hr). Lateral guidance is effected by coils on each side of the guideway which are cross-connected in a figure-eight shape. Thus when the vehicle is centered there is no induced currents in these coils. The vehicles also incorporate an active secondary suspension system.

The propulsion uses an LSM system. Direct-current voltage is brought to inverters spaced 4 km along the guideway which produce a variable frequency and voltage for the guideway-mounted LSM coils.

7-4 TECHNICAL ISSUES

Some of the engineering and operational issues in Mag-Lev design are listed below. In this section we discuss only a few of these.

Technical Issues in Magnetic Transportation

- Aerodynamics
- Communication and control
- Cost
- Curve negotiation, tilting, grade climbing ability
- Emergency operation: egress, loss of magnet, loss of LSM, fire
- Guideway flexibility
- Guideway repair and inspection
- Headway, capacity
- Magnet design, cryogenics, helium management
- Magnetic field in cabin
- Maintenance
- Noise
- On-board power
- Power: utility interface, substations
- Primary and secondary suspension
- Ride quality
- Route selection
- Speed, acceleration, braking
- Station-guideway compatibility
- Switching

Aerodynamic Forces

As speed increases, the magnetic drag force decreases in eddy-current-repulsion-based systems, whereas aerodynamic forces increase dramatically. It was for this reason that Robert Goddard (1909) proposed an evacuated tube form of Mag-Lev transportation. [See also Guderjahn et al. (1969).]

Aerodynamic forces on moving vehicles close to the ground plane are usually divided into two parts: form drag and skin drag. The form drag is proportional to the frontal area (A), whereas the skin drag is proportional to the vehicle length (L), cross-sectional perimeter p, and the viscosity of air (μ).

Form drag is analogous to moving a flat plate in air normal to its frontal area. Ideally, the flow should move from the front face to the back face with no drag. In reality, however, the flow separates in the form of viscous eddies, resulting in a higher fluid pressure on the front of the object and a lower pressure on the back. Form drag is usually expressed as follows,

$$F_1 = \tfrac{1}{2}\rho V^2 A C_{D1} \tag{7-4.1}$$

where V is the average speed of the vehicle, ρ is the density of air, and C_{D1} is a dimensionless geometry-dependent form drag coefficient. Aerodynamic-drag-producing eddies are usually found near edges, so that guideway sidewalls, protruding guideway coils, or openings or cutouts may become a source of aerodynamic drag (Barrows et al., 1992).

The skin friction drag force is usually written as

$$F_2 = \tfrac{1}{2}\rho V^2 p L C_{D2} \tag{7-4.2}$$

where C_{D2} is called a skin friction coefficient, L is the vehicle length, and p is the cross-section perimeter. Barrows et al. (1992) quote a C_{D2} value of 0.004. Thus, total drag increases as the length increases, but the specific drag force per vehicle mass usually decreases, which favors slender vehicles.

The two expressions (7-4.1) and (7-4.2) are usually combined into one expression:

$$F = F_1 + F_2 = \tfrac{1}{2}\rho V^2 A C_D \tag{7-4.3}$$

According to Rhodes and Mulhall (1981), a lower bound on the dimensionless C_D is 0.2, but typical Mag-Lev vehicles might be in the $C_D \approx 0.3$ range.

Using $V = 500$ km/sec (134 m/sec) and $A = 10\ \text{m}^2$ $\rho = 1.2\ \text{kg/m}^3$, the aerodynamic drag force is on the order of 32 kN. The power is given by

$$P = F \cdot V \tag{7-4.4}$$

In the example just cited, the power in the aerodynamic losses amounts to 4.3 MW to move a vehicle carrying on the order of 100 passengers at a speed of 500 km/hr.

Special Mag-Lev Aerodynamic Effects In Section 7-2 we saw that flat, channel, and box-beam guideways have been proposed for Mag-Lev systems. According to Barrows et al. (1992), $C_{D1} = 0.15$ for a channel and $C_{D1} = 0.11$ for a box-beam guideway. However, the flat or box-beam guideways subject the moving vehicles to lateral cross-wise aerodynamic forces while the channel offers some protection.

Another Mag-Lev specific aerodynamic effect is the generation of eddies and turbulence in and around the moving vehicle magnets and guideway magnets. Barrows et al. (1992) quote results from the German EML vehicle TR-07—namely, that a large contribution to the drag comes from the magnets, and that significant reductions may be possible with careful design—although there are presently no general guidelines, and empirical methods of wind tunnel experiments and experience are needed (see Hammitt, 1973).

Grade-Climbing Capability

If T is the maximum thrust and D is the total aerodynamic and magnetic drag, then the relation between acceleration A, grade angle α, and peak thrust force is given by

$$A = \frac{(T - D)}{m} - g \sin \alpha \tag{7-4.5}$$

For example, for a Mag-Lev system competitive with the TR07, we have $D = 60$ kN, $T = 120$ kN, $m = 100 \times 10^3$ kg. For zero grade ($\alpha = 0$), we have $A = 0.6\ \text{m/sec}^2 = 0.061$ G. For zero acceleration ($A = 0$), we have $\sin \alpha = 0.06$ and $\alpha = 3.5°$. Thus, to better the performance of the German TR07, designers must either decrease the mass, decrease the aerodynamic drag, or increase the peak thrust of the propulsion motor. For example, decreasing the mass to 60×10^3 kg would increase the grade to $\alpha \approx 5°$. If, in addition, the drag is reduced to 40 kN, the grade would increase to $\alpha = 7.7°$ and the peak acceleration at zero grade would be 0.136 G or 1.33 m/sec^2, which is double the original performance.

Guideway Banking and Turns

While Mag-Lev enthusiasts have always spoken of 500 km/sec as an operating cruise speed, engineers have to face the reality of building a guideway that must have turns in order to fit into an existing natural and man-made landscape. A turn at 100 m/sec with a circular radius of 1000 m produces a lateral acceleration of 10 m/sec, which is slightly greater than the acceleration under gravity. Thus, to avoid lateral forces on the passengers the vehicle must enter the turn in a banked state. One can show that the required angle of banking for the net force on the passengers to be normal to the floor of the vehicle is given by (Figure 7-21)

$$\tan\theta = \frac{v^2}{Rg} \tag{7-4.6}$$

where v is the tangential velocity, R is the radius of the turn, and $g = 9.8\ m/s^2$ is the acceleration of gravity. It is easy to see that if $v = 100$ m/sec and $R = 2$ km, then $\theta \approx 30°$. This is a very large bank angle, typical of bobsled tracks. However, several U.S. teams in the NMI design studies have proposed a combination of tilt mechanism in the vehicle and in the guideway. For example, one could design in a vehicle tilt capability of 10° and 15° in the guideway.

Another design issue is the transition into and out of a banked section of guideway. It is believed by some designers that too fast a roll rate might be unsettling to some of the passengers.

Another guideway design problem is to design a jog in the right of way as shown in Figure 7-22. This involves both positive and negative roll angles. The relation between jog distance, radius of curvature R, and turn angle θ is given by

$$\Delta = 2R(1 - \cos\theta) \tag{7-4.7}$$

Propulsion Power

Almost all EDL Mag-Lev design incorporate a linear synchronous motor concept (LSM) in which an active magnetic field in the guideway interacts with the superconducting coils in the vehicle. The

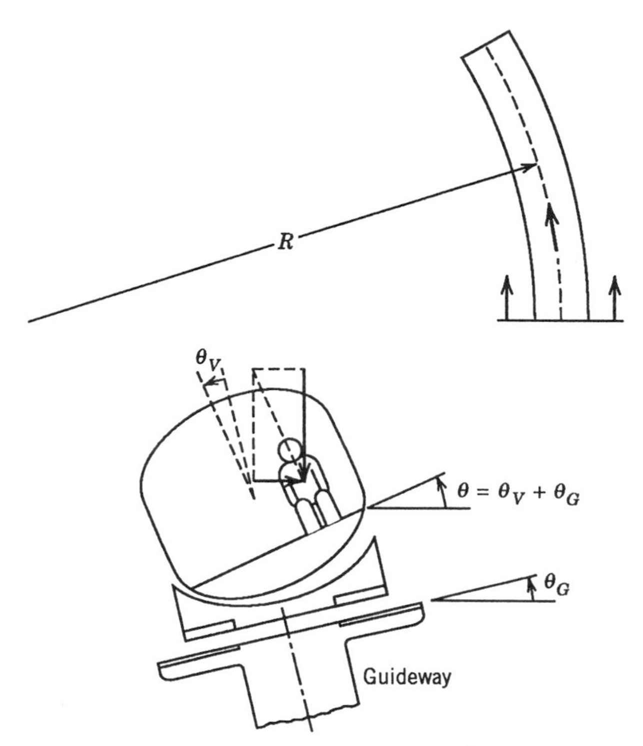

Figure 7-21 Tilt geometry for guideway and vehicle under a circular turn of radius R.

guideway must then be equipped with a sequence of propulsion coils which are sequentially activated to produce a traveling magnetic wave. In most designs, a block of guideway coils are activated (e.g., 2–4 km). A 2-km block with a vehicle traveling at 100 m/sec must be energized for 20 sec. The guideway coils are fed alternating currents with varying voltage, frequency, and phase. Each block is fed by ac or dc voltage power from a local area substation. Each of the substations are, in turn, fed by the regional electric power station. The NMI design concept of the Foster–Miller team described above uses a radically different scheme in which only a few sets of guideway propulsion coils are activated from the track-side dc power using local commutation solid-state switches.

Vehicle – Guideway Interaction

It is estimated that 75–80% of the construction costs of a high-speed transport system will be in the guideway. Guideway costs and design

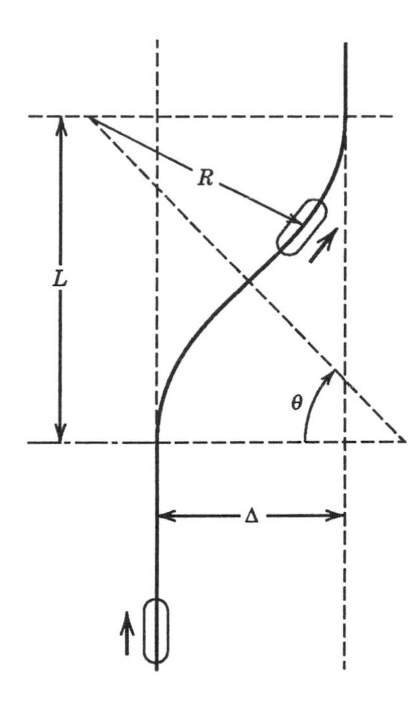

Figure 7-22 Dual-turn geometry for a "jog" or lateral change in guideway path.

are of critical importancc rcgardless of whether the vehicle is suspended by steel wheels or by magnetic forces. However, because of the importance of dynamic forccs at these high speeds, the nature of thc magnetic stiffness between the guideway and the vehicle greatly affects the ride quality and the structural design [see, e.g., Popp, (1982*b*)]. Given the importance of vehicle–guideway interaction, the U.S. Department of Transportation NMI studies sponsored two projects on this topic. In this section we review some of the highlights of the NMI studies on Mag-Lev vehicle–guideway interactions and their implication for design (Daniels et al., 1992 and Wormley et al., 1992). Dynamic tests and simulation of the Japanese EDL two-car vehicle MLU001, under guideway irregularities have been reported by Yoshioko and Miyamoto (1986).

Although much of the French TGV is an at-grade guideway, in the United States it is expected that a good percentage of a potential Mag-Lev route will be elevated in order to take advantage of existing interstate highway rights of way. Thus we focus our attention on vehicles on flexible guideway beams supported at discrete points (Figure 7-23). For reference we consider an example presented in a U.S. DOT/NMI report by a team headed by an engineering company Parsons Brinkerhoff Quade & Douglas, Inc. (Daniels et al., 1992). In their study they examined five guideway configurations and a baseline

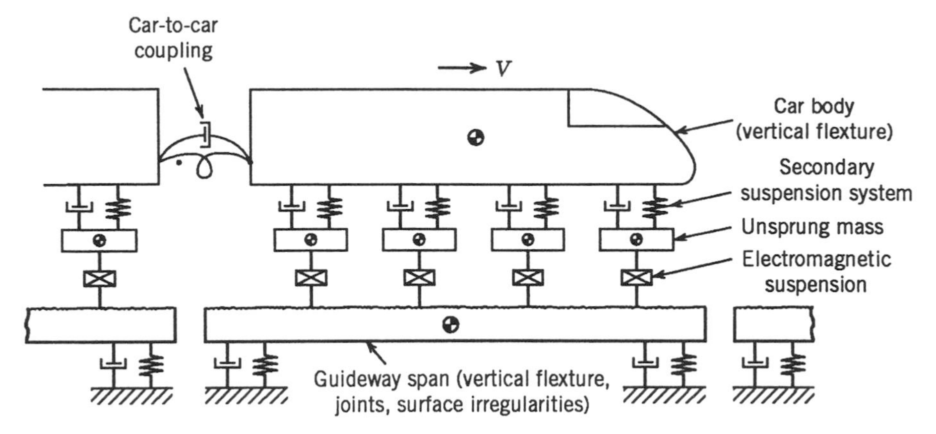

Figure 7-23 Sketch of guideway–vehicle interaction for a Mag-Lev vehicle on a multi-span system. [From Daniels et al., (1992).]

EML and EDL vehicle system. These two systems are dynamically very different. An EML system such as the TR07 or HSST has a distributed magnetic load on the guideway and a much higher stiffness than does an EDL system. Because of higher fields, current EDL systems concentrate magnets in bogies at each end of the vehicle, thus producing more concentrated loads on the guideway and requiring a stiffer fuselage to avoid vehicle structural bending. With larger air gaps, the EDL vehicle has vertical and pitch natural frequencies in the 1.5- to 2.5-Hz range, whereas an EML vehicle of equal mass has vertical and pitch frequencies in the 10- to 13-Hz range. Earlier studies of EDL designs suggested that a secondary suspension could be avoided. But except for the Magneplane design, both EML and EDL systems now envision secondary suspension systems to achieve suitable ride quality. Table 7-3 gives the major parameters of analysis for the Parsons–Brinkerhoff NMI study for one of the guideway geometries.

In choosing a Mag-Lev guideway design, one is faced with several competing design goals:

- Ride quality; minimize passenger level accelerations—this leads to a stiff guideway
- Minimize guideway mass (i.e., lower cost)—this, however, leads to higher flexibility
- Minimize dynamic stresses as well as static deflections—this sometimes leads to lower flexibility

TABLE 7-3 Mag-Lev Design Specifications

Speed	134–150 m/sec	
Passengers per vehicle	50–120	
Vehicle length	30–38 m	
Vehicles per consist	1–8	
Thrust per mass	1–3.3	
Maximum power per vehicle	4–5 MW	
Peak power/pass kW/seat	60–133	
Maximum thrust	60–120 kN	
Minimum curve radius	400 m	
Frontal dimensions	3.5 m × 3.5 m	
Braking rate	0.16 G	
Liftoff speed	5–10, 20–50 m/sec	70 m/sec (magneplane)
Magnet gap	40–75 mm	150 mm (magneplane)
Guideway span length	25–30 m	9.1 m (magneplane)
Magnet mass per vehicle	7–12 Mg	

Results of numerous simulations using linear vibration models of Mag-Lev systems were performed by Daniels et al. (1992). Two graphs illustrating the conflicting design demands are shown in Figure 7-24 for the reinforced concrete box-beam cross section with a 21-m span. When one looks at the vehicle body peak acceleration versus speed, the EDL system shows a resonance near 250–300 km/hr and decreases near 500 km/hr (Figure 7-24*a*). The acceleration in the EML vehicle however increases with speed.

In contrast, the dynamic loads on the guideway as measured by the midspan peak acceleration (Figure 7-24*b*) show the EML system to be below 0.3 G at resonances of up to 500 km/hr, which is a desired construction industry limit. The EDL vehicle shows much higher guideway accelerations, with a resonance effect around 325–350 km/hr. It should be noted that the crossing frequency at 250 km/hr is 3.3 Hz, which is in the range of the low EDL natural frequencies. However, even at 500 km/hr (6.6 Hz for a 21-m span) the EML vehicle frequency is still much higher (10–13 Hz).

As noted in Chapter 5, almost all of these analyses of vehicle–guideway dynamics neglect nonlinear effects. Nonlinear effects can originate from several sources:

- Nonlinear magnetic suspension forces
- Nonlinear active control forces
- Aerodynamic forces
- Multi-vehicle interaction

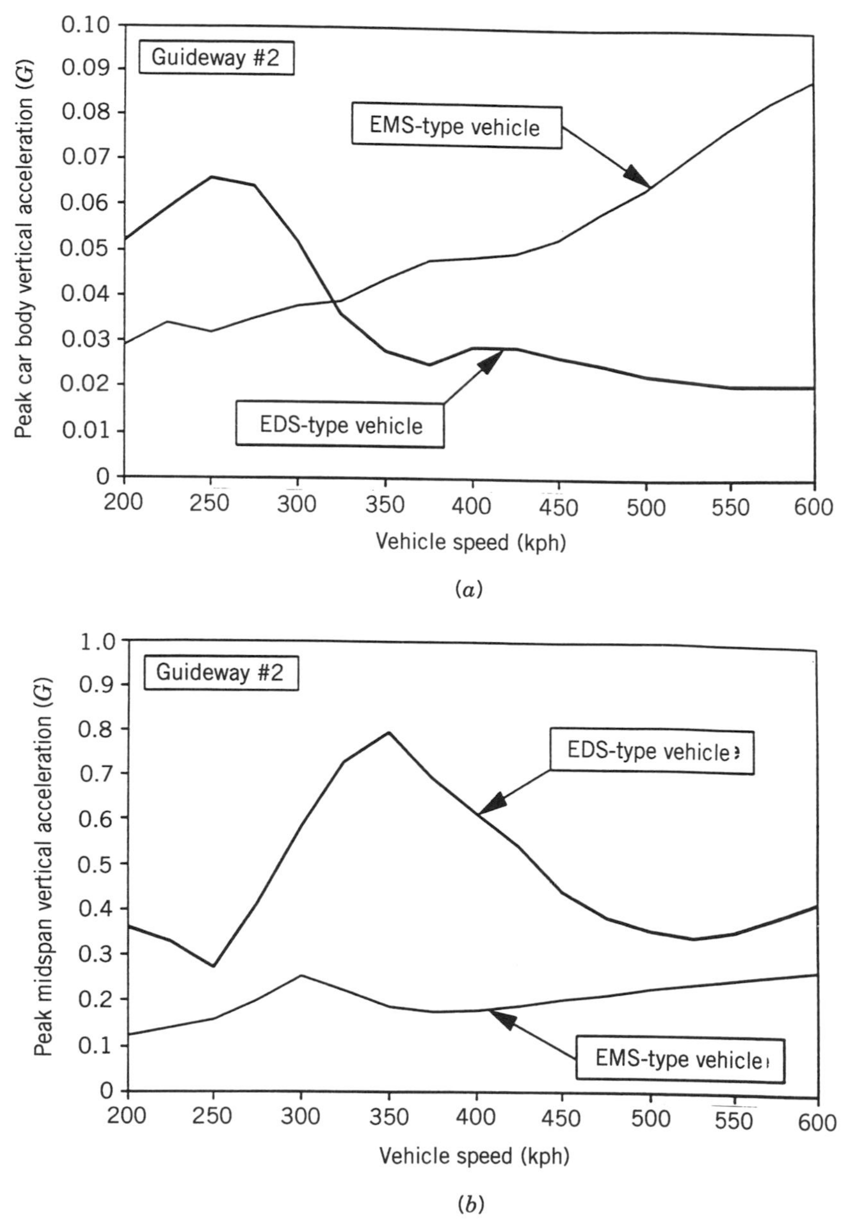

Figure 7-24 (*a, b*) Comparison of peak vertical acceleration of EML and EDL Mag-Lev vehicles on a 21-meter-long concrete box guideway. [From Daniels et al. (1992).]

- Magnetic drag interaction
- Guideway span bearing pads

Nonlinear dynamic effects include numerous behavior not predicted by linear simulation codes:

- Dynamic and static instabilities
- Limit cycle behavior
- Quasiperiodic oscillations
- Chaotic oscillations
- Nonlinear damping
- Mode interactions (e.g., lateral and yaw)
- Parametric oscillations

Also not performed in these studies are off-design vehicle dynamics such as guideway impact recovery or the loss of one or more levitation magnets. Some studies have looked at response to wind gusts.

Finally, it should be noted that other guideway design issues are of importance in a public sector application of Mag-Lev. Some of these include:

- Switching
- Passenger emergency egress
- Ice and snow buildup
- Inoperable vehicle removal
- Maintenance
- Effect of reinforcement bars on magnetic field
- Effects of magnetic fields and forces on corrosion and fatigue
- Earthquake response

Noise

To the surprise of many skeptics, wheeled railroad engineers have succeeded in pushing the speed of steel-wheeled vehicles to 300 km/hr (186 miles/hr) and beyond. However, the generation of high levels of acoustic noise by these vehicles continues to be a problem. The promise of low noise and high speeds is one of the reasons Japanese Mag-Lev proponents have used to justify continued Mag-Lev development.

An example of the limits that noise can place on high-speed rail can be found in a *Nikkei Weekly* article of December 14, 1992 by M. Matsumoto. The article states that although new Japanese Railway high-speed bullet trains (WIN 350 and Star 21) have met speed trial specifications of 300–350 km/hr, excessive noise will force operation at lower speeds.

As part of the recent U.S. National Mag-Lev Initiative, a comparative study of noise from Mag-Lev and wheel-rail vehicles was made by the firm of Harris Miller Miller &A Hanson, Inc. (Hanson et al. 1992). We will report only a few data from this excellent report. Other sources of noise data include the U.S. Office of Technology Assessment (OTA) Report (U.S. OTA, 1991) and technical advertising data from HSST Corporation (Japan) and Transrapid International (Germany).

A comparison of noise for various steel-wheel systems and the TR07 EML vehicle at a distance of 25 m from the track is given in Figure 7-25 (source: Harris–Miller report and Transrapid International). A comparison of everyday noise levels is given in Table 7-4, taken from both the OTA and Harris–Miller reports. One can clearly see that at 200 km/hr (124 miles/hr) the TR07 is 10 dB quieter than an American Amtrak vehicle and even the Shinkansen. But now wheel technologies such as TGV and ICE seem to have closed the gap

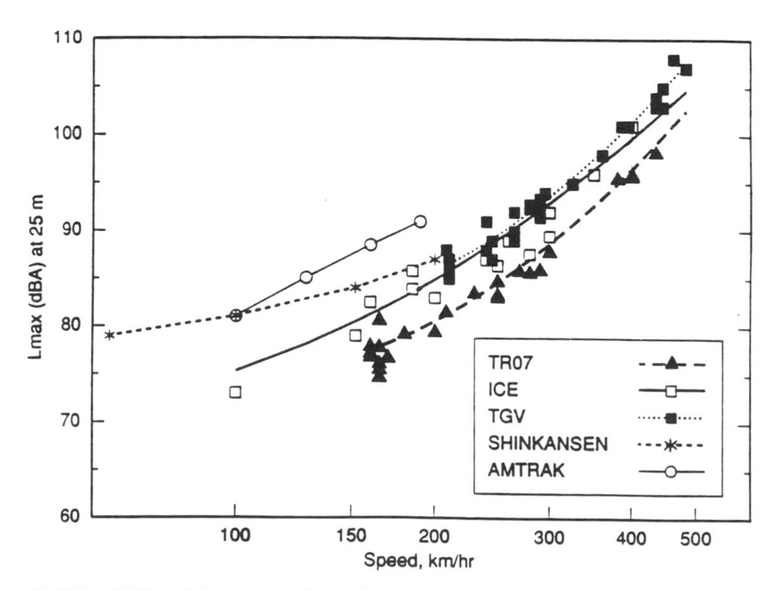

Figure 7-25 Off-guideway noise from Mag-Lev and electrically powered high-speed rail systems. [From Hanson et al. (1992).]

TABLE 7-4 Noise Characteristics of Transportation and Other Activities

Activity	Sound Level (dB)
Whispering	30
Light auto traffic at 100 ft	50
Conversational speech	60
Vacuum cleaner at 10 ft	69
Freight train at 50 ft	75
Shinkansen at 150 miles / hr at 82 ft	**80**
Alarm clock at 2 ft	80
Riding inside a city bus	83
Transrapid at 185 miles / hr at 82 ft	**84**
Heavy truck at 50 ft	90
TGV at 185 miles / hr at 82 ft	**91**
Jet takeoff at 2000 ft	105
Jet takeoff at 200 ft	120
Threshold of physical pain	130

Source: Office of Technology Assessment, 1991, based on U.S. General Accounting Office data.

between steel-wheel and EML Mag-Lev noise levels. At 400 km/hr the TR07 appears to be 5 dB quieter than the TGV. However, as speeds approach 500 km/hr, this gap seems to disappear. This is because aerodynamic or aeroacoustic sources begin to dominate at higher speeds. However, these different reports do not report noise data for the Japanese EDL vehicle tests in Miyazaki. This vehicle may have noise characteristics that are significantly different from those of the EML TR07 because of a larger gap and different guideway coil layout geometry.

A misunderstanding regarding levitated vehicles is that if there is no physical contact between vehicle and guideway, there can be no mechanical noise. However, mechanical or structural noise results from vibration excitation forces on the guideway and the vehicle. In magnetically levitated systems, large magnetic forces can and do have time-varying components that can excite the vehicle and guideway structure. These frequency sources include:

- Guideway-span crossing frequencies and harmonics
- Stator-pole passing frequencies (high in EML systems, lower in EDL systems)
- Line frequencies (e.g., LSM frequencies)

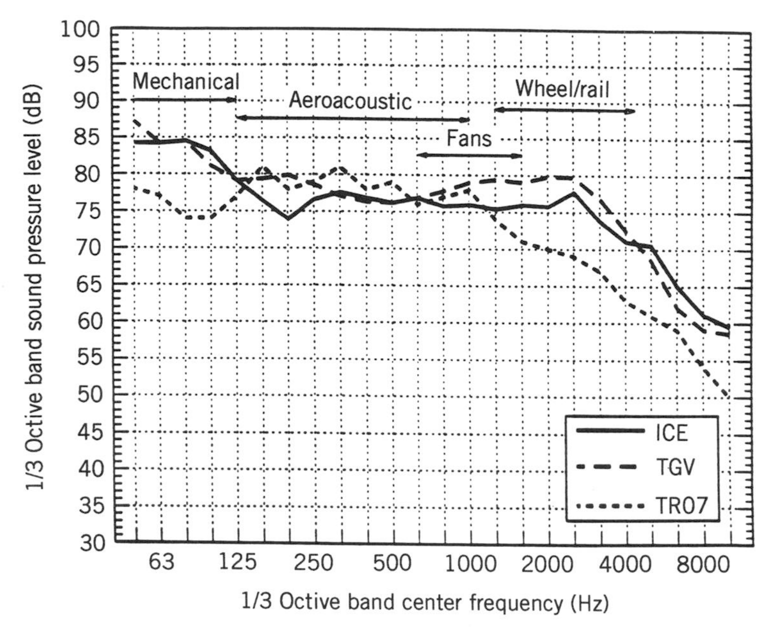

Figure 7-26 Noise spectra from Mag-Lev and high-speed rail systems at 290 km/hr. [From Hanson et al. (1992).]

As the Harris–Miller report points out, in ferromagnetic materials (EML systems), magnetostriction effects can cause vibration. Also, nonlinearities can create both harmonics and subharmonics, and in chaotic vibrations a single frequency source can create a wide spectrum of frequencies.

With regard to observed frequency content in Mag-Lev noise, measurements on the TR07 seem to indicate lower noise levels at higher frequencies (> 1250 Hz) than do measurements on steel-wheel vehicles (see Figure 7-26).

Aerodynamic noise is still an empirical subject with some broad principles. Noise can be generated at the nose and tail sections of the vehicle, along the vehicle surface, and between the vehicle–guideway gap. The Harris–Miller report identifies some basic aeroacoustic mechanisms:

- Fluid shear forces as in boundary layers and gaps
- Transition from laminar to turbulent flow
- Flow separation and reattachment
- Vortex generation, separation, and reattachment

- Flow over cavity resonances
- Flow past edges
- Turbulent boundary layer along the body surface

For Mag-Lev designs, these mechanisms suggest that the manner in which the guideway coils (for levitation, guidance, or LSM propulsion) are attached to the structure may significantly affect the radiated aeroacoustic noise. Also, some designers want to open up the bottom of the guideway to avoid snow and debris accumulation. However, open sections might increase radiated noise.

The Harris–Miller conclusions point out that public acceptance of a new transportation system may depend heavily on human environmental effects such as noise. The choices between sheet or discrete coil guideways, ladder or null-flux coils, or "U"-shaped or box-beam guideways should be examined not only for the electromagnetic or mechanical trade-offs but also for the affect the design will have on noise, especially at speeds above 350 km/hr.

7-5 TECHNICAL IMPACT OF HIGH-T_c SUPERCONDUCTORS ON MAG-LEV

Prototype superconducting Mag-Lev vehicles and small models all use the low-temperature, liquid helium (4.2 K)-based niobium–titanium or niobium–tin superconductors. These materials have been successfully used in magnets over the past 20 years for fusion studies, magnetohydrodynamics or MHD, and (in more recent applications) medical imaging magnets. In early 1987, new discoveries were made of superconducting materials that exhibited zero resistance to current flow at temperatures below 95 K which permitted their use at liquid nitrogen temperatures (77 K) (see Chapter 3). At a symposium on magnetic transportation in Buffalo, New York, sponsored by The New York State Institute of Superconductivity, two proponents of Mag-Lev expressed differing views on the use of high-T_c superconductors in Mag-Lev. Dr. Henry Kolm of Magneplane International, Inc., one of the early pioneers of Mag-Lev research, expressed the belief that conventional superconductors operating at liquid helium temperatures are still the best materials for Mag-Lev transportation vehicles. However, Dr. Yoshihiro Kyotani, the former head of the Japanese Mag-Lev project, said that if the new high-temperature superconductors were available in wire form, he would recommend their use in the Japanese Mag-Lev program. This debate continues in the Mag-Lev community. However, in the last two years, 1991–1993, significant progress has been made in high-temperature superconductivity bulk and wire such that a re-examination of its use in Mag-Lev is warranted.

In Table 7-5, I have tried to give a qualitative assessment of the potential impact of high-T_c superconductors on superconducting Mag-Lev transportation. At the present time the new superconductors such as BSSCO and YBCO have begun to demonstrate sufficient current-carrying capacity at high enough magnetic fields. Besides an increase in temperature, high magnetic fields and high current densities can destroy the superconducting state. High-temperature superconductor wire with 10^4- to 10^5-A/cm^2 capability does now exist in short lengths (1–10 m). The assessment of impact on Mag-Lev transportation, however, assumes that such wire will be available in the future, although that future may be 3–8 years away.

TABLE 7-5 Effect of High T_c Mag-Lev Design Issues

Design Issue	Major	Minor	None
Magnet			
Lift, guidance forces		×	
Cryogenics	×		
Thermal stability	×		
Wire manufacturing and design	×		
Stress effects		×	
Fatigue		×	
Vehicle			
Structure		×	
Shielding	×		
Suspension		×	
Aerodynamics			×
Dynamics and control			×
Guideway			
Continuous versus discrete track		×	
Active coil design			×
Wayside power		×	
Vehicle–guideway communication			×
Curves, elevation, banking		×	
Other			
Service	×		
Maintenance	×		
Capital costs		×	
Operating costs		×	
Human factors		×	

In brief, we believe that the new high-T_c superconductor could impact two areas of design for Mag-Lev transportation: (1) magnet design and (2) maintenance and service.

Cryostat Design

The major impact of the use of the new high-critical-temperature (T_c) superconductors will involve new cryostat designs for the magnets. Because of the almost infinite availability of liquid nitrogen vis-à-vis liquid helium, one can design a cryostat without the need to collect the gas and recool it. Elimination of the compressor and helium refrigerator would save significant weight. In addition, it would be expected that additional weight savings would be achieved in the cryostat itself.

It is a basic law of physics that stored magnetic energy requires a minimum amount of mass to resist the self-induced magnetic stresses in the magnet. This theorem is known as the *Virial Theorem* [see, e.g., Moon (1984).] However, it has been shown that for magnets that have length scales below 3–5 m, the need to minimize heat transfer effects in liquid helium magnets results in a magnet mass many times the minimum required by the Virial Theorem. Thus, at liquid helium temperatures, it becomes uneconomical and mass-inefficient to build magnet/cryostats below 1 m in size. At liquid nitrogen temperatures, however, it might be more mass-efficient to build magnetic/cryostats on a scale of 0.5 m or less. The consequences of a larger number of smaller magnets for Mag-Lev vehicles are discussed below.

Thermal Stability

Another important area of impact on magnet design is the thermal stability of the superconducting magnets. In low-temperature superconductors such as niobium–titanium wire, a local rise in temperature above the superconducting transition will lead to resistive heating which can produce a propagating normal conductor zone inside a magnet. To avoid such thermal instabilities, magnet designers have had to surround the superconducting filaments with copper to shunt away current when the material becomes resistive, or they have provided sufficient convective cooling using liquid helium. Because the new superconductors have heat capacity properties about an order of magnitude higher than those of either Nb–Ti or Nb_3Sn, they are less sensitive to such thermal instability. On the other hand, the low-temperature conductors have a rather high normal zone velocity; thus if a quench does occur, it will spread rapidly throughout the whole magnet, thus distributing the magnet energy uniformly in the form of heat. The new superconductors, on the other hand, have quench velocities more than an order of magnitude lower than those of their low-temperature counterparts; thus if a thermal quench is initiated, it will dump the whole magnet energy into one zone of the magnet and perhaps permanently damage the magnet. Thus, the drastically different thermal properties of the new superconductors, as well as liquid nitrogen vis-à-vis liquid helium, will require new thinking about how to design the magnet for safe reliable operation in the face of small thermal disturbances.

By far the greatest impact of the use of liquid-nitrogen-cooled superconductors on Mag-Lev is in the ease of maintenance and service of the cryostat-magnets. One can handle liquid nitrogen as easily as one pours coffee out of a thermos bottle. Liquid helium, although routinely used in modern research laboratories, requires special technical expertise to handle it safely and efficiently, not to mention the need for refrigeration and gas and liquid storage for cryostats. The use of liquid nitrogen in Mag-Lev vehicles will substantially increase the acceptance of this technology by operating companies.

Another important design parameter in Mag-Lev vehicles is the magnetic field levels in the passenger compartment. Although there are no clear studies showing any danger in working near low, static magnetic fields, standards for long-time human exposure have been decreasing in the last two decades. One of the design problems is to screen out stray fields from the passenger compartment. Some designers have used magnetic iron, but this has a severe weight penalty. The new superconductors could be used as a thick film on a lightweight structure to screen out magnetic fields through the Meissner effect when the fields are below the Type II threshold of around 100 G.

Smaller Magnet Design

Another consequence of the new materials on magnetic exposure criteria is that they permit the use of smaller magnets. The use of a spatial array of small magnets (less than 0.5 m in length), with reversed currents in neighboring coils, can lead to a shorter decay length of the field as one moves away normal to the magnets. This means that one can use a thinner, more lightweight shield, thereby saving additional weight on the vehicle.

The consequences of a larger number of smaller superconducting magnets with opposite polarity can be summarized below.

1. Small coils permit easier manufacture, especially if the wind and react method is required.
2. A larger number of smaller coils should give a better lift distribution under the vehicle.
3. A lower magnetic pressure should require lower currents per turn in the magnets, as well as lower fields.
4. Lower currents imply that a J_c value of $\sim 10^4$ A/cm^2 might lead to an acceptable magnet design; this is an order of magnitude below conventional wisdom regarding useful current densities for high-power application of superconductors.
5. Lower currents and lower fields, as well as the shorter spatial wavelength due to the larger number of coils, should lead to lower fields in the passenger compartment.

Mag-Lev with Superconducting Permanent Magnets

Is superconducting magnetic transportation at liquid nitrogen temperatures possible without wire or tape wound coils?

Conventional superconducting Mag-Lev vehicles have been built and tested for over 20 years using coil wound Nb–Ti magnets at liquid helium temperatures. These coils, however, require an intricate cryostat design that must transmit lift and guidance forces to the vehicle and, at the same time, minimize heat leaks. The Japanese design also employs helium collection and liquifier systems on board, which constitutes a weight penalty.

At least two obstacles obstruct the use of high-T_c coils for Mag-Lev: (1) low J_c and brittleness of high-T_c material and (2) difficulty of manufacture.

However, two developments in materials processing of high-T_c materials have improved the feasibility of liquid nitrogen Mag-Lev. First, there has been rapid improvements in the bulk material critical current levels in magnetic fields. For example, the Japanese Superconducting Research Center at ISTEC in Tokyo have claimed a J_c value of $\sim$ 35,000 A/cm^2 in $YBa_2Cu_3O_7$ in a 1-T field for a bulk material prepared by a "quench–pulverize–melt–growth" process (see Chapter 3). Related processes have been under study at ATT Laboratories, the University of Houston, and the Catholic University of America.

The second development is the improvement of magnetization properties of YBCO material using melt–growth methods. This is shown in Figure 3-19. These properties allow one to *magnetically charge* a bulk material to a given magnetization level and then retain a significant remnant magnetization when the applied field is removed. Conventional sintered YBCO materials retain only low fields on the order of 10–20 G, whereas melt–growth materials have retained over 1000 G. ISTEC scientists in Japan have claimed that one can process a YBCO material to retain up to 1.5–2.0 T. Significant progress in remnant fields has also been made at the University of Houston (Weinstein and Chen, 1992) with trapped fields of over 4 T in YBCO. This raises the interesting prospects of developing superconducting permanent magnets (SPMs) (see also Chapter 3).

Proposals have been made to use permanent magnets instead of superconducting coils to generate the magnetic pressure needed to levitate vehicles (Polgreen, 1966). The problem has been that the room-temperature conventional magnets have a severe weight penalty and generate fields on the order of 0.5 T. However, recent discoveries raise the possibility of employing superconducting permanent magnets —that is, bulk forms that retain superconducting circulating currents after the applied field is removed. It is thought that fields as high as 2.0–4.0 T might be achieved. Because the magnetic pressure is proportional to the square of the magnetic field, these superconducting permanent magnets at liquid nitrogen temperatures can generate more than ten times the levitation pressure of room-temperature rare earth permanent magnets, thereby decreasing the weight penalty of the magnets. This means that vehicles could be levitated with the new superconductors without the need for wire or tape forms of these new materials (Figure 7-27). This is a *radically new concept* still in the exploratory stage.

In this concept, small bulk pellets (e.g., 4–6 cm in diameter) of YBCO or other high-T_c material would be placed in a magnetizing field at liquid nitrogen temperature and removed while maintaining this temperature. They would then be assembled in small cryostat modules to be installed under the Mag-Lev vehicle. This process could be automated. No on-board current supply would be required. The multiplicity of cryostats would ensure redundancy in case of a cryostat failure.

There is a basis theorem of electromagnetics which states that the field generated by a current filament is identical to a field pattern generated by a dipole distribution. Thus, one can arrange a set of

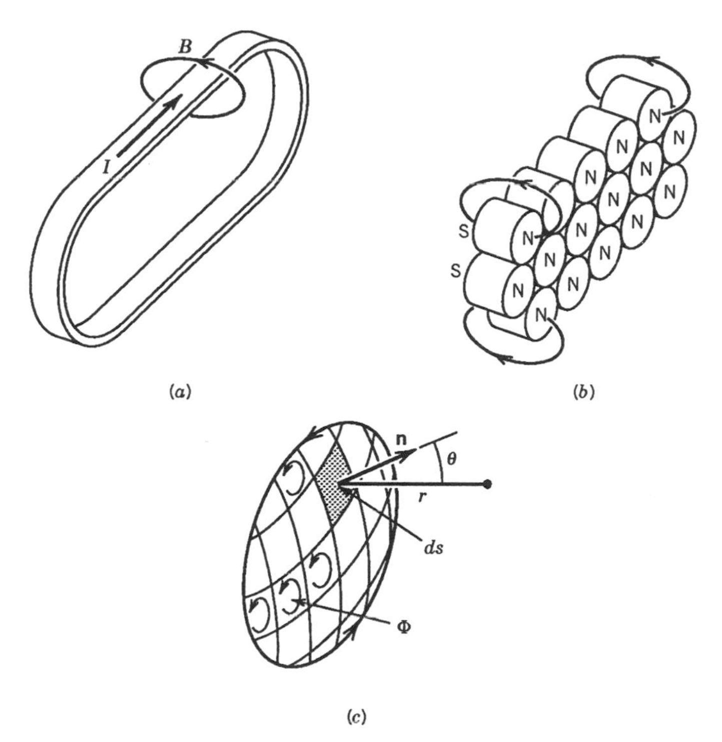

Figure 7-27 (*a*) Conventional Coil-wound magnet. (*b*) Sketch of superconducting permanent magnet array. (*c*) Magnetic shell.

SPM dipoles in a planar array and produce the magnetic field of an equivalent coil wound magnet. This principle has been stated in a theorem in Jeans (1925):

> A current flowing in any closed circuit produces the same magnetic field as a certain magnetic shell known as the "equivalent magnetic shell." This shell may be taken to be any shell having the circuit for its boundary, its strength being uniform and proportional to that of the current.

The field potential of a magnetic shell is also found in Jeans (1925). When the magnetization of the shell is normal to the shell thickness and of strength Φ, then the field potential is given by

$$\mathbf{H} = -\nabla\Omega$$

$$\Omega = \int\int \Phi \frac{\partial}{\partial n}\left(\frac{1}{r}\right) dS$$

When Φ is uniform, we obtain

$$\Omega = \Phi\alpha$$

where α is the solid angle subtended by the magnetic shell at the field point—for example,

$$\alpha = \iint \frac{\cos\theta}{r^2}\, dS$$

where θ is the angle between the normal at dS and the line joining dS to the field point. Any element of the magnetic shell of strength Φ behaves like a magnetized particle of strength $\Phi\, dS$ with its axis normal to dS.

However, this material can also be used as a magnetic field shield. Thus, it might be possible to use melt–growth bulk material both as a source of field and as a field-shaping and field-containment material to achieve high fields under the vehicle and near earth level fields (1–2 G) in the passenger space.

Hybrid Ferromagnetic – Superconducting Mag-Lev

Finally there is a theoretical study by Braunbeck (1939a) that claims that the magnetic instability of ferromagnetic attractive forces may be stabilized by the presence of superconducting or diamagnetic material. A so-called "mixed-mu" Mag-Lev system has been described by Rhodes (1982) (see also Rhodes and Mulhall, 1981), in which a superconducting screen is placed between an electromagnet and a ferromagnetic rail. Thus, it is not inconceivable that a hybrid levitation system using bulk or thick-film YBCO as a superconducting screen might obviate the need for feedback control in an EML design.

7-6 THE FUTURE OF MAG-LEV

A wise sage once said, "Time erases most visions of the future." So it is guaranteed that these words written in January 1994 will likely be irrelevant in a year hence. The history of Mag-Lev transportation has had its high and low points. Earlier technical advances and promised political decisions evaporated in the face of periodic recessions and cheap airplane fares and subsidized automobile travel. At one of the high points, Canada had signed an agreement with Germany to build a Mag-Lev Transrapid line from Montreal to Toronto in the 1970s. But within two years the Germans had to withdraw because technology and economics were not secure enough. And, commitments by both Japan and Germany to build revenue Mag-Lev service in their own countries have also had to face economic and technical realities. Part of that reality is the nature of macrotechnology in public works. The railroad took nearly a century to mature from its first beginnings in the nineteenth century, and the time from the Wright brother's experiments in flight to the first passenger jet service was nearly half a century and included the tragedy of the first Comets. Even the time from the first Boeing 707 to the 747 jumbo jet was more than a decade. This is in sharp contrast to electronic and computer microtechnology in which development to product time can be a few years. It has only been 20 years from the first prototype Mag-Lev tests at high speeds to the present threshold of potential public operation. And given history, we should not expect anything less than another decade or two before the vision of Mag-Lev dreamers of the 1960s and 1970s will be realized.

In spite of these sobering thoughts, the development of Mag-Lev in the United States is at a high point in early 1994. The National Mag-Lev Initiative was established in 1990 and a major transportation bill passed in 1991 which called for a $750 million prototype Mag-Lev line. In late 1993, the final NMI report from the U.S. Department of Transportation concluded that U.S. Mag-Lev technology could leapfrog existing Japanese and German systems and outperform current high-speed rail systems such as TGV and ICE. They also concluded that the two application corridors that were most economically viable were the northeast corridor (Boston to Washington) and the California corridor (San Francisco to San Diego). In a minor setback, the U.S. Congress refused funds to begin prototype development. Instead they approved a 20 million dollar research and development program in Mag-Lev transportation. This represents the largest annual U.S. allocation for Mag-Lev research.

In another development, a joint study by the states of New York and Massachusetts, conducted by the EBASCO Corporation of New York, recommends development of a Mag-Lev system between Boston, Albany, and New York City and west to Buffalo using existing interstate highway rights of way (*MAGLEV News*, 1993). A principle factor in this study was the cost of new land acquisition for high-speed rail alignments to avoid steep grades and sharp turns. The ability of a Mag-Lev system to negotiate high grades and to bank into turns using existing highway grades was a key technical advantage to the economic problem of funding suitable rights of way in dense population areas.

So Mag-Lev has moved from the laboratory to the legislature. The creation of new transportation systems in technologically developed societies presents a paradox. The very advanced state of those societies from whom the technology emanates also generates sophisticated opposition to new technology as a threat to quality of life because of noise, pollution, and visual aesthetics. The next decade will see if Mag-Lev transportation can overcome these fears and fulfill the early promise of levitation scientists and engineers.

Mag-Lev will also have to compete with new initiatives in the United States to build an "information superhighway." Communication technologists have often claimed that new communication networks will obviate the need for new transportation investment. However, history has shown that every new advance in communication, from the telegraph to television, has been met with an increased demand for travel. Thus, wise public planners should see the need for investments in both new transportation as well as communication technologies.

REFERENCES

Adler, R. J., and Anderson, W. W. (1988). "Magnetic Suspension of Superconductors at 4.2 K," *Appl. Phys. Lett.* **53**, 2346–2347.

Allaire, P., ed. (1992). *Proceedings of the Third International Symposium on Magnetic Bearings*, Technomic Publishing Co., Lancaster, PA.

Allaire, P. E., Humphris, R. R., and Lewis, D. W. (1992). "The Father of Magnetic Bearings: Professor Jesse W. Beams," in *Electromagnetic Forces and Application. Proceedings of the Third International ISEM Symposium*, Sendai, Japan, January 1991, J. Tani and T. Takagi, eds., Elsevier, Amsterdam, pp. 155–158.

Allen, L. H., Broussard, P. R., Claassen, J. H., and Wolf, S. A. (1988). "Temperature and Field Dependence of the Critical Current Densities of Y–Ba–Cu–O," *Appl. Phys. Lett.* **53**, 1338–1340.

Anderson, P. W., and Kim, Y. B. (1964). "Hard Superconductivity: Theory of the Motion of Abrikosov Flux Lines," *Rev. Mod. Phys.* **Jan.**, 39–45.

Arkadiev, V. (1945). "Hovering of a Magnet over a Superconductor," *J. Phys. (Moscow)* **9**(2), 148.

Arkadiev, V. (1947). "A Floating Magnet," *Nature* **160**, 330.

Atherton, D. L. and Eastham, A. R. (1974). "Electromagnetic Dynamic Levitation Development in Canada," *High Speed Ground Transport Journal*, **8**(2), 101–110.

Atherton, D. L., Belanger, P. R., Eastham, A. R., et al. (1980). "High-Speed Magnetically Levitated Transport Development in Canada," *J. Adv. Transp.* **14**, 73–105.

Azukizawa, T., Morishita, M., M., Kasahara, N., Akoshi, M., and Ogiwara, H. (1992). "Characteristics of Magnets Suspension System Using High T_c Superconductors," in *Proceedings of the International Symposium on Nonlinear Phenomena in Electromagnetic Fields*, January 1992, ISEM-Nagoya.

Baiko, A. V., Voevodskii, K. E., and Kochetkov, V. M. (1980). "Vertical Unstable Stability of Electrodynamic Suspension of High Speed Ground Transport," *Cryogenics* **20**, 271–276.

Balachandran, U., Shi, D., dos Santos, D. I., Graham, S. W., Patel, M. A., Tani, B., Vandervoort, K., Claus, H., and Poeppel, R. B. (1988). "120 K Superconductivity in the (Bi, Pb)–Sr–Ca–Cu–O System," *Physica C* **156**, 649–651.

Barrows, T., McCullum, D., Mark, S., Castellino, R. C. (1992). *Aerodynamic Forces on Maglev Vehicles*. U.S. Dept. of Transportation Report for National Maglev Initiative DOT/FRA/NMI-92/21, December.

Basinger, S. A., Hull, J. R., and Mulcahy, T. M. (1990). "Amplitude Dependence of Magnetic Stiffness in Bulk High-Temperature Superconductors," *Appl. Phys. Lett.* **57**,

Beams, J. N., Spitzer, D. M., Jr., and Wade, J. P., Jr. (1962). *Rev. Sci. Instr.* **33**, 151.

Bean, C. P. (1964). "Magnetization of High-Field Superconductors," *Rev. Mod. Phys.* **36**, 31–39.

Bean, C. P. (1972). "Magnetization of Hard Superconductors," *Phys. Rev. Lett.* **8**(6), 250–253.

Beaugnon, E., and Tournier, R. (1990). "Levitation of Water and Organic Substances in High Static Magnetic Fields," *Nature*.

Beloozerov, V. N. (1966). "Confinement of a Superconducting Sphere by a System of Ring Currents," *Soviet Phys. Tech. Phys.* **11**(5), 631–635.

Bobrov, E. S. (1989). "Mechanical Characteristics of High Temperature Ceramic Superconductors," in *Applied Electromagnetics in Materials*, K. Miya, ed., Pergamon Press, Elmsford, NY, pp. 347–357.

Boerdijk, A. B. (1956). *Phillips Res. Rep.* **11**, 45.

Borcherts, R. H. (1975). "Repulsion Magnetic Suspension Research—U.S. Progress to Date," *Cryogenics* **15**, 385–393.

Borcherts, R. H., and Davis, L. C. (1972). "Force on a Coil Moving Over a Conducting Surface Including Edge and Channel Effects," *J. Appl. Phys.*, **43**, 2418–2427.

Borcherts, R. H., and Reitz, J. R. (1971). "High-Speed Transportation via Magnetically Supported Vehicles. A Study of the Magnetic Forces," *Trans. Res.* **5**, 197–209.

Borcherts, R. H., Davis, L. C., Reitz, J. R., and Wilke, D. F. (1973). "Baseline Specifications for a Magnetically Suspended High-Speed Vehicle," *Proc. IEEE* **61**(5), 569–577.

Bourke, R. D. (1964) "A Theoretical and Experimental Study of a Superconducting Magnetically-Supported Spinning Body," NASA Contractor Report CR-108, October 1964.

Brandt, E. H. (1988). "Friction in Levitated Superconductors," *Appl. Phys. Lett.* **53**, 1554–1556.

Brandt, E. H. (1989). "Levitation in Physics," *Science* **243**, 349–355.

Brandt, E. H. (1990a). "Fluctuation, Melting, De-Pinning, Creep, and Diffusion of the Flux-Line Lattice in High-T_c Superconductors," International Conference on Low Temperature Physics LT19, Brighton, England, *Physica B*.

Brandt, E. H. (1990b). "La Lévitation," *Recherche* **224**.

Brandt, E. H. (1990c). "Rigid Levitation and Suspension of High-Temperature Superconductors by Magnets," *Am. J. Phys.* **58**, 43–49.

Braun, M., Buszka, P., Motylewski, T., Przydrozny, W., and Sliwa, C. (1990). "Vibration Frequency and Height of a Magnet Levitating over a Type II Superconductor," *Physica C* **171**, 537–542.

Braunbeck, W. (1939a). "Freies Schweben Diamagnetischer Körper in Magnetfeld," *Z. Phys.* **112**, 764–769.

Braunbeck, W. (1939b). "Freischwebende Körper in electrischen and magnetischen Feld," *Z. Phys.* **112**(11–12), 753–763.

Brechna, H. (1973). *Superconducting Magnet Systems*, Springer-Verlag, New York.

Briggs, B. M., and Oman, H. (1990). "Superconducting Magnetically Levitated Carrier—A Commuting Solution, *25th IECEC* **3**, 419–424.

Brown, W. F., Jr. (1966). *Magnetoelastic Interactions*, Springer-Verlag, New York.

Buchhold, T. A. (1960). "Applications of Superconductivity" *Sci. Am.* **202**(3).

Burke, H. E. (1986). *Handbook of Magnetic Phenomena*, Van Nostrand Reinhold Company, New York.

Cai, Y., Chen, S. S., and Rote, D. M. (1992). "Dynamics and Controls in Maglev Systems," U.S. Department of Energy Report, Argonne National Laboratory ANL-92/43.

Cai, Y., Chen, S. S., Mulcahy, T. M., and Rote, D. M. (1992). "Dynamic Stability of Maglev Systems," U.S. Department of Energy Report, Argonne National Laboratory ANL-92/21.

Celani, F., Fruchter, L., Giovannella, C., Messi, R., Pace, Saggese, S. A., and Sparvieri, N. (1989). "Torque Measurements of Textured $YBa_2Cu_3O_{7-x}$ Sintered Pellets," *IEEE Trans. Magn.* **25**, 2348–2351.

Chang, P.-Z. (1991). "Mechanics of Superconducting Magnetic Bearings," Ph.D. Dissertation, Cornell University, Ithaca, NY.

Chang, P.-Z., and F. C. Moon. (1991). "Low-Temperature Measurements of Levitation Forces in $YBA_2CU_3O_7$; 4.2 K to 77 K," Cornell University Report.

Chang, P.-Z., Moon, F. C., and Hull, J. R. (1989). "Drag Force and Magnetic Stiffness in Bulk High-Temperature Superconductors," Argonne National Laboratories Report, December.

Chang, P.-Z., Moon, F. C., Hull, J. R., and Mulcahy, T. M. (1990). "Levitation Force and Magnetic Stiffness in Bulk High-Temperature Superconductors," *J. Appl. Phys.* **67**, 4358–4360.

Chang, P.-Z., Chiu, M., Moon, F. C., Jin, S., and Tiefel, T. H. (1992). "Grain-Size Dependence of Levitation Forces in Bulk $YBa_2Cu_3O_7$ Superconductors," Cornell University Report, Mechanical and Aerospace Engineering.

Chen, P. C., Oliversen, R. J., and Hojaji, H. (1992). "Advanced Technology Lunar Telescopes," in *HTS Materials, Bulk Processing, Bulk Applications, Proceedings of the 1992 TCSUH Workshop*, Houston, Texas, February 1992, C. W. Chu, W. K. Chu, P.-H. Hor, and K. Salama, eds., World Scientific, Singapore, pp. 513–524.

Chu, D. (1982). "On the Instabilities Inherent in Electrodynamic Magnetic Levitation," M. S. Dissertation, Cornell University, Ithaca, NY.

Chu, D., and Moon, F. C. (1983). "Dynamic Instabilities in Magnetically Levitated Models," *J. Appl. Phys.* **54**(3), 1619–1625.

Chu, W. K., Ma, K. B., McMichael, C. K., and Lamb, M. A. (1992). "Applications of High Temperature Superconductors on Levitation Bearings and Other Levitation Devices," in *Applied Superconductivity*, Special Issue, *Proceedings of the 3rd International Conference World Congress on Superconductivity*, September 1992, Part II, Munich, pp. 1259–1264.

Clem, J. R. (1979). "Theory of AC Losses in Type-II Superconductors with a Field-Dependent Surface Barrier," *J. Appl. Phys.* **50**, 3518–3530.

Coffey, H. T. (1974). "Magnetic Suspension Studies for High-Speed Vehicles," *Advances in Cryogenic Engineering*, Vol. 19, Plenum Press, New York, pp. 137–153.

Coffey, H. T., Chilton, F., and Hoppie, L. O. (1972). "Magnetic Levitation of High-Speed Ground Vehicles," in *Proceedings of the 5th International Applied Superconductivity Conference*, IEEE Publication No. 72CH0682-5-TABSC, May 1–3, 1972, pp. 62–75.

Coffey, H. T., Solinski, J. C., Colton, J. D., and Woodbury, J. R. (1974). "Dynamic Performance of SRI Maglev Vehicle," *IEEE Trans. Magn.* **10**(3),

Connor, K. A., and Tichy, J. A. (1988). "Analysis of an Eddy Current Journal Bearing," *Trans. ASME* **110**,

Crandall, S. H., Karnopp, D. C., Kurtz, E. F., Jr., and Pridmore-Brown, D. C. (1968). *Dynamics of Mechanical and Electromechanical Systems*, McGraw–Hill, New York.

Crapo, A. D., and Lloyd, J. D. (1990). "Homopolar DC Motor and Trapped Flux Brushless DC Motor Using High Temperature Superconductor Materials," presented at the 1990 Applied Superconductivity Conference, Snowmass Village, Aspen, CO.

Cullity, B. D. (1972). *Introduction to Magnetic Materials*, Addison–Wesley, Reading, MA.

Culver, W. H., and Davis, M. H. (1957). "An Application of Superconductivity to Inertial Navigation," Rand Corporation Report RM-1852, January 7, 1957.

Culver, W. H., and Davis, M. H. (1960). Rand Corporation Report R-363, 19 April 1960.

Dalgleish, E. G., and Riches, E. E. (1986). "A Review of Birmingham MAGLEV After One Year in Public Service," in *Proceedings of the International Conference on Maglev and Linear Drives*, May 14–16, 1986, Vancouver, B.C., Canada, IEEE Publication No. 86CH2276-4, p. 105.

Danby, G. T., Jackson, J. W., and Powell, J. R. (1972). "Force Calculations for Hybrid (Ferro-Nullflux) Low-Drag Systems," in *Proceedings of the 5th International Applied Superconductivity Conference* IEEE Publication No. 72CH0682-5-TABSC, pp. 113–119.

Daniels, L. E., Ahlbeck, D. R., Stekly, Z. J., and Gregorek, G. M. (1992). "Influence of Guideway Flexibility on Maglev Vehicle/Guideway Dynamic Forces," U.S. Department of Transportation, Federal Railway Administration, National Maglev Initiative, DOT/FRA/NMI-92/09, July.

Davis, L. C. (1990). "Lateral Restoring Force on a Magnet Levitated Above a Superconductor," *J. Appl. Phys.* **67**, 2631–2636.

Davis, L. C., and Wilke, D. F. (1971). "Analysis of Motion of Magnetic Levitation Systems: Implications," *J. Appl. Phys.* **42**(12), 4779–4793.

Davis, L. C., Logothetis, E. M., and Soltis, R. E. (1988). "Stability of Magnets Levitated Above Superconductors," *J. Appl. Phys.* **64**, 4212–4218.

Delprette, C., Genta, G., Mazzocchetti, L., Rava, E., Ricca, A., Ripamonti, G., Somtini, L., Tonoli, A., Varesi, E., and Zannella, S. (1992). "High Speed Asynchronous Motor with Superconducting Bearings," in *Proceedings of the Third International Symposium on Magnetic Bearings*, P. E. Allaire, Alexandria, VA, ed., Technomic Publ., Lancaster, PA, pp. 287–296.

Dill, J. F., Rao, D. K., and Decher, R. (1990). "A Feasibility Study for the Application of High Temperature Superconducting Bearings to Rocket Engine Turbopumps," presented at the Conference on Advanced Earth-to-Orbit Propulsion Technology, May 15–17, NASA Marshall Space Flight Center.

Downer, J., Anastas, G., Goldie, J., Gondhalekar, V., and Torti, R. (1990). "A Superconducting Large-Angle Magnetic Suspension," *25th IECEC* **3**, 482–487.

Dreimann, K. (1989). "M-Bahn MAGLEV Transit System Experience, Status, Application," in *Proceedings of the 11th International Conference on Magnetically Levitated Systems and Linear Drives*, July 7–11, 1989, Yokohama, Japan, pp. 101–104.

Earnshaw, S. (1842). "On the Nature of Molecular Forces which Regulate the Constitution of Luminiferous Ether," *Trans. Cambridge Philos. Soc.* **7**, 97–112.

Eastham, A. R., and Atherton, D. L. (1975). "Superconducting Maglev and Linear Synchronous Motor Development in Canada," *IEEE Trans.* **MAG-11**, 627–632.

Eyssa, Y. M., and Huang, X. (1990). "High Pressure Superconducting Radial Magnetic Bearing," in *Proceedings of the 25th IECEC*, Reno, pp. 454–459.

File, J., Martin, G. D., Mills, R. G., and Upham, J. L. (1968). "Stabilized, Levitated Superconducting Rings," *J. Appl. Phys.* **39**, 2623.

Fink, H. J., and Hobrecht, **X. X.** (1971). "Instability of Vehicles Levitated by Eddy Current Repulsion—Case of an Infinitely Long Current Loop," *J. Appl. Phys.* **42**(9), 3446–3450.

Foster–Miller, Inc. (1992). "Maglev System Concept Definition Report," U.S. Department of Transportation, Federal Railway Administration, National Maglev Initiative, DOT/FRA/ORD-92/01.

Frazier, R. H., Gilinson, P. J., Jr., and Oberbeck, G. A. (1974). *Magnetic and Electric Suspensions*, MIT Press, Cambridge, MA.

Fukuyama, H., Seki, K., Takizawa, T., Aihara, S., Murakami, M., Takaichi, H., and Tanaka, S. (1991). "Superconducting Magnetic Bearing Using MPMG YBaCuO," presented at the ISTEC Meeting, October 1991, Tokyo, Japan in *Adv. in Superconductivity IV*, H. Hayakawa, N. Koshizuka (eds.), Springer-Verlag, Tokyo, 1992, 1093–1096.

Gafka, D., and Tani, J. (1992). "Chaotic Behavior, Strange Attractors and Bifurcations in Magnetic Levitation Systems," presented at the International Symposium on Nonlinear Phenomena in Electromagnetic Fields, January 1992, Japan, ISEM-Nagoya.

Geary, P. J., (1964). *Magnetic and Electric Suspensions*, British Scientific Instrument Research Association.

Glatzel, K., Khurdok, G., and Rogg, D. (1980). "Development of the Magnetically Suspended Transportation System in the Federal Republic of Germany," *IEEE Trans. Veh. Technol.* **Vt-19**, 3–17.

Goddard, R. (1909). "The Limit of Rapid Travel" [Editorial], *Sci. Am.* **20**.

Goldstein, H. (1950). *Classical Mechanics*, Addison–Wesley, Reading, MA.

Golkowski, Cz and Moon, F. C. (1994). "Levitation Forces in Bulk YBCO in a High Magnetic Field and Low Temperatures," *J. Appl. Physics*, to appear.

Guderjahn, C. A., Wipf, S. L., Fink, H. J., Boom, R. W., McKenzie, K. E., Williams, D., and Downey, T. (1969). "Magnetic Suspension and Guidance for High Speed Rockets by Superconducting Magnets," *J. Appl. Phys.* **40**, 2133–2140.

Hagedorn, P. (1988). *Non-Linear Oscillations*, 2nd edition, Oxford University Press, New York.

Haldar, P., and Motowidlo, L. (1992). "Recent Developments in the Processing of High J_c Silver Clad Mono- and Multi-filament $(Bi, Pb)_2Sr_2Ca_2Cu_3O_{10}$ Wires and Tapes," *J. Metals*.

Hammitt, A. G. (1973). *The Aerodynamics of High Speed Ground Transportation*. Western Periodicals Co., North Hollywood, CA.

Hanson, C., Abbot, P. D., and Dyer, I. (1992). "Noise Sources of High Speed Maglev Trains," U.S. Department of Transportation, Federal Railway Administration, National Maglev Initiative Report, May.

Harding, J. T. (1961). "Force and Torque on a Superconducting Ellipsoid in an Axially Symmetric Magnetic Field," NASA Jet Propulsion Laboratory Technical Release No. 32-242.

Harding, J. T. (1965a). "Force on a Superconducting Sphere in a Magnetic Field: The General Case," NASA, Jet Propulsion Laboratory Technical Release No. 32-806.

Harding, J. T. (1965b). "Levitation of Superconducting Rotors," *Int. Adv. Cryogenic Eng.*, **10B**, 137.

Harding, J. T., and Lawson, D. B. (1968). "Superconducting Gyroscope: Drift Data and Mathematical Model," *AIAA J.* **6**(2), 305.

Harding, J. T., and Tuffias, R. H. (1960). "The Cryogenic Gyro," NASA, Jet Propulsion Laboratory Technical Release No. 34-100.

Hayes, W. F. (1979). "Conceptual Design Study of a High Speed MAGLEV Guided Ground Transportation System," *Natl. Res. Counc. Can. Div. Mech. Eng. Q. Bull.*, No. 3, 35–37.

He, J. L., Coffey, H. T., Rote, R. M., and Wang, Z. (1991). "Publications on Maglev Technologies," U.S. Department of Energy Report, Argonne National Laboratory, No. ANL/ESD/TM-29, December.

Hein, R. A. (1974). "Superconductivity: Large-Scale Applications," *Science* **185**, (4147) 211–222.

Hellman, F., Gyorgy, E. M., Johnson, D. W., Jr., O'Bryan, H. M., and Sherwood, R. C. (1988). "Levitation of a Magnet Over a Flat Type II Superconductor," *J. Appl. Phys.* **63**, 447–450.

Hikihara, H. and Moon, F. C. (1994). *Chaotic Levitated Motions of a Magnet Supported by a Superconducting Bearing*, Cornell University Report, Sibley School of Mechanical and Aerospace Engineering.

Hoard, R. W. (1980). *The Effects of Strain on the Superconductivity Properties of Niobium-Tin Conductors*, Ph.D. dissertation, University of Washington,

Lawrence Livermore Laboratory Report No. UCRL-53069 Livermore, California.

Hockney, R., D. Eisenhaure, and T. Hawkey (1988). "Magnetic Bearings for an Optical Disk Read/Write Head," *23rd IECEC* **2**,

Hojaji, H., Michael, K. A., Barkatt, A., Thorpe, A. N., Ware, M. F., Talmy, I. G., Haught, D. A., and Alterescu, S. (1989). "A Comparative Study of Sintered and Melt–Grown Recrystallized $YBa_2Cu_3O_x$," *J. Mater. Res.* **4**, 28–32.

Hojaji, H., Barkatt, A., Michael, K. A., Hu, S., Thrope, A. N., Ware, M. F., Talmy, I. G., Haught, D. A., and Alterescu, S. (1990). "Yttrium Enrichment and Improved Magnetic Properties in Partially Melted Y–Ba–Cu–O Materials," *J. Mater. Res.* **5**, 721.

Homer, G. J., Randle, T. C., Walters, C. R., Wilson, M. N., and Bevir, M. K. (1977). "A New Method for Stable Levitation of an Iron Body Using Superconductors," *J. Phys. D. Appl. Phys.* **10**, 879–886.

Hoppie, L. O., Chilton, F., Coffey, H. T., and Singleton, R. C. (1972). "Electromagnetic Lift and Drag Forces on a Superconducting Magnet Propelled Along a Guideway Composed of Metallic Loops," in *Proceedings of the 5th International Applied Superconductivity Conference*, May 1–3, 1972, IEEE Publication No. 72CH0682-5-TABSC, pp. 113–119.

Huang, X., and Eyssa, Y. M. (1990). "A Passive Bearing System Using Superconducting Magnets," in *Proceedings of the Intersociety Energy Conversion Engineering Conference*, Reno, NV, pp. 464–468.

Huebener, R. P. (1979). *Magnetic Flux Structures in Superconductors*, Springer-Verlag, Berlin.

Hull, J. R. (1990). "Materials Research Issues in Superconducting Levitation and Suspension as Applied to Magnetic Bearings," *25th IECEC* **3**, 425–431.

Hull, J. R., Mulcahy, T. M., Lynds, L., Weinberger, B. R., Moon, F. C., and Chang, P.-Z. (1990). "Phenomenology of Forces Acting Between Magnets and Superconductors," in *Proceedings of the Intersociety Energy Conversion Engineering Conference*, Reno, NV, pp. 432–437.

Huseyin, K. (1978). *Vibrations and Stability of Multiple Parameter Systems*, Noordhoff International Publications, Alphenaan den Rijn, The Netherlands.

Iannello, V. (1990). "Superconducting Meissner Effect Bearings for Cryogenic Turbomachines," preprint, NASA Goddard contractors meeting.

Ishigaki, H., Itoh, M. Hida, A., Endo, H. Oya, T., Ohyama, T., and Minemoto, T. (1990). "Measurement of Repulsive Force of High T_c Materials Due to Meissner Effect and Its Two Dimensional Distribution," in *Proceedings of the 1990 Applied Superconductivity Conference*, Snowmass Village, Aspen, CO.

Itoh, M., Ishigaki, H., and Hida, A. (1989). Fabrication of Y–Ba–Cu–O Superconductor for Magnetic Bearing, *IEEE Trans. Magn.* **25**, 2518–2521.

Iwamoto, M., Yamada, T., and Ohno, E. "Magnetic Damping Force in Electrodynamically Suspended Trains," in *1974 Digest of Inter-Mag Conference of IEEE*, Toronto, Canada, 458–461.

Iwasa, Y. (1973). "Electromagnetic Flight Stability by Model Impedance Simulation," *J. Appl. Phys.* **44**(2), 858–862.

Jackson, J. D. (1962). *Classical Electrodynamics*, John Wiley & Sons, New York.

Jayawant, B. V. (1981). *Eelectromagnetic Levitation and Suspension Techniques*, Edward Arnold, London.

Jayawant, B. V. (1988). "Review Lecture on Electromagnetic Suspension and Levitation Techniques," *Proc. R. Soc. Lond.* **A416**, 245–320.

Jeans, J. (1925) *The Mathematical Theory of Electricity and Magnetism*, 5th edition, Cambridge University Press, Cambridge, England.

Jin, S., Tiefel, T. H., Sherwood, R. C., van Dover, R. B., Davis, M. E., Kammlott, G. W., and Fastnacht, R. A. (1988). "Melt-Textured Growth of Polycrystalline $YBa_2Cu_3O_{7-d}$ with High Transport J_c at 77 K," *Phys. Rev. B* **37**, 7850–7853.

Jin, S., Tiefel, T. H., Nakahara, S., Grabener, J. E., O'Bryan, H. M., Fastnacht, R. A., and Kammlott, G. W. (1990). "Enhanced Flux Pinning by Phase Decomposition in Y–Ba–Cu–O," *Appl. Phys. Lett.* **56**, 1287–1289.

Johansen, T. H., Bratsberg, H., and Yang, Z. J. (1990). "Thickness Dependence of the Levitation Force in Superconducting $YBa_2Cu_3O_x$," in *Proceedings of The 2nd World Congress on Superconductors*, Houston, TX.

Johansen, T. H., Bratsberg, H., Yang, Z. J. Helgesen, G., and Skjeltorp, A. T. (1990). "A Pendulum Feedback System to Measure the Lateral Force on a Magnet Placed Above a High-T_c Superconductor," *Rev. Sci. Instrum.* **61** (2) 3827–3829.

Johansen, T. H., Yang, Z. J., Bratsberg, H., Helgesen, G., and Skjeltorp, A. T. (1991). "The Lateral Force on a Magnet Placed Above a Planar $YBa_2Cu_3O_x$ Superconductor," *Appl. Phys. Lett.* **58** (2), 1–3.

Johnson, W. G., Jr. and Dress, D. A. (1989). *The 13-Inch Magnetic Suspension and Balance System Wind Tunnel*, NASA TM-4090.

Kim, Y.-K., Katsurai, M., and Fujita, H. A. (1989). "A Superconducting Actuator Using the Meissner Effect, *Sensors and Actuators* **20**, 33.

Kitaguchi, H., Takada, J., Oda, K., Osaka, A., Miura, Y., Tomii, Y., Mazaka, H., and Takano, M. (1989). "Magnetic Suspension of a Bi–Pb–Sr–Ca–Cu–O Superconductor Due to the Meissner Effect," *Physica C* **157**, 267–271.

Kitamura, T., Hasegawa, T., Takeshita, F., Yamamoto, K., Murase, S., and Ogiwara, H. (1991). "Development of Silver Sheathed Bi-based Superconducting Coil," in *Advances in Superconductivity IV, Proceedings of the 4th International Symposium on Superconductivity*, H. Hayakawa and

N. Koshizuka, eds., Tokyo, October 1991, Springer-Verlag, Tokyo, 1992, pp. 599–601.

Kolm, H. H., and Thornton, R. D. (1972). "Magneplane: Guided Electromagnetic Flight," in *Proceedings of the 5th International Applied Superconductivity Conference*, IEEE Publication No. 72CH0682-5-TABSC, May 1–3, 1972, pp. 76–85.

Kolm, H. H. and Thornton, R. D. (1973). Electromagnetic Flight, *Sci. Am.* **229**, 17–25.

Komori, M. (1991). "Drag Pressures of a Set of Alternating-Polarity Magnets Over a Superconducting Tile," *J. Appl. Phys.* **70**(4), 2226–2229.

Komori, M., and Kitamura, T. (1991). "Static Levitation in a High-T_c Superconductor Tile on Magnet Arrangements," *J. Appl. Phys.* **69**(10).

Kozorez, V. V., Kolodeev, I. D., Krjukov, M. I., Ljashenko, A. M., Roshkovan, V. M., and Cheborin, O. G. (1976). "On Potential Well of Magnetic Interaction of Ideal Current Circuits," *Doklady Akad. Nauk. Ukr. SSR Ser A*, **3**, 247–248.

Kramer, E. J. (1975). "Scaling Laws for Flux Pinning in Hard Superconductors," *J. Appl. Phys.*, **44**(3), 1360–1370.

Kuroda, M., Tanada, N., and Kikushima, Y. (1992). "Chaotic Vibration of a Body Levitated by Electromagnetic Force," in *ISEM-Nagoya International Symposium on Nonlinear Phenomena in Electromagnetic Fields*, Japan, January 1992.

Kyotani, Y. (1975). "Development of Superconducting Levitated Trains in Japan," *Cryogenics* **15**, 372–376.

Kyotani, Y. (1986). "MAGLEV," *Cryog. Eng.* **B21**, 14–16.

Kyotani, Y. (1988). "Recent Progress by JNR on MAGLEV," *IEEE Trans. Magn.* **24**, 804–807.

Laithewaite, E. R. (1977a). *Transport Without Wheels*, Elek Science Publishers, London.

Lamb, M. A., Ma, K.-B., McMichael, C. K., Meng, R. L., Hor, P. H., Weinstein, R., Chen, I., and Chu, W.-K. (1992). "Characterization of Non-contact Vibration Absorbers Using $YBa_2Cu_3O_{7-\delta}$," *Proceedings of the 1992 TCSUH Workshop of HTS Materials, Bulk Processing, and Bulk Applications*, Houston, Texas, C.-W. Chu, W.-K. Chu, P.-H. Hor, and K. Salama, eds.,

Lee, D. F., Chaud, X., and Salama, K. (1992). "Processing and Transport Current Denisty of Melt-Textured YBCO Composites," *Proceedings of the* 1992 *TCSUH Workshop of HTS Materials*, C. W. Chu, W. K. Chu, P.-H. Hor, and K. Salama, eds. World Scientific Publ. 337–342.

Ma, K.-B., Liu, J-R., McMichael, C., Bruce, R., Mims, D., and Chu, W.-K. (1991). "Spontaneous and Persistent Rotations of Cylindrical Magnets

Levitated by Y–Ba–Cu–O Superconductors," *J. Appl. Phys.* **70**(7), 3961–3963.

Ma, K.-B., McMichael, C. C. and Chu, W.-K. (1992). "Applications of High Temperature Superconductors in Hybrid Magnetic Bearings," *Proceedings of the 1992 TCSUH Workshop of HTS Materials, Bulk Processing, and Bulk Applications*, Houston, Texas, C.-W. Chu, W.-K. Chu, P-H. Hor, and K. Salama, eds., pp. 425–429.

Magneplane International (1992). "System Concept Definition Report," U.S. Department of Transportation, National Maglev Initiative Report, September 1992.

Marek, A. (1990). "Levitation Stability of Superconducting Rings in a Central Magnetic Field," *Proc. 25th IECEC* **3**, 438–443.

Marinelli, M., Morpurgo, G., and Olcese, G. L. (1989). "Diamagnetism of Powder and Bulk Superconducting YBCO Measured with a New Levitometer." *Physica C* **157**, 149–158.

Marinescu, M., Marinescu, N., Tenbrink, J., and Krauth, H. (1989). "Passive Axial Stabilization of a Magnetic Radial Bearing by Superconductors," in *Proceedings of IEEE Meeting CH2731-8/89* (AB-9).

Martini, G., Rivetti, A., Pavese, F. (1990). "A Self Rotating Magnet Levitation Above a YBCO Specimen," in *Advances in Cryogenic Engineering*, Vol. 35, R. W. Fast, ed., Plenum Press, New York.

McCaig, M. and Clegg, A. G. (1987). *Permanent Magnets*, 2nd edition, John Wiley & Sons, New York.

McMichael, C., Ma, K.-B., Lamb, M. A., Lin, M. W., Chow, L., Meng, R. L., Hor, P. H. and Chu, W.-K. (1992). "Practical Adaptation in Bulk Superconducting Magnetic Bearing Applications," *Appl. Phys. Lett.* **60**, 1893–1895.

Meisenholder, S. G., and Wang, T. C. (1972). "Dynamic Analysis of an Electromagnetic Suspension System for a Suspended Vehicle System," U.S. Federal Railway Administration Report, No. FRA-RT-73-1.

Meissner, W., and Ochsenfeld, R. (1933). "Ein neuer Effekt bei Eintritt der Supraleitfähigkert," *Naturwissenschaften* **21**, 787–788.

Mikhalevich, V. S., Kozoriz, V. V., and Shabiij, K. (1991). "Magnetic Levitation Based on Magnetic Potential Well (MPW) Effect," Society of Automotive Engineers, SAE Paper 911G26, 4pp.

Miller, L., and Ruoss, W. (1989). "Performance Analysis of the Transrapid 07," in *Proceedings of the 11th International Conference on Magnetically Levitated Systems and Linear Drives*, July 7–11, 1989, Yokohama, Japan, IEE Japan, pp. 85–92.

Montgomery, B. (1980). *Solenoid Magnet Design*, John Wiley & Sons, New York.

Moon, F. C. (1974). "Laboratory Studies of Magnetic Levitation in the Thin Track Limit," *IEEE Trans. Magn.* **MAG-10**(3), 439–442.

Moon, F. C. (1977). "Vibration Problems in Magnetic Levitation and Propulsion," in *Transport Without Wheels*, E. R. Laithwaite, ed., Elek Science Publishers, London, pp. 122–161.

Moon, F. C. (1978). "Nonconservative Instabilities in Electrodynamic Magnetic Levitation of Vehicles," in *Proceedings of the International Seminar on Superconductive Magnetic Levitated Train*, Miyazaki, Japan, November 1978, pp. 108–112.

Moon, F. C. (1979). "Buckling of a Superconducting Ring in a Toroidal Magnetic Field," *J. Appl. Mech.* **46**(1), 151–155.

Moon, F. C. (1980). "Magnetoelastic Instabilities in Superconducting Structures and Earnshaw's Theorem," *Mechanics of Superconducting Structures*, ASME, Applied Mechanics Monograph, Vol. 41, pp. 77–90.

Moon, F. C. (1984). *Magneto-Solid Mechanics*, John Wiley and Sons, New York.

Moon, F. C. (1988). "Chaotic Vibrations of a Magnet Near a Superconductor," *Phys. Lett. A* **132**(5), 249–251.

Moon, F. C. (1990a). "Magnetic Flexure System for Determining Superconductive Properties of a Sample," U.S. Patent Number 4,931,732, June 5, 1990.

Moon, F. C. (1990b). "Magnetic Forces in High-T_c Superconducting Bearings," *Appl. Electromagn. Mater.* **1**, 29–35.

Moon, F. C. (1992a). *Chaotic and Fractal Dynamics*, John Wiley & Sons, New York.

Moon, F. C. (1992b). "Levitation Studies in High T_c Superconductors at Lower Temperatures and High Fields," in *Advances in Superconductivity, Proceedings of the 4th International Symposium on Superconductivity*, H. Hayakawa and N. Koshizuha, eds., October 1991, Tokyo, Springer-Verlag, Berlin, 1992, pp. 1049–1054.

Moon, F. C. and Chang, P.-Z. (1990). "High-Speed Rotation of Magnets on High T_c Superconducting Bearings," *Appl. Phys. Lett.* **56**, 397–399.

Moon, F. C. and Hull, J. (1990). "Materials Issues in Superconducting Levitation and Suspension as Applied to Magnetic Bearings" *Proceedings of the Intersociety Energy Conversion Engineering Conference*, Reno, Nev.

Moon, F. C. and Li, G.-X., (1986). "Chaotic Dynamics in Magnetic and Magneto-Mechanical Systems." IUTAM Symposium on Electro-magneto-mechanical Interactions in Deformable Solids and Structures, Tokyo, Oct., 1986.

Moon, F. C., and Raj, R. (1989). "Superconducting Rotating Assembly," U.S. Patent Number 4,886,778, Dec. 12, 1989.

Moon, F. C., Weng, K.-C., and Chang, P.-Z. (1989). "Dynamic Magnetic Forces in Superconducting Ceramics." *J. Appl. Phys.* **66**, 5643–5645.

Moon, F. C., Wang, J. D., and Raj, R. (1987). "High-T_c Superconducting Magnetic Bearings for Levitation of High Speed Rotors," Cornell University Report.

Moon, F. C., Yanoviak, M. M., and Ware, R. (1988). "Hysteretic Levitation Forces in Superconducting Ceramics," *Appl. Phys. Lett.* **52**, 1534–1536.

Moon, F. C., Chang, P.-Z., Hojaji, H., Barkatt, A., and Thorpe, A. N. (1990). "Levitation Forces, Relaxation and Magnetic Stiffness of Melt–Quenched $YBa_2Cu_3O_x$," *Jpn. J. Appl. Phys.* **29**, 1257–1258.

Moon, F. C., Chang, P.-Z., Golkowski, C., and Yang, Z. (1992), in *HTS Materials, Bulk Processing and Bulk Applications*, C. W. Chu, W. K. Chu, P.-H. Hor and K. Salama (eds.), *Proc. 1992 TCSUH Workshop*, Houston, World Scientific, Singapore, 442–452.

Moon, F. C., Golkowski, Cz., and Kupperman, D. (1993). "Superconducting Bearings for High Load Applications," *Appl. Superconductivity* **1**(7–9), 1175–1184.

Murakami, M. (1989). "The Quench and Melt Growth Process," *Supercurrents* **8**, 41–47.

Murakami, M., Morita, M. Doi, K., and Miyamoto, K. (1989a). "A New Process with the Promise of High J_c in Oxide Superconductors," *Jpn. J. Appl. Phys.* **28**(7), 1189–1194.

Murakami, M., Morita, M., and Koyama, N. (1989b). "Flux Creep in High J_c $YBa_2Cu_3O_7$ Crystals," *Jpn. J. Appl. Phys.* **28**(10), L1754–L1756.

Murakami, M., Morita, M., Koyama, N. (1989c). "Magnetization of a $YBa_2Cu_3O_7$ Crystal Prepared by the Quench and Melt Growth Process," *Jpn. J. Appl. Phys.* **28**, L1125–L1127.

Murakami, M., Gotoh, S., Koshizuka, N., Tanaka, S., Matsushita, T., Kambe, S., and Kitazawa, K. (1990). "Critical Currents and Flux Creep in Melt Processed High T_c Oxide Superconductor," *Cryogenics* **30**, 390–396.

Murakami, M., Kondoh, A., Fujimoto, H., Sakai, N., Takaichi, H., Yamaguchi, K., Takata, T., Takamuku, K., Ogawa, N., Hirabayashi, I., Koshizuka, N., and Tanaka, S. (1991). "Melting Processing of Bulk YBCO Superconductors with High J_c," presented at the ASME Winter Annual Meeting, December 1991, Atlanta, Paper No. 91-WA-AES-2.

Nagaike, T., and H. Takatsuka (1989). "Present Status and Prospect of HSST," in *Proceedings of the 11th International Conference on Magnetically Levitated Systems and Linear Drives*, Yokohama, Japan, IEE Japan, pp. 29–35.

Nelson, D. R. (1988). "Vortex Entanglement in High-T_c Superconductors," *Phys. Rev. Lett.* **60**, 1973–1976.

Nemoshkalenko, V. V., Ivanov, M. A., Nikitin, B. G., Pogorelov, Y. G., and Klimenko, G. A. (1990a). "Continuous Range of Stable Equilibrium Positions in the System of Magnet and High-T_c Superconductor," *Solid State Commun.* **74**, 637–639.

Nemoshkalenko, V. V., Brandt, E.H., Kordyuk, A. A., Nikitin, B. G. (1990b). "Dynamics of a Permanent Magnet Levitating Above a High-T_c Superconductor," *Physica C* **170**, 481–485.

Neumüller, H.-W., Assmann, H., Kress, B., and Ries, G. (1991), "Preparation and Electrical Characterization of Melt Textured 2212 Bi–Sr–Ca–Cu–Oxide Layers on Ag-Tape," in *Advances in Superconductivity IV, Proceedings of the 4th International Symposium on Superconductivity* October 1991, Tokyo, H. Hayakawa and N. Koshizuka, eds., pp. 553–558.

O'Connor, L. (1992). "Active Magnetic Bearings Give Systems a Lift," *Mech. Eng.* **115**(10), 52–57.

O'Neill, G. K., and Kolm, H. (1978). "Mass Driver for Lunar Transport and as a Reaction Engine,' *J. Astron. Sci.* **XXV**(4).

Ooi, B. T. (1976). "Levitation, Drag and Transverse Forces in Finite Width Sheet Guideways for Repulsive Magnetic Levitation," *High-Speed Ground Transp. J.* **19**, 369–373.

Orlando, T. P., and Delin, K. A. (1991). *Foundations of Applied Superconductivity*, Addison–Wesley, Reading, MA.

Oyama, T., Murakami, M., Fujimoto, H., Shiohara, Y., Koshizuka, N., and Tanaka, S. (1990). "Large Levitation Force Due to Flux Pinning in MPMG Processed Y–Ba–Cu–O Superconductors with Ag Doping," in *Proceedings of the 2nd International Conference in Magnetic Bearings*, Tokyo.

Pao, Y.-H. (1978). "Electromagnetic Forces in Deformable Continua," *Mech. Today* **4**, 209–305.

Parker, R. J. (1990). *Advances in Permanent Magnetism*, John Wiley & Sons, New York.

Permag Corp. (1986). *Magnet Catalog P5A*, Permag Corp., U.S.A.

Peters, P. N., Sisk, C., and Decher, R. (1990). "Self-Stable Suspended Rotors Utilizing Superconducting Niobium," *25th IECEC* **3**, 444–447.

Phillips, J. C. (1989). *Physics of High-T_c Superconductors*, Academic Press, New York.

Polgreen, G. R. (1966). *New Applications of Modern Magnets*, Macdonald, London.

Politis, C., and Stubhan, F. (1988). "Is the Magnetic Suspension in High-Temperature Superconductors a General Phenomenon?" *Mod. Phys. Lett. B* **2**, 1119–1123.

Pollard, M. G., and Riches, E. E. (1985). "Birmingham Maglev: Development for the Future," in *International Conference of Maglev Transport '85*, pp. 123–136.

Poole, C. P., Jr., Datta, T., and Farach, H. A. (1988). *Copper Oxide Superconductors*, John Wiley & Sons, Inc., New York.

Popp, K. (1982a). "Mathematical Modeling and Control System Design of MagLev Vehicles," *Dynamics of High-Speed Vehicles*, W. O. Schiehlen (ed.), CISM Courses and Lectures No. 274, 333–363. Springer-Verlag.

Popp, K. (1982b). "Stochastic and Elastic Guideway Models," *Dynamics of High-Speed Vehicles*, W. O. Schiehlen (ed.), CISM Courses and Lectures No. 274, 13–38, Springer-Verlag, Wien.

Powell, J. R. (1963). "The Magnetic Road: A New Form of Transport," in *ASME Railroad Conference*, Paper 63-RR4.

Powell, J. R., and Danby, G. T. (1967). "A 300-MPH Magnetically Suspended Train," *Mech. Eng.* **89**, 30–35.

Powell, J. R., and Danby, G. R. (1968). "Dynamically Stable Cryogenic Magnetic Suspensions for Vehicles in Very High Velocity Transport Systems," *Recent Adv. Eng. Sci.* **5**(I), 159–182.

Powell, J. R., and Danby, G. T. (1969a). "Electromagnetic Inductive Suspension and Stabilization System for a Ground Vehicle," U.S. Patent 3,470,828, filed November 21, 1967, Granted October 7, 1969.

Powell, J. R., and G. T. Danby (1969b). "Magnetically Suspended Trains: The Application of Superconductors to High-Speed Transport," *Cryog. Indust. Gases*, 19–24.

Rao, D. K., and Bupara, S. S. (1992). "Development of a Passive Superconducting Bearing to Support Heavy Rotors at High Speeds," in *HTS Materials, Bulk Processing, and Bulk Applications, Proceedings of the 1992 TCSUH Workshop*, Houston, Texas. C. W. Chu, W. K. Chu, P.-H. Hor, and K. Salama, eds., pp. 436–441.

Rebhan, E., and Salat, A. (1967). "Equilibrium and Stability of Normal and Superconducting Current Loops," *Z. Naturforsch. A* **22** 1920–1926.

Reitz, J. R. (1970). "Forces on Moving Magnets due to Eddy Currents," *J. Appl. Phys.* **41**, 2067–2071.

Reitz, J. R., and Borcherts, R. H. (1975). "U.S. Department of Transportation Program in Magnetic Suspension (Repulsion Concept)," *IEEE Trans. Magn.* **MAG-11**(2) 615–618.

Reitz, J. R., and Davis, L. C. (1972). "Force on a Rectangular Coil Moving Above a Conducting Slab," *J. Appl. Physics.* **43**(4), 1547–1553.

Reitz, J. R. and Milford, F. J. (1960). *Foundations of Electromagnetic Theory*, Addison-Wesley, Reading, MA.

Rhodes, R. G. (1982). "Maglev Research in the U.K.," in *Second International Seminar on Superconductive Magnetic Levitated Trains*, Miyazaki, Japan, November 1982 pp. 31–37.

Rhodes, R. G. and Eastham, A. R. (1971). "Magnetic Suspension for High Speed Trains," *Hovering Craft and Hydrofoil* **11**, 12–26.

Rhodes, R. G. and Mulhall, B. E. (1981). *Magnetic Levitation for Rail Transport*, Clarendon Press, Oxford.

Rigney, T. K., and Trivedi, A. N. (1992). "Application of High-Temperature Superconductors in Turbomachinery Bearings for Space Systems," in *HTS Materials, Bulk Processing, and Bulk Applications, Proceedings of the 1992 TCSUH Workshop*, Houston, Texas, C. W. Chu, W. K. Chu, P.-H. Hor, and K. Salama, eds., pp. 430–435.

Rivetti, A., Martini, G., Goria, R., and Lorefice, S. (1987). "Turbine Flowmeter for Liquid Helium with the Rotor Magnetically Levitated," *Cryogenics* **27**, 8–11.

Rogg, D. (1985). "Development of Magnetic Levitation Transport Systems in the Federal Republic of Germany; Survey, Present State, Prospect and Reasons," *International Conference on Maglev Transport '85*, Tokyo, Japan, 1-11, Published by Institute of Electrical Engineers of Japan.

Rogg, D. (1986). "The Research and Development Program, 'Magnetically Suspended High Speed Transport Systems' in the Federal Republic of Germany," 31-39, *International Conference on Maglev and Linear Drives*, Vancouver, British Columbia, 1986, IEEE, New York, 86 CH2276-4.

Rudback, N. E., Hayes, W. F., Fife, A. A., Eastham, A. R., and Audette, M. (1985). "An Overview of Canadian Maglev Research and Development," *Proceedings of the International Conference on Maglev Transport '85*, Tokyo, Institute of Electrical Engineer's, Japan, pp. 13–20.

Sakamoto, T., and Eastham, A. R. (1991). "Static Stability of the Split-Guideway Electrodynamic Levitation System," in *Electromagnetic Forces and Application*, *Proceedings of the Third International ISEM Symposium*, Sendai, Japan, pp. 351–354.

Salama, K., Selvamanickam, V., Gao, L., and Sun, K. (1989). "High Current Density in Bulk $YBa_2Cu_3O_x$ Superconductor," *Appl. Phys. Lett.* **54**(23), 2352–2354.

Sato, K.-I., Shibuta, N., Mukai, H., Hikata, T., Masuda, T., Ueyama, M. Kato, T., and Fujikami, J. (1991). "Bismuth Superconducting Wire and Application," in *Advances in Superconductivity IV*, *Proceedings of the 4th International Symposium on Superconductivity*, October 1991, Tokyo, H. Hayakauva and N. Koshizuka, eds., pp. 559–564.

Scanlan, R. M. (1979). *Superconducting Materials*, Lawrence Livermore Laboratory, Preprint. UCRL-83492.

Scharf, B. R., Eidelloth, W., Barnes, F. S., and Dugas, M. (1989). "Suspension in Magnetic Recording Using High T_c Superconductors," *IEEE Trans. Magn.* **25**, 3230–3232.

Schrieffer, J. R. (1964). *Theory of Superconductivity*, Addison–Wesley, Reading, MA.

Schweitzer, G., ed. (1988). *Magnetic Bearings*, Springer-Verlag, Berlin.

Scott, W. T. (1959). "Who Was Earnshaw?" *Am. J. Phys.* **27**, 418.

Seckel, E. (1964). *Stability and Control of Airplanes and Helicopters*, Academic Press, New York.

Shapira, Y., Huang, C. Y., McNiff, E. J., Jr., Peters, P. N., Schwartz, B. B.,and Wu, M. K. (1989). "Magnetization and Magnetic Suspension of $YBa_2Cu_3O_x$–AgO Ceramic Superconductors," *J. Magn. Magn. Mater.* **78**, 19–30.

Simon, I. (1953). "Forces Acting on Superconductors in Magnetic Fields," *J. Appl. Phys.* **24**(1).

Simon, R., and Smith, A. (1988). *Superconductors*, Plenum Press, New York.

Slemon, G. R., (1975). "Canadian MAGLEV Project on High-Speed Interurban Transportation," *IEEE Trans. Magn.* **MAG-11**, 1478–1483.

Smythe, N. R. (1968). *Static and Dynamic Electricity*, 3rd edition. McGraw–Hill, New York.

Sommerfeld, A. (1952). *Electrodynamics*, Academic Press, New York.

Stoker, J. J. (1950). *Nonlinear Vibrations*, Interscience, New York (reissued by John Wiley & Sons, 1993).

Stratton, J. A. (1941). *Electromagnetic Theory*, McGraw–Hill, New York.

Sugiura, T., Hashizume, H., and Miya, K. (1991). "Numerical Electromagnetic Field Analysis of Type-II Superconductors," *Proceedings of the International Symposium on the Application of Electromagnetic Forces*.

Takahata, R., Ueyama, H., and Yotsuya, T. (1992). "Load Carrying Capacity of Superconducting Magnetic Bearings," *Electromagnetic Forces and Applications*, J. Tani and T. Takagi, (eds.), ELsevier Scientific Publisher, B.V.

Takaichi, H., Murakami, M., Kondoh, A., Koshizuka, N., Tanaka, S., Fukuyama, H., Seki, K., Takizawa, T., and Aihara, S. (1992). "The Application of Bulk YBaCuO for a Practical Superconducting Magnetic Bearing," in *Proceedings of the Third International Symposium on Magnetic Bearings*, July 1992, Alexandria, VA, P. E. Allaire, ed., Technomic Publishers, Lancaster, PA, pp. 307–316.

Tamura, H., Matsunaga, J.-I., and Xu, Z. (1992). "Experimental Study on Chaotic Vibrations of a Magnetically Levitated Body," in *ISEM-Nagoya International Symposium on Nonlinear Phenomena in Electromagnetic Fields*, January 1992, Japan.

Tanigaki, K., Ebbesen, T. W., Saito, S., Mizuka, J., Tsai, J.-S., Kubo, Y., and Kuroshima, S. (1991). "A New Type of Molecular Superconductor Based on C_{60} Doped with Cesium and Rubidium," *Advances in Superconductivity IV, Proceedings of the 4th International Symposium on Superconductivity* October 1991, Tokyo, H. Hayakawa and N. Koshizuka, eds., pp. 203–206.

Tenney, F. H. (1969). "On the Stability of Rigid Current Loops in an Axisymmetric Field," Plasma Physics Laboratory Report No. MATT – 693, Princeton, NJ.

Terentiev, A. N. (1990). "Disappearance of Friction in a Levitated $YBa_2Cu_3O_{7-x}$ Superconductor in a Variable Magnetic Field," *Physica* **166**, 71–74.

Terentiev, A. N., and A. A. Kuznetsov (1990). "Rotation of Levitating $YBa_2Cu_3O_{7-x}$ Superconductor in a low frequency magnetic field," *SPCT* **3**(12).

Terentiev, A. N., and Kuznetsov, A. A. (1992). "Drift of Levitated YBCO Superconductor Induced by Both a Variable Magnetic Field and a Vibration," *Physica C* **195**, 41–46.

Tichy, J. A., and Connor, K. A. (1989). "Geometric Effects on Eddy Current Bearing Performance," *J. Tribol.* **111**.

Tinkham, M. (1975). *Introduction to Superconductivity*, McGraw–Hill, New York.

U.S. Department of Transportation (1993a). *Compendium of Executive Summaries from the Maglev System Concept Definition Final Reports*, National Maglev Initiative DOT/FRA/NMI-93/02, March 1993.

U.S. Department of Transportation (1993b). *Final Report on the National Maglev Initiative*, Report No. DOT/FRA/NMI-93/03, September, 1993.

U.S. Office of Technology Assessment (1991). *New Ways: Tiltrotor Aircraft and Magnetically Levitated Vehicles*, OTA-SET-507. U.S. Government Printing Office, Washington, D.C.

Urankar, L. (1976). "Intrinsic Damping in Basic Magnetic Levitation Systems with a Continuous Sheet Track," *Siemens Forsch. Entwicklungsber* **5**(2), 110–119.

Vasil'ev, S. V., Kim, K. I., Matin, V. I., and Mikirtichev, A. A. (1977). "Repulsive-Type Magnetic Levitation Systems for High Speed Transportation (Survey of Foreign Investigations)," *Izv. Vyssh. Ucheb. Zaved. Elektromekh.* 882–888.

Waldron, R. D. (1966). "Diamagnetic Levitation Using Pyrolytic Graphite," *Rev. Sci. Instr.* **37**(1), 29–35.

Wang, J., Yanoviak, M. M., and Raj, R. (1989). "Type II Magnetic Levitation on Sinter-Forged YBCO Superconductor," *J. Am. Ceram. Soc.* **72**, 846–848.

Webb, W. W. (1971). "Mechanisms Determining the Critical Current in Hard Superconductors," *J. Appl. Phys.* **42**(1), 107–115.

Weeks, D. E. (1989). "Levitation Properties of the $YBa_2Cu_3O_x$ and Tl–Ba–Ca–Cu–O Superconducting Systems," *Appl. Phys, Lett.* **55**, 2784–2786.

Weeks, D. E. (1990). High T_c Superconducting Levitation Motor with a Laser Commutator, *Rev. Sci. Instrum.* **61**, 195–200.

Weh, H. "Magnetic Levitation Technology and Its Development Potential," in *Proceedings of the 11th International Conference on Magnetically Levitated Systems and Linear Drives*, July 7–11, 1989, Yokohama, Japan, IEE Japan, pp. 1–9.

Weinberger, B. R., Lynds, L., and Hull, J. R. (1990a). "Magnetic Bearing Using High-Temperature Superconductors, Some Practical Considerations," *Supercond. Sci. Technol.* **3**, 381–388.

Weinberger, B. R., Lynds, L., Van Valzah, J., Eaton, H. E., Hull, J. R., Mulcahy, T. M., and Basinger, S. A. (1990b). "Characterization of Composite High Temperature Superconductors for Magnetic Bearing Applications," in *Proceedings of the Applied Supercondivity Conference*, Snowmass Village, Aspen, CO.

Weinberger, B. R., Lynds, L., Hull, J. R., and Balachandran, U. (1991). "Low Friction in High Temperature Superconducting Bearings," *Appl. Phys. Lett*.

Weinstein, R., and Chen, I. (1992). "Persistent Field: Techniques for Improvement and Applications," presented at the TCSUH Workshop on HTS Materials, Bulk Processing, and Bulk Applications, Houston, February 1992.

Weinstein, R., Chen, I. G., Liu, J., Narayanan, R., Ren, Y. R., Xu, J., Obot, V., and Wu, J. (1992). "Materials Characterization and Applications for High-T_c Superconducting Permanent Magnets," in *Proceedings of the Third World Congress on Superconductivity*, Munich, September 1992, Pergamon Press, Elmsford, NY.

Weldon, W. F. (1992). "The Potential for Application of High Temperature Superconductors to Electromagnetic Launchers," in *HTS Materials, Bulk Processing, Bulk Applications, Proceedings of the 1992 TCSUH Workshop*, C. W. Chu, W. K. Chu, P. H. Hor, K., Salama, eds., pp. 525–529.

"West Germany Okays World's First Commercial MAGLEV Line," *New Technol. Week* July 5, 1988, p. 3.

Wilke, D. F. (1972). "Dynamics, Control and Ride Quality of a Magnetically Levitated High Speed Ground Vehicle," *Transp. Res.* **6**, 343–369.

Williams, R., and Matey, J. R. (1988). "Equilibrium of a Magnet Floating Above a Superconducting Disc," *Appl. Phys. Lett.* **52**(9), 751–752.

Wilson, M. N. (1983). *Superconducting Magnets*, Clarendon Press, Oxford, England, 1983.

Wolfshtein, D., Seidel, T. E., Johnson, D. W., Jr., and Rhodes, W. W. (1989). "A Superconducting Magnetic Levitation Device for the Transport of Light Payloads," *J. Supercond.* **2**, 211.

Wong, J. Y., Howell, J. P., Rhodes, R. G., and Mulhall, B. E. (1976). "Performance and Stability Characteristics of an Electrodynamically Levitated Vehicle Over a Split Guideway," *Trans. ASME J. Dyn. Syst. Meas. Control Ser. G* **98**(3), 277–285.

Woods, L. C., Cooper, R. K., Neil, V. K., and Taylor, C. E. (1970). "Stability Analysis of a Levitated Superconducting Current Ring Stabilized by Feedback and Eddy Current," *J. Appl. Phys.* **41**(8), 3295–3305.

Woodson, H. H., and Melcher, J. R. (1968). *Electromechanical Dynamics*, Parts I, II, and III, John Wiley & Sons, New York.

Wormley, D. N., Thornton, R. D., Yu, S.-H., and Cheng, S. (1992). "Interactions Between Magnetically Levitated Vehicles and Elevated Guideway Structures," U.S. Department of Transportation, Federal Railway Administration, National Maglev Initiative DOT/FRA/NMI-92/93, July 1992.

Yamamura, S. (1976). "Magnetic Levitation Technology of Tracked Vehicles: Present Status and Prospects," *Trans. IEEE* **MAG-12**, 874–878.

Yang, Z. J. (1992). "Thickness Dependence of Levitation Forces Acting on Magnets Over a Thin Superconducting Sheet," *Jpn. J. Appl. Phys.* **31**, L938–L941.

Yang, Z. J. and Moon, F. C. (1992). "Amplitude-Dependent Magnetic Stiffness of Melt-Quenched YBCO Superconductors," Cornell University Report, Sibley School of Mechanical and Aerospace Engineering, Ithaca, N.Y.

Yang, Z. J., Johansen, T. H., Bratsberg, H., Helgesen, G., and Skjeltorp, A. T. (1989). "Vibration of A Magnet Levitated Over a Flat Superconductor," *Physica C* **160**, 461–465.

Yang, Z. J., Johansen, T. H., Bratsberg, H., Helgesen, G., and Skjeltorp, A. T. (1990a). "Investigation of the Interaction Between a Magnet and a Type-II Superconductor by Vibration Method," *Physica C* **165**, 397–403.

Yang, Z. J., Johansen, T. H., Bratsberg, H., Helgesen, G., and Skjeltorp, A. T. (1990b). "Comment On Lateral Restoring Force on a Magnet Levitated Above a Superconductor," *J. Appl. Phys.* **68**, 3761–3762.

Yoshioka, H., and Miyamoto, M. (1986). "Dynamic Charcteristics of Maglev Vehicle MLU001-Guideway Irregularity Test," in *Proceedings of the International Conference on Maglev and Linear Drives*, May 1986, Vancouver, B.C., Canada, IEEE 86CH2276-4, pp. 89–94.

Yotsuga, T., Shibayama, M., Takahata, R. (1991). "Characterization of High-Temperature Superconducting Bearing," Preprint CEC '91 meeting, from R. Takahata, Koyo Seiko R & D Center, Osaka, Japan.

AUTHOR INDEX

The page numbers in *italic* refer to References.

SUBJECT INDEX